Non-Asymptotic Theory of Waves in Excitable Media Across Nature and Technology

Gennadiy Vladimiroxich Zhizhin
Russian Academy of Natural Sciences, Russia

IGI Global
Scientific Publishing
Publishing Tomorrow's Research Today

Vice President of Editorial Melissa Wagner
Director of Acquisitions Mikaela Felty
Director of Book Development Jocelynn Hessler
Production Manager Mike Brehm
Cover Design Phillip Shickler

Published in the United States of America by
IGI Global Scientific Publishing
701 East Chocolate Avenue
Hershey, PA, 17033, USA
Tel: 717-533-8845
Fax: 717-533-7115
Website: https://www.igi-global.com E-mail: cust@igi-global.com

Library of Congress Cataloging-in-Publication Data

Names: Zhizhin, G. V. (Gennadiĭ Vladimirovich), author.
Title: Non-asymptotic theory of waves in excitable media across nature and technology / Gennadiy Vladimirovich Zhizhin.
Description: Hershey, PA : Engineering Science Reference, [2025] | Includes bibliographical references and index. | Summary: "Non-Asymptotic Theory of Waves in Excitable Media Across Nature and Technology discusses the occurrence of various types of waves in chemical, physical, and biological processes. These waves involve mass movements in space and are described using mathematical models. The narrow reaction zone method, proposed by Zel'dovich, is one such model used to describe combustion waves. However, observations indicate that the chemical reaction zone is not always negligible. Isothermal waves, such as those observed in Belousov-Zhabotinsky reactions, also have finite-length reaction zones. New mathematical methods are needed to study waves in excitable media, considering the full extent of kinetic processes. The book also discusses waves in biological populations and the limitations of asymptotic methods in analyzing them. Non-asymptotic methods have been developed to study waves in various excitable media, such as radical polymerization waves and waves of biological populations. Complex waves, like solid-flame combustion and natural fires, involve multiple kinetic processes and require specific models. The book also discusses the propagation of waves in excitable media related to gas dynamics, such as detonation waves. Various mathematical models and methods have been developed to analyze these waves and their characteristics"-- Provided by publisher.
Identifiers: LCCN 2023029654 (print) | LCCN 2023029655 (ebook) | ISBN 9781668477458 (hardcover) | ISBN 9781668477465 (paperback) | ISBN 9781668477472 (ebook)
Subjects: LCSH: Wave-motion, Theory of. | Waves--Mathematical models. | Approximation theory. | Nonlinear systems. | Mathematical physics.
Classification: LCC QA927 .Z475 2023 (print) | LCC QA927 (ebook) | DDC 531/.113015114--dc23/eng/20231004
LC record available at https://lccn.loc.gov/2023029654
LC ebook record available at https://lccn.loc.gov/2023029655

British Cataloguing in Publication Data
A Cataloguing in Publication record for this book is available from the British Library.

Table of Contents

Preface

In the world around us, various kinds of chemical, physical and biological processes are constantly taking place, accompanied by a significant movement of masses in space. Briefly describing these processes, we say that a wave, a wave of something, a wave of chemical reaction, a wave of insects, a combustion wave, a wave of mammals, a wave of bacteria, a wave of microbes, a wave of detonation, an earthquake wave, etc., is running through space. Trying to explain the observed phenomena of this type, we are forced to build models of these phenomena that could reveal the main features of this type of mass movement in various media (excitable media) in which these movements are formed. Physical, chemical and biological models of the corresponding waves are built. Having described this phenomenon with equations, we then need to find a mathematical solution to these equations, from which we could predict the wave speed, its intensity, geometric dimensions, etc. Often this turns out to be a very difficult task. We have to resort to additional assumptions and simplifications. One of these simplifying assumptions was the narrow reaction zone method proposed by Academician Ya.B. Zel'dovich (Zel'dovich, 1944), who played a great role in describing combustion waves in relation to the tasks of military equipment. According to this method, all kinetic processes in the wave are concentrated in a narrow layer of asymptotically zero extent (Zel`dovich and al., 1980). The idea of limiting schematization of a combustion wave with a zero length of the chemical reaction zone apparently goes back to M. Faraday. The theory describing combustion waves in this approximation is called the asymptotic combustion theory. However, observations of various types of waves indicate that even in highly activated processes one cannot purely visually consider the chemical reaction zone equal to zero. For example, in the work of A.G. Gaydon, H.G. Wolfgrad, (Gadon, Wolfgrad, 1959) photographs of a flame in a mixed ethylene-air mixture with excess and lack of air are given. These photographs show how different a gas flame can look, and it is difficult to isolate the area of the flame where the actual chemical reactions take place. Apparently, in the brightest places of the flame, chemical reactions occur. But firstly, whether chemical reactions occur only in these places, and, in processes, in comparison with combustion waves, and secondly, what area in space is occupied by at least these brightest places remains unclear. Taking into account that the glow of the flame is the result, first of all, of high temperature, it is possible, in order to exclude the influence of temperature, to refer to the results of observations of waves of chemical reactions under isothermal conditions, i.e. in cases where the cause of flame propagation is the diffusion of molecules. Widely known examples of such waves are autowave processes in Belousov-Zhabotinsky reactions. Moreover, these reactions are accompanied by a change in the color of the liquid and are easily observed. From photographs of these processes (Zhabotinsky, Zaikin, 1973) it is clearly seen that the zones of chemical reactions in spirals have a finite length. Direct experimental proof of the distribution of the zone of chemical reactions in a flame can be the results of measurements of the concentration distributions of components in a flat gas flame, obtained using the method of molecular beam mass spectrometry (Westmoreland et al., 2006). It follows from these measurements that the change in the concentrations of the components occurs almost over the entire length of the combustion wave. In this regard, the need to create new mathematical methods for studying waves in excitable media is fully justified. Since in less activated processes, in comparison with combustion waves, deviations from the approximations of the narrow reaction zone method will be even more significant. Isothermal waves of a chemical reaction are characteristic not only of the Belousava-Zhabotinsky reaction. It was observed in mixtures of hydrocarbons with air (Neiman, 1938), carbon disulfide with oxygen (Voronkov and Semyonov, 1939), and fluorine with hydrogen (Fedotov and Chaikina, 1972). Naturally, the narrow reaction zone method is inapplicable to all these reactions. Such reactions form what is said to be a cold flame. Consequently, here it is impossible to determine the wave speed and wave structure relatively simply. Asymptotic methods are also inapplicable in the analysis of waves of biological populations. In these waves, as well as in cold flame waves, it is necessary to take into account kinetic processes over the entire wavelength. The simplest model of a wave of a biological population, which simultaneously describes the simplest model of a cold flame, was considered in 1937 (Kolmogorov, Petrovsky, Piskunov, 1937). The merit of this work is that it considered the non-stationary diffusion equation in general form, and it was shown that the stationary solution of this equation is possible only at a wave speed equal to 2. However, an analytical wave solution was not obtained in this work. Therefore, after the work of Kolmogorov, Petrovsky, Piskunov, none of the researchers posed and thus solved the problem of determining the speed and structure of a wave of biological populations, based on those equations that corresponded to the models of biological populations they adopted. At best, referring to the work of Kolmogorov, Petrovsky, Piskunov, they set the wave speed according to the solution obtained in it, as if forgetting that the wave speed in the stationary solution is an eigenvalue of the system of equations and it must be determined from the given system of equations under study and correspond to the solution as an own function of the problem (Berezovskaya, Karev, 1999). Thus, asymptotic methods, despite their effectiveness in some cases, cannot be the basis for the fundamental study of waves in excitable media in many cases. In numerous works of the author, carried out at different times, in the study of waves in various excitable media, a non-asymptotic theory of waves in excitable media was developed and applied. The beginning of such studies was laid during the development of a fundamentally new type of polymerization reactor - a reactor with a standing cylindrical wave of radical polymerization. Subsequently, this reactor was called the frontal polymerization reactor (Zhizhin, 1992). A preliminary analysis of the results of experimental studies of radical polymerization waves showed that these waves are characterized by an extended zone of chemical reactions and asymptotic calculation methods are not applicable here. For the calculation and design of the reactor, it was necessary to develop adequate methods corresponding to the features of radical polymerization. These methods were created (Zhizhin, 1982, 1984, 1985). Operational frontal radical polymerization reactors with cylindrical and spherical reaction fronts have been created, and these reactors have been tested in plants. This made it possible to sharply increase the productivity of the polymerization reactors.

The developed non-asymptotic methods make it possible to take into account kinetic processes at a wavelength after the wave front (semi-infinite reaction zone method) and over the entire length wave (infinite reaction zone method). The semi-infinite reaction zone method is used in cases where transport processes predominate before the wave front, i.e. kinetic processes develop very slowly, but in the vicinity of the front, the waves accelerate sharply and slowly weaken after it. This method is of great importance for the calculation of waves in chemical-technological processes, for which the process of "fading" of chemical reactions is essential. The infinite reaction zone method is used when kinetic processes along the entire wavelength cannot be neglected, i.e., for example, for cold flame waves and biological population waves.

Turning to the analysis of waves of biological populations, it should be noted that the simplest population model of the wave by A. N. Kormogorov et al. is too simple and does not take into account the characteristics of living organisms. These features include primarily reproduction and death. In addition, living organisms need to eat, living organisms pollute the environment with waste products, which also affects the kinetic processes in the wave. the death of living organisms in a wave can be natural, or it can be the result of the destruction of organisms by other organisms. In these cases, the diffusion equation becomes so much more complicated and supplemented by other equations that the classical result of 1937 becomes inapplicable. However, the non-asymptotic methods developed by the author in the analysis of radical polymerization waves remain applicable. With their help, in all the listed cases, approximate analytical solutions were obtained and expressions for the speed and structure of the waves were determined (Zhizhin, 2004 a, 2019). Non-asymptotic methods made it possible to analytically describe the waves of biological populations of two types: waves of conquest and solitary waves. A wave of conquest running through space fills it with individuals. A solitary wave represents an antinode of individuals running through space. The existence of solitary waves of biological populations was noted earlier (Ivanov, 1984; Keller and Segel, 1970; Kennedy and Aris, 1980). Living organisms also have features that consist in the interaction of organisms at a distance. An example of such an interaction is the distribution of pheramones (odorous substances) by some organisms to convene other similar organisms into a common flock. These substances are used, for example, by locusts and bark beetles. Non-asymptotic methods make it possible to describe population waves in such cases as well (Zhizhin, 2012, 2014; 2019; Zhizhin, Selikhovkin, 2012).

In nature and technology, there are waves in excitable media with the simultaneous passage of several different kinetic processes. One example of such complex waves can be a wave of solid-flame combustion with a variable surface of chemical interaction. A model of combustion wave in condensed mixtures, in which one of the reactants and reaction products are refractory substances, was constructed (Aldushin et al., 1972). It was believed that a refractory reagent is particles distributed in a low-melting reagent, a chemical reaction occurs on the surface of a refractory reagent, slowing down the reaction, since the low-melting reagent must diffuse through the product layer to enter into the reaction. This model was later called the ideal solid-flame combustion model (Merzhanov, 2000). In this model, the change in the interface area of the reagents, which naturally decreases as the refractory reagent burns out, was not taken into account. This was the contradiction of the adopted model. It was assumed that the refractory reagent is consumed in the reaction, but neither the change in particle size nor their number was taken into account. To eliminate this shortcoming and obtain an approximate analytical solution, the semi-infinite reaction zone method was applied (Zhizhin, 2004 b).

Undoubtedly, natural fires are among the complex types of waves in excitable media. They are diverse and cause enormous harm to nature and people. Despite the urgent need to determine the patterns of fire propagation, a mathematical description encounters serious obstacles. General mathematical models of fires were developed in the works of Grishin A.M. (Grishin, 2002, 2004). However, these models are too general and represent an eclectic combination of equations of various processes containing too many unknown parameters and equations. Of the various types of natural fires, the most suitable for a mathematical description is a ground (forest or steppe) fire. In this paper, the model of a ground fire is considered by the methods of the non-asymptotic theory of combustion. It was found that the length of the zone of chemical reactions in the ground fire wave is the internal degree of freedom of the ground fire.

In the works of B.V. Novozhilov (Novozhilov, 1965, 1966, 1973, 2004) studied non-stationary modes of combustion of gunpowder and solid propellants with small deviations from the stationary state. The non-stationary heat equation was reduced to an infinite system of ordinary differential equations. It was found that in a certain range of parameters there are regimes of chaotization of the powder burning rate. In this paper, following the principles of the non-asymptotic theory of waves in excitable media, a model of gunpowder combustion is constructed at any distance from stationary combustion regimes. It was found that the modes of chaotization of the burning rate of gunpowder are a consequence of the methods used in 2004 in the vicinity of stationary modes. They do not exist if the system is not limited to a neighborhood of equilibrium positions.

The propagation of waves in excitable media is often associated with dynamic processes, in particular with gas dynamic effects. In the classical Chapman-Jouguet theory of detonation, it is assumed that a stationary adiabatic detonation wave is a complex of a shock wave, and a combustion zone located behind it (a zone of chemical reactions). Moreover, it is considered that at the end of the combustion zone, the gas velocity is equal to the local sound velocity (Chapman-Jouguet condition). In addition, it is assumed that at the end of the chemical reaction zone, the depth of the chemical reaction is equal to unity (everything that could burn was burned), but the adiabatic coefficient of the gas remains constant. It is easy to see that such assumptions contradict each other. If a chemical reaction has taken place in the gas, then the adiabatic coefficient will change, since the composition of the gas will change, especially if everything that could burn has burned down. This is especially important when determining the Mach number at the end of the chemical reaction zone, which is a function of the adiabatic coefficient. In addition, if we assume that at the end of the chemical reaction zone, the Mach number is equal to unity, i.e. If a flow crisis sets in, which, as is known (Vulis, 1950), occurs at a finite distance from the beginning of the flow, then there is no reason to believe that when a flow crisis is reached, everything that can burn will necessarily burn out. Despite the obviousness of these contradictions, their significance has not been discussed in literature. In this work, the theory of detonation is developed taking into account the dependence of the specific heat capacities of the gas and its adiabatic coefficient on the composition of the gas, which changes during a chemical reaction. A model of a detonation wave in a dusty gas is considered.

As you know, the conditions of life on planet Earth are greatly influenced by mass ejections from the Sun. In this paper, the features of gas motion in prominences on the Sun are considered, taking into account the various effects exerted on gas particles in these flows (Zhizhin, 2021).

The topics outlined in the preface made up the content of the monograph, each of the topics is considered in a separate chapter.

The first chapter of the monograph examines the principles of the asymptotic theory of combustion, which has become widespread and is briefly called the narrow reaction zone method. According to this method, the length of the radical polymerization zones of chemical reactions in the combustion wave is asymptotically assumed to be zero. The chapter proves that this method contradicts the known equations of heat conduction and diffusion, written taking into account the laws of conservation of energy and mass.

The second chapter reviews some observed experimental data on the propagation of waves in excitable media: insect waves, isothermal waves of chemical reactions, radical polymerization waves accompanied by the release of a large amount of thermal energy, combustion waves of activated chemical reactions. It is noted that many researchers, when analyzing waves in excitable media, are guided by the narrow reaction zone method, although experimental studies using modern methods show that changes in the concentration of reagents occur along the entire length of the wave.

The third chapter deals with non-asymptotic methods for describing waves in exciting media. A system of equations for a combustion wave in a condensed phase with an arbitrary source and a reaction of order m is considered. Solutions with a semi-infinite and infinite reaction zone are obtained. A cold flame with an autocatalytic reaction considered. A waves of logistic population wave are considered. The dependence of the wave propagation velocity on the conditions of its location is proved. A system of waves in a one-dimensional approximation is also considered. The features of the interaction of waves are investigated.

The fourth chapter is devoted to an important example in practical terms of applying non-asymptotic methods for calculating radical polymerization waves, in which the chemical reaction zone occupies a significant part of the wave. Moreover, the reaction waves here can take the form of a cylinder and a sphere. The corresponding reactor designs were developed and tested in laboratory and factory conditions. It is shown that the calculation of reactors using the methods of non-asymptotic theory coincides well with the experimental data obtained on these reactors. The procedure for industrial start-up of continuous frontal polymerization reactors with stopped polymerization wave of cylindrical and spherical symmetry is described.

In chapter 5 the infinite reaction zone method is used to solve mathematical problems about the propagation of the carbon disulfide reaction in the form of a cold flame. It is shown that this method males it possible to obtain approximate analytical solutions for both one-component and two-component cold flame models. Calculations were carried out using the obtain power ed solution and results were compared with the available experimental data for of a cold flame.

In chapter 6 non-asymptotical research methods are applied to the analysis of diffusion-thermal combustion waves. At the beginning a diffusion-thermal combustion flame is considered for an arbitrary source form. Then a power low is adopted for the dependence of the reaction rate on the mass fraction of the reagent (non-adiabatic combustion flame).

In chapter 7, based on the concepts of non-asymptotic wave theory, detonation waves in gas are considered taking into account the change in composition during a chemical reaction.

In chapter 8 the theory of nonequilibrium detonation waves in gas with condensed chemically inert condensed particles and a variable adiabat coefficient of gas, changing its composition during the reaction, is developed. It is proven that the presence of condensed particles in a gas flow can ensure the achievement of supersonic speeds in a detonation wave after a shock wave, and, thereby, significantly increase the conversion of chemical reactions and the momentum imparted by the flow to the aircraft.

Chapter 9 is devoted to the analysis of solid-flame combustion waves. A model for a combustion wave in condensed mixtures with autoinhibition by a high-melting reaction product is considered with allowance for a reduction in the particle surface of high-melting reactant during interaction with the low-melting reactant.

In chapter 10 an essentially non-stationary model of gunpower combustion is constructed. Regimes with a critical burning rate of gunpowder have been found. The condition for the existence of combustion of gunpowder with supercritical velocities is investigated.

In chapter 11 self-replating nonlinear waves in various biological populations are considered as waving attractors in excitable media. It is shown that there are two types of waves a wave of conquest and a solitary wave. In all cases considered formulas for calculating the velocity of the wave and its profile were obtained.

In chapter 12 the processes of biocenosis in the wave mode are considered. The biocenosis wave has the form of a solitary wave for each of the population included in the biocenosis. Approximated analytical solutions for the profiles of population concentrations are wave population velocities are obtained.

In chapter 13 using the methods of non-asymptotic theory of waves, waves of boot-beetles that spread through the forests of planet Earth, and cause enormous damage to forests, under mining and destroying trees, are considered. Mathematical models of boot-beetles' waves are constructed, analytical solutions are obtained, and numerical studies are carried out. A description of field studies of the state of Russian forests susceptible to bark beetle attacks is provided.

In chapter 14 using the methods of non-asymptotic theory of waves in excitable media, a mathematical model of a ground fire has been constructed and studied. A qualitative study of the corresponding system of differential equations has been carried out, an approximate analytical solution has been obtained. It has been established that the length of the zone of the chemical reaction zone is the internal degree of freedom of the ground fire wave.

In chapter 15 the motion of gas in prominences on the Sun, emanating from the chromosphere and returning to the San`s chromosphere is investigated. The study continuous transition of the gas velocity through the speed of sound in prominences emanating from the solar chromosphere.

REFERENCES

Aldushin, A.P. et al. (1972). Propagation of an exothermic reaction front in condensed mixtures during the interaction of components through a layer of a refractory product. *Physics of combustion and explosion*. 8(2), 202-212.

Berezovskaya, F. S., & Karev, G. P. (1999). Traveling wave bifurcations in population models with taxis. *Успехи физических наук*, *169*(9), 1011–1024. DOI: 10.3367/UFNr.0169.199909d.1011

Chapman, D. L. (1899)... *Philosophical Magazine*, *47*, 347.

Fedotov, V. G., & Chaikina, A. M. (1972). Investigation of the cold flame of the reaction of fluorine with hydrogen in a flow. *Reports of the Academy of Sciences of the USSR*, *203*(2), 406–408.

Grishin, A.M. (2002). General mathematical models of forest and peat fires and their application. *Mechanical advances*. 12, 41-89.

Grishin, A.M. (2004). General mathematical model of steppe fires. *Ecological systems and devices*. 2, 25 - 29.

Jouguet, E. (1905). *Jorn. De Mathematique*. 347.

Kolmogorov, A. N., Petrovsky, I. G., & Piskunov, N. S. (1937). An investigation of the equation of diffusion coupled with an increase in the amount of matter, and its application to a biological problem. *Bull. Moscow State University Series A.Mathematics and Mechanics*, *1*(6), 1–26.

Merzhanov, A. G. (2000). *Solid flame combustion*. ISMAN Publishing House.

Neiman, M.B. (1938). The role of peroxides in the formation of cold and hot hydrocarbon flames. *Advances in chemistry*. 7(7), 341-384.

Novozhilov, B. V. (1965). Criterion of stability of the stationary mode of combustion of gunpowder. *Journal of Applied Mechanics and Technical Physics*, *4*, 157–160.

Novozhilov, B. V. (1966). Nonlinear fluctuations in the burning rate of gunpowder. *Journal of Applied Mechanics and Technical Physics*, *5*, 31–41.

Novozhilov, B. V. (1973). *Unsteady combustion of solid rocket propellants*. Nauka.

Novozhilov, B. V. (2004). Chaotization of the non-stationary burning rate of gunpowder. *Chemical Physics*, *5*, 68–74.

Voronkov, V. G., & Semenov, N. N. (1939). Propagation of a cold flame in combustible mixtures containing 0.03% carbon disulphide. *Journal of Physical Chemistry, 13*(12), 1695–1727.

Vulis, L. A. (1950). *Thermodynamics of gas flows*. Gosenergoizdat.

Westmoreland, P. R., Law, M. E., Cool, T. A., Wang, J., McIlroy, A., Taatjes, C. A., & Hansen, N. (2006). Analysis of flame structure by molecular-beam mass spectrometry using electron-impact and synchrotron-photon ionization. *Combustion, Explosion, and Shock Waves, 42*, 672–677.

Zel'dovich, Ya. B. (1944). *Theory of combustion and detonation of gases*. Publishing House of the Academy of Sciences of the USSR.

Zeldovich, Ya. B., Barenblatt, V. B., Librovich, G. M., & Makhviladze, G. M. (1980). *Mathematical theory of combustion and explosion*. Science.

Zhabotinskii, A. M., & Zaikin, A. N. (1973). Autowave Processes in a Distributed Chemical System. *Journal of Theoretical Biology, 40*(1), 45–61. DOI: 10.1016/0022-5193(73)90164-1 PMID: 4723554

Zhizhin, G. V. (1982). Wave structure of frontal polymerization. *Reports of the Academy of Sciences of the USSR, 263*(6), 1399–1402.

Zhizhin, G. V. (1984). On the calculation of the wave structure of frontal polymerization. *Kinetics and Catalysis, 2*, 292–298.

Zhizhin, G. V. (1985). Determination of the coordinates of the location of the front of radical polymerization. *Kinetics and Catalysis, 4*, 1002–1005.

Zhizhin, G.V. (1992). *Macrokinetics in frontal polymerization reactors*. Petersburg: Polytechnic.

Zhizhin, G. V. (2004a). *Self-regulating waves of chemical reactions and biological populations*. Science.

Zhizhin, G.V. (2004 b). Wave Model of Ideal Solid-Flame Combustion with Variable Chemical Interaction Surface. *Physics of combustion and explosion*. 40(1), 95-102.

Zhizhin, G. V. (2012). Mathematical model of a solitary wave a population of beetles-bark beetles. *Proceedings of the Petersburg State Forestry Academy, 196*, 153–162.

Zhizhin, G. V. (2014). Mathematical model of flight of the gregarious locust. *Biosphere, 6*(2), 112–117.

Zhizhin, G. V. (2019). *Attractors and Higher Dimensions in Population and Molecular Biology: Emerging Research and Opportunities*. IGI Global. DOI: 10.4018/978-1-5225-9651-6

Zhizhin, G. V. (2021). Synergetics in Gas Streams. *JNSST*, *15*(3), 127–144.

Zhizhin, G. V., & Selikovkin, A. V. (2012). *Mathematical modeling of the development and distribution of insect populations-stem pests in the forests of Russia*. State Forestry University.

Chapter 1
The Asymptotic Theory:
The Waves of Burning

ABSTRACT

The assumptions of the asymptotic method of a narrow reaction zone in a gas heating wave are considered. Using the known laws of heat and mass transfer in a gaseous medium in which chemical reactions can take place, it is proved that the assumptions of the narrow reaction zone method contradict the laws of conservation of energy and mass. In the three-dimensional phase space, corresponding to the system of equations of heat conduction and diffusion, there are no solutions with some approximation that correspond to the narrow reaction zone method. The assumptions of the asymptotic method of a narrow reaction zone in a gas heating wave are considered.

INTRODUCTION

The most common example of wave propagation in excitable media is the propagation of combustion waves. In this case, everything that can burn can be an excitable medium: gas, wood, grass, plants, polymers, various building materials, etc. During combustion, as a rule, thermal energy is released. They want thermal energy and is the very first type of energy that a person has learned to use, the science of combustion has arisen quite recently. Only in the 18th century, Lomonosov and Lavoisier rejected the erroneous view of alchemists, who claimed that substances capable of burning contain a special "substance of fire" - phlogiston. In 1773, Lavoisier came to the correct understanding of combustion as a chemical combination of substances with atmospheric oxygen. Systematic studies of combustion processes were started only at the end of the 19th century, when catastrophic explosions in mines prompted scientists from a number of countries to take up a thorough study of the flame propagation regimes in pipes. By the visual description of combustion

DOI: 10.4018/978-1-6684-7745-8.ch001

was proposed by M. Faraday (Faraday, 1861). In middle of the last century, new incentives for the development of the science of combustion appeared, associated with the creation of military equipment. During combustion, complex physical, mechanical and chemical phenomena occur. When starting to study combustion, it is necessary to identify the main inherent in this phenomenon, leaving for a while sides factors that can be taken into account later. Therefore, the first researchers of the combustion process greatly simplified the problem, built idealized models, which, in their opinion, described the characteristic aspects of the combustion process. This chapter is devoted to the presentation of the most common simplified combustion model.

COMBUSTION WAVES IN GASES

Let a certain vessel contain a mixture of gases capable of an exothermic chemical reaction. The proportion of active particles in a mixture is determined by the temperature and activation energy of a chemical reaction by the Arrhenius law. If the gas temperature is low (for example, room temperature), then the proportion of active particles is small. As the temperature rises, the rate of a chemical reaction increases, and with it, the rate of heat release increases. This will lead to further spontaneous heating of the gas due to its chemical energy. This process accelerates rapidly. If there is heat loss to the environment, a stationary regime can be established. By heating the gas at one end of the pipe, the mixture can be ignited locally. The released heat will be transferred to neighboring, cold layers of gas, which will lead to the propagation of the reaction along the pipe. If the mixture has reacted at one end of the pipe, say, at the right, then not only the temperature will change along its length, but also the concentration of the initial and final products. Temperature unevenness leads to the process of heat conduction, i.e. to the transfer of the type from already reacted volumes of gas to a fresh, cold mixture. Due to the variability of concentrations, the phenomenon of diffusion arises - the molecular transfer of a substance from areas with a high concentration of it to those places of the gas where it is in short supply. The initial substances diffuse to the right, the final ones - into the unburned mixture. Diffusion and heat conduction processes smooth out temperature and concentration distributions. Between the fresh mixture and the combustion products, a zone appears where the final products are present, and the temperature is increased despite the fact that the reaction does not take place here. At the left end of this zone, called the heating zone, the temperature is close to the initial one, and at the right end, it is close to the combustion temperature. The concentration of the initial substance falls from left to right and at the end of the heating zone is many times less than the initial one. Schematic distribution of concentrations and temperatures along the length of the pipe is shown in Figure 1.

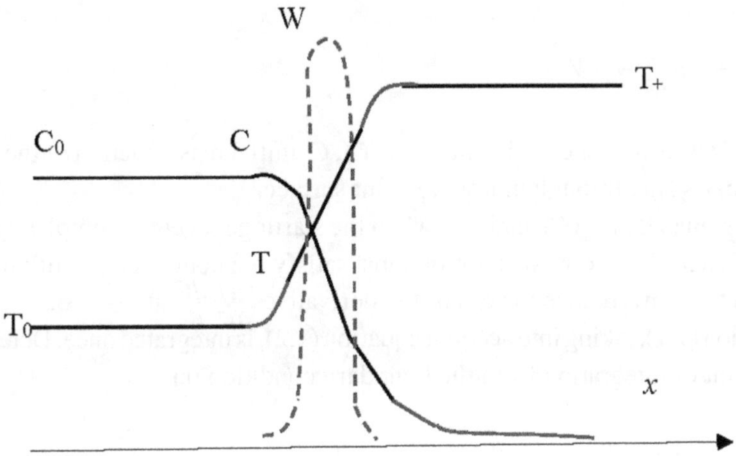

Where does a chemical reaction take place? We know that its speed depends on the concentration of the initial substances and temperature. On the left, a high concentration, but low temperature, on the right, on the contrary. But the reaction rate depends very strongly (exponentially) on temperature and much weaker on concentration. Therefore, it can be argued that the maximum rate of a chemical reaction is achieved in the place where the temperature is highest equal to T_+. The dotted line in Figure 1.1 shows the reaction rate along the flame front. The curve has a sharp maximum at T_+. In the narrow reaction zone method (Zel'dovich and Frank-Kamenetsky, 1938), it is assumed that the chemical reaction zone is a narrow front. To the left of it is the heating zone of the initial mixture due to the heat generated in the reaction zone.

In a stationary one-dimensional formulation, the equation describing the temperature distribution in a gas combustion wave has the form

$$c_p \rho u \frac{dT}{dz} = \frac{d}{dz}\lambda\frac{dT}{dz} + \rho Q W. \tag{1.1}$$

There T is the temperature of gas, c_p is specific heat capacity of gas at constant pressure, ρ is gas density, λ is gas thermal conductivity, Q is specific thermal effect, W is chemical reaction rate, z is wave coordinate associated with the reaction front,

u is combustion wave velocity, $W = k_0\varphi'(C)\phi'(T), k_0$ is reaction rate constant, C is mass fraction of reactant, $z=ut-x$, t is time, x is spatial coordinate.

The equation for changing the mass fraction of a reagent in a stationary wave in coordinates associated with the wave, in the one-dimensional approximation, has the form

$$\rho u\frac{dn}{dz} = \frac{d}{dz}D\rho\frac{dn}{dz} - \rho W. \tag{1.2}$$

There D is diffusion coefficient, $n= C/C_0$, C_0 initial mass fraction of the reagent, $\rho u=m$ is mass flux through unit wave front surface.

It is assumed that $\varphi'(C)$ vanishes when the starting reagent is completely burned out ($C=0$). The desired wave solution must satisfy the boundary conditions, which consist in the convergence to zero of the derivatives $\frac{dT}{dz},\frac{dC}{dz}$ at $z\to\pm\infty$.

Equation (1.1), taking into account equation (1.2), is integrated once. Determining the constant of integration from the boundary condition on

$$z = -\infty, T = T_0, \frac{dT}{dz} = 0, \frac{dn}{dz} = 0, n = 1,$$

we obtain the equation

$$\lambda\frac{dT}{dz} = c_p\rho uT + \rho uQn - Q\rho D\frac{dn}{dz} - (c_pT_0 + Q)\rho u. \tag{1.3}$$

We write equations (1.1), (1.3) and the boundary conditions for them in a dimensionless form, using the values accepted in the combustion theory (Zel'dovich et al., 1980),

$$w^2\frac{d^2\eta}{d\zeta^2} - w^2\frac{d\eta}{d\zeta} + \frac{\varphi(n)\phi(\eta)}{C_0} = 0, \tag{1.4}$$

$$\frac{d\eta}{d\zeta} = \eta + \frac{n - 1 - Le\,dn/d\zeta}{C_0}, \tag{1.5}$$

$$\eta = \frac{T - T_+}{T_+ - T_0}, \zeta = \frac{z\rho u c_p}{\lambda}, Le = \frac{\rho c_p D}{\lambda}, w = u\left(\frac{\rho c_p}{\lambda k_0}\right)^{1/2}, \varphi(n) = \varphi'(C_0 n),$$

$$\phi(\eta) = \phi'[\eta(T_+ - T_0) + T_0], C_0 = c_p\frac{T_+ - T_0}{Q}.$$

The system of equations (1.4), (1.5) can be represented as an autonomous system of three first-order differential equations

$$\frac{d\eta}{d\zeta} = y, \tag{1.6}$$

$$\frac{dy}{d\zeta} = y - \frac{\varphi(n)\phi(\eta)}{C_0 w^2}, \tag{1.7}$$

$$\frac{dn}{d\zeta} = \frac{C_0(\eta - y) + n - 1}{Le}, \tag{1.8}$$

$$\phi(\eta) = \exp\left(-\frac{\beta}{1 + \eta/(1+\sigma)}\right), \sigma = \frac{T_0}{T_+ - T_0}, \beta = \frac{E}{R_0 T_+}, \frac{d\eta}{d\zeta}(\pm\infty) = \frac{dn}{d\zeta}(\pm\infty) = 0,$$

$$-1 \le \eta \le 0, \eta(-\infty) = -1, \eta(+\infty) = 0, 0 \le n \le 1, n(-\infty) = 1, n(+\infty) = 0, \varphi(0) = 0,$$

R_0 is universal gas constant, E is activation energy of a chemical reaction.

The system of equations (1.6) - (1.8) has a singular point b - the equilibrium position with coordinates η=0, y=0, n=0. This equilibrium position in represents the final position in the wave, when all the reactant has burned out and the temperature has reached its maximum possible value. Solutions of the system of equations (1.6) - (1.8) can be represented by autonomy by trajectories in the phase space. At least one of these trajectories must correspond to the desired wave solution. The initial position in the wave solution corresponds in the phase space to the point a with coordinates η=-1, y=0, n=1. Point a is not an equilibrium position, since the derivatives in it are not equal to zero. The wave solution must correspond to a continuous trajectory leaving point a entering the point b in Figure 2.

Figure 2. Phase space of the system (1.6) - (1.8)

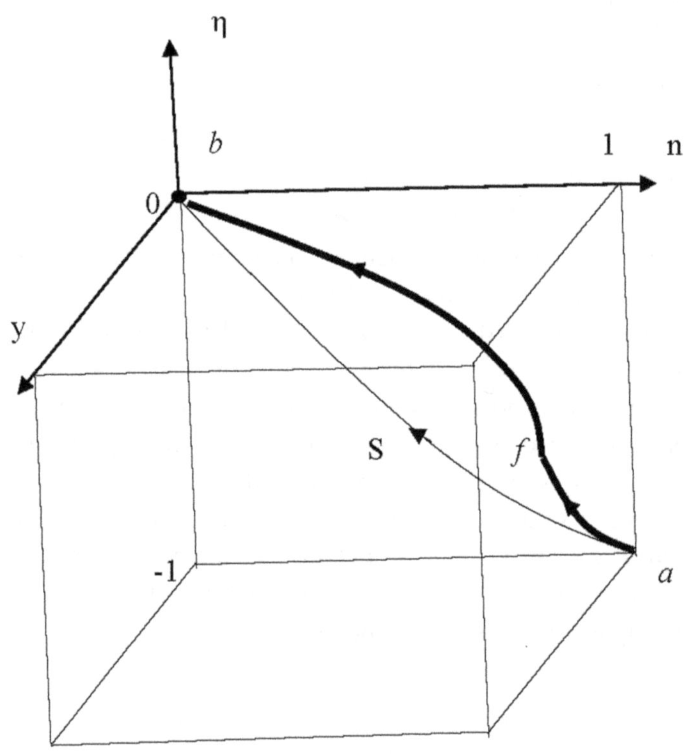

However, due to the method of a narrow wave response zone, it consists of a heating zone adjacent to the initial station, followed by a chemical reaction zone. Moreover, the chemical reaction zone has an infinitesimal extent. According to this assumption, the trajectory corresponding to the heating zone is located on the plane at $n = 1$. It ends at some point f on this plane, corresponding to the reaction front, with some temperature T_f. After it, a trajectory corresponding to the chemical reaction zone should be located. This trajectory is located in the open part of the phase space, and it must end in the equilibrium position b. In this zone, according to the assumption of the narrow reaction zone method, the concentration n quickly disappears, while its duration is zero. Such assumptions are in no way consistent with the phase space in Figure 2. Consequently, these assumptions do not agree with the original equations describing the change in temperature and concentration in the wave (and there are no other physically justified equations). The correctness of the introduction of some delta functions as narrow reaction zone models in Figure 2 is not visible in any way. It is also unclear from what conditions one can determine the coordinates of the point *f* and the value of the wave velocity. All the reasoning that was previously used in the narrow reaction zone method for such determinations is devoid of physical justification within the framework of the accepted heat conduction and diffusion equations.

CONCLUSION

Thus, the ideas of the narrow reaction zone method put forward by physicists purely logistically (Zel`dovich, Frank-Kamenetsky, 1938) do not agree with the known equations of heat conduction and diffusion with sources constructed in accordance with the laws of conservation of energy and mass. Numerous attempts to reconcile these ideas with conservation laws have not been successful (Novikov and Ryazantsev 1965; Berman, Ryazantsev, 1972; Bachelis and Melamed 1965, 1966; Barenblatt 1978; Zel'dovich et al. 1980). At the same time, many researchers of combustion waves did not use the narrow reaction zone method (Williams, 1971; Warnatz et al., 2001; Lewis, Pease, Taylor, 1956). In this regard, in the following chapters, when studying waves in various excitable media, we will not use the representations of the narrow reaction zone method, but will develop a non-asymptotic theory of waves, assuming that the length of the zone of kinetic processes in waves is different from zero.

REFERENCES

Bachelis, R. D., & Melamed, V. G. (1965). On the non-uniqueness of stationary solutions for the system of equations of combustion theory with piecewise constant rate constant, thermal conductivity and diffusion coefficients. *Reports AN USSR, 163*(6), 1338–1341.

Bachelis, R. D., & Melamed, V. G. (1966). On the non-uniqueness of the stationary solution of the system of equations of the theory of combustion. *Applied Mathematics and Mechanics, 30*(2), 368–374.

Barenblatt, G. I. (1978). *Similarity, self-similarity, intermediate asymptotics.* Gidrometioizdat.

Berman, V. S., & Ryazantsev, Yu. S. (1972). On the analysis of the problem of the thermal propagation of a flame by the method of matched asymptotic expansions. *Applied Mathematics and Mechanics, 36*(4), 659–666.

Faraday, M. (1861). *The chemical history of the candle.* Royal Institution of Great Britain. DOI: 10.1038/scientificamerican03301861-194

Lewis, B., Pease, R. N., & Taylor, H. S. (1956). *Combustion Processes.* Princeton University Press. DOI: 10.1515/9781400877027

Novikov, S. S., & Ryazantsev, Yu. S. (1965). On the existence and uniqueness of the solution of the system of equations of the thermal theory of combustion. *Journal of Applied Mechanics and Technical Physics, 4*(4), 86–88. DOI: 10.1007/BF01565821

Warnatz, J., Mass, U., & Dibble, R. W. (2001). *Combustion. Physical and chemical fundamentals, modeling and simulations, experiments, pollutant formation.* Springer.

Williams, F. A. (1964). *Combustion Theory.* University of California.

Zel'dovich, Ya. B., Barenblatt, G. I., Librovich, V. B., & Makhviladze, G. M. (1980). *Mathematical theory of combustion and explosion.* Nauka.

Zeldovich, Ya. B., & Frank-Kamenetsky, D. A. (1938). Theory of thermal propagation of the flame. *Journal of Physical Chemistry, 12*(1), 100–105.

Chapter 2
Real Wave Structures in the Excitable Media

ABSTRACT

The available observational data and modern experimental studies of the structure of waves in various excitable media, including biological populations and chemical mixtures, are analyzed. The results of experimental studies of both isothermal waves of chemical reactions and non-isothermal waves of chemical reactions with different activation energies, including strongly activated chemical reactions, are considered. The analysis is caused by the need to critically review the widely used asymptotic method for constructing models and the calculations of wave processes based on this method in nature and technology.

INTRODUCTION

Describing the behavior of animals, R. Chauvin noted their social nature, the desire to form packs (Chauvin, 1963). This desire is especially strong in insects, which tend to spread over the territory in the form of waves. Combustion processes, for example, fires or detonation waves, also spread through the territory in waves. In this chapter, we will consider some observational or experimental data on the propagation of waves in excitable media, and the theoretical description of these waves will be referred to subsequent chapters.

WAVES OF INSECTS.

A striking example of wave propagation in excitable media is the flight of gregarious locusts. The famous entomologist B.P. Uvarov in the 20s of the last centuries showed (Bei-Bienko, 1970), that the phenomenon of so-called phase variability is

DOI: 10.4018/978-1-6684-7745-8.ch002

typical for locusts. Depending on the density of the group and the concentration of food products, either single forms or gregarious forms of larvae are formed. The solitary form forms when there is enough food. This form is characterized by discreet coloration and calm behavior. With a lack of food and high crowding, a gregarious form of locusts arises (excited). This form is characterized by bright coloring, great mobility and aggressiveness. Locust individuals unite in large communities and fly over the Earth for long distances in search of food. Following the concept of S.S. Chetverikov, the process of locust spread can be considered as the spread of living matter in space in the form of "waves of life". During the flight, a swarm of locusts periodically stops, descending to the Earth and eating vegetation over large areas. Part of the locust individuals die for various reasons in flight however, this loss is compensated by joining the swarm of locust individuals located in the overflying area. Thus, the structure of the "wave of life" is a stationary soliton - a solitary wave, i.e. the graph of the concentration of individuals in a flock has the form (Zhizhin, 2014) of a curve with a maximum. It is obvious that in this case it is simply not possible to speak about the approximation of a narrow reaction zone.

Figure 1. Concentration profile in a solitary wave of a locust population

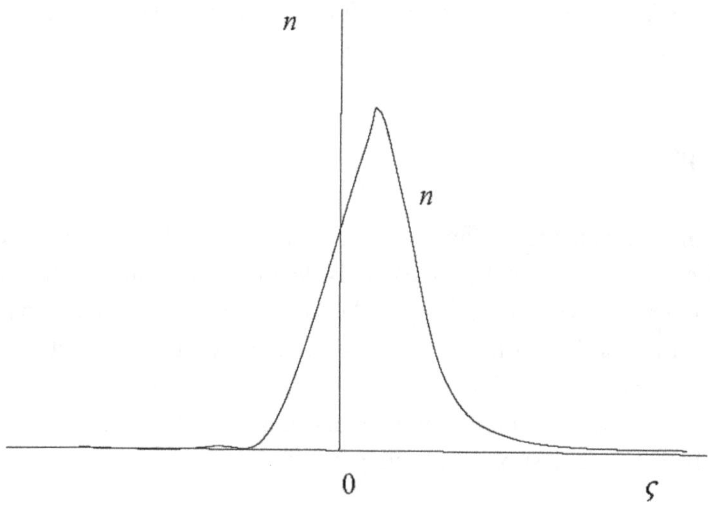

It should be noted that there is another form of flight of gregarious locusts, which is not yet amenable to mathematical description. This is the flight of locusts over very long distances without stopping and without feeding the swarm of locusts from the environment. For example, in 1954, these insects from Northwest Africa reached the British Isles, covering a distance of 2400 km. At the same time, the flight of locusts took place at an altitude of 2000m. Apparently, in this case, the non-stationary mode of flight of the gregarious locust is realized.

Beetles - bark beetles cause enormous damage to forests on Earth, biting into the bark of trees, causing them to dry out. Detailed information about forest insects and their stages of development is contained in the monograph (Zhizhin and Selikhovkin, 2012). These insects also spread in waves (Zhizhin, 2011).

Let us now turn to the analysis of data on the propagation of waves of chemical reactions.

THE WAVES OF ISOTHERMAL CHEMICAL REACTIONS.

Let us first consider isothermal chemical reactions, i.e. in cases where the cause of the propagation of a wave of a chemical reaction is the diffusion of molecules. Widely known examples of such waves are autowave processes in Belousov-Zhabotinsky reactions. Moreover, these reactions are accompanied by a change in the color of the liquid and are easily observed. Figure 2 shows one of the photographs of the self-oscillatory oxidation reaction with malonic acid bromate with the formation of spiral waves (Zaikin, Zhabotinsky, 1977).

Figure 2. Spiral waves of malonic acid oxidation

Figure 2 clearly shows that the chemical reaction zones have a finite extent.

THE WAVES OF RADICAL POLYMERIZATION

Let us now consider data on the propagation of waves of exothermic chemical reactions. An important example of a wave in excitable media, which is of practical importance, is the propagation of polymerization waves. The first experiments that confirmed the possibility of carrying out the polymerization reaction using the propagation of reaction waves initiated by a thermal pulse were carried out under the guidance of Academician N.S. Enikolopyan (Chechilo, Khvilivitsky, Enikolopyan, 1972). The study was carried out on a liquid mixture of methyl methacrylate (MMA) with a low molecular weight initiator (benzoyl peroxide) at pressures up to 5 10* Pa. Observations of the propagation of the polymerization reaction were carried out using copper-constantan micro thermocouples immovably installed over the volume of a vertical cylindrical ampoule. The upper section of the ampoule was heated by electric current. The measurements showed the practical absence of a temperature gradient in the cross section of the ampoule during the passage of the polymerization wave. This indicates that adiabatic propagation of the polymerization wave occurred under the experimental conditions and that the polymerization wave can be considered flat. Based on a comparison of the diffusion and thermal conductivity coefficients, a conclusion was made about the thermal nature of the polymerization wave propagation. Figure 3 shows a potentiometric record of measuring the temperature of the mixture over time at two different points along the axis of the ampoule, which correspond to the propagation of the MMA polymerization wave in the presence of 0.4 g of benzoyl peroxide per 100 ml of the mixture at a pressure of 3.5 10^8 Pa and an initial temperature of 295 K.

Figure 3. Change in temperature of the mixture of MMA, benzoyl peroxide in time in two sections of the ampoule

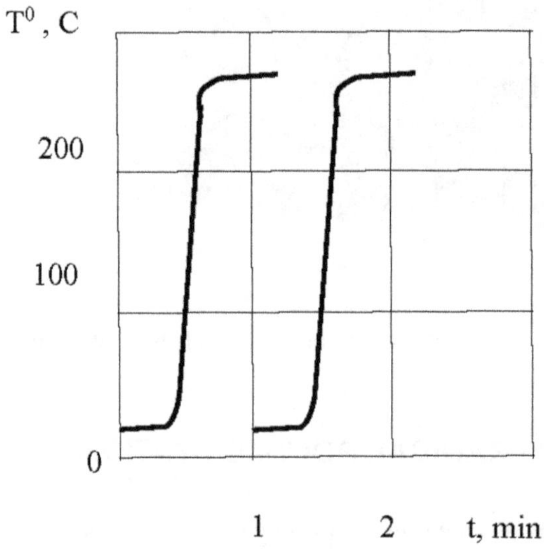

The experimentally discovered phenomenon of layer-by-layer propagation of the polymerization reaction was later called frontal polymerization.

The effect of the type of initiator on the structure of the MMA radical polymerization wave was studied (Chchilo and Enikolopyan, 1975). Figure 2.4 shows graphs of the structure of the polymerization wave at a pressure of 3.5 10^8 Pa, an initial temperature of 295 K and the same concentration of 0.0165 mol / l of various initiators: dicyclohexylperoxydicarbonate (CPC), benzoyl peroxide (PB), ditretbutyl peroxide (DTBP) (Bagdasaryan, 1966).

Figure 4. Dependence of temperature on the spatial coordinate in the polymerization wave with various initiators with a concentration of 0.0165 mol/l: 1 - CPC, 2 - PB, 3 - DTBP

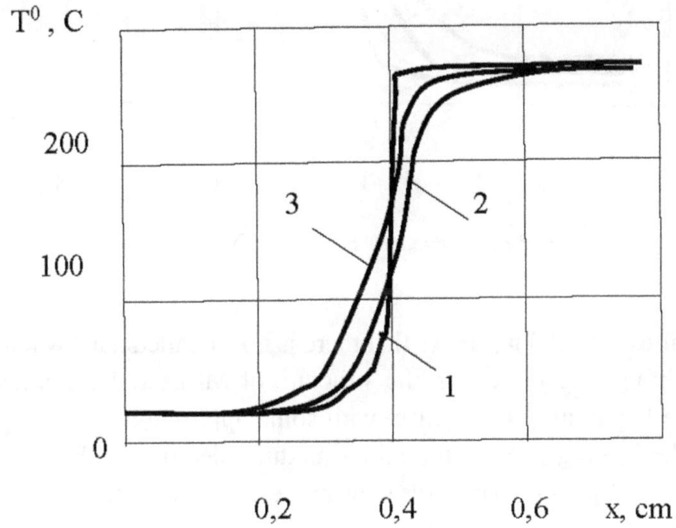

The effect of pressure was studied (Chechilo, Enikolopyan, 1976) on the structure of the radical polymerization wave (Figure 5).

Figure 5. Dependence of temperature on the spatial coordinate in the polymerization wave at a CPC concentration of 0.165 mol/l, T_0=295 K and various pressures

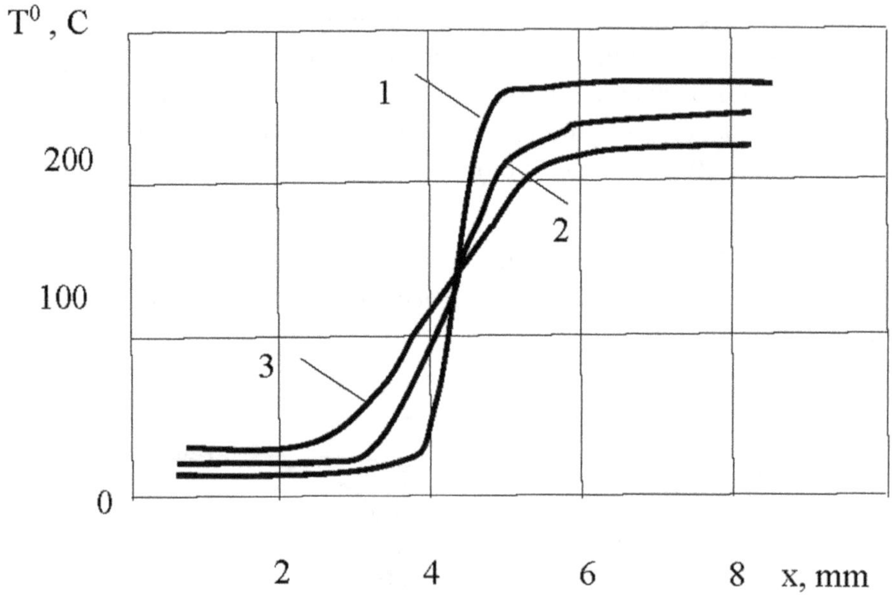

1 - P = 2,5 10^8 Pa, 2 – P= 3 10^8 Pa, 3 – P=3,5 10^8 Pa.

From Figures 2.3 - 2.5 it follows that there is no so-called narrow reaction zone asymptotically tending to zero in the structure of MMA radical polymerization waves. At the beginning of the wave, with some approximation, it is possible to distinguish the heating zone of the initial mixture, determined by the Michelson solution (Mikhelson, 1930). Here, the main process is the transfer of heat by thermal conduction. This zone by no means fills the entire profile of the radical polymerization wave. In the rest of the wave, chemical reactions take place. In addition, in the polymerization wave one can clearly see an extended zone at the end of the wave, which corresponds to the slow decay of the reaction.

In the considered experimental works on radical polymerization, the reaction medium in the ampoules was immobile, and the polymerization wave moved along the immobile medium. An industrial implementation of such a physicochemical phenomenon would correspond to a batch frontal polymerization reactor. Of significant practical interest for the continuous production of polymers is the continuous frontal polymerization reactor. In such a reactor, under stationary conditions, the polymerization wave must be motionless, and the medium must move towards the wave. However, in this process, a significant role is played by the hydrodynamic features of the movement of the medium associated with a sharp change in viscosity during the reaction and sticking of the polymer to the reactor walls. Attempts to implement such a process in a tubular reactor revealed a sharp curvature of the wave profile and, moreover, the formation of a breakdown of the reactor, i.e., a jet of the initial mixture along the axis of the pipe (Butakov, Zanin, 1978; Malkin, Zhirkov, 1984). The author found a way out of this situation. He proposed to create a reactor in which the walls would not have a significant effect on the polymerization process. So, under his leadership, reactors of a special design with a stationary cylindrical reaction front and a stationary spherical polymerization reaction front were created and tested (Zhizhin, 1995). It is obvious that the features of a flat reaction front, associated with a large length of the reaction zone, are also preserved in the cylindrical and spherical front of the polymerization reaction.

COMBUSTION WAVES OF ACTIVATED CHEMICAL REACTIONS

Let us turn to the analysis of data on the structure of combustion waves with activated chemical reactions, i.e. with high activation energies.

In the book of Gaydon A.G., Wolfgard X.G. (Gaydod, Wolfgard, 1959) shows photographs of a flame in a mixed ethylene-air mixture with an excess and a lack of air, as well as an ethylene flame in air in the absence of preparation of the initial mixed mixture of ethylene with air.

This shows how different a gas flame can look. In this case, it is difficult to single out the region of the flame where the actual chemical reactions take place. Apparently, in the brightest places of the flame, chemical reactions occur. But firstly, whether chemical reactions take place only in these places, and, secondly, what area in space is occupied by at least these brightest places, remains unclear.

The results of measurements of the concentrations of components in a flat gas flame obtained using the method of molecular beam mass spectrometry (Westmoreland et al., 2006) can serve as direct experimental proof of the distribution of the zone of chemical reactions in a flame. Figure 6 shows the results of these measurements for a flame in an ethylene-oxygen stoichiometric mixture.

Figure 6. Temperature profiles and mole fractions of the main components in a flat flame of SRRS at a pressure of 2.666 kPa, measured by molecular beam mass spectrometry with electron impact ionization

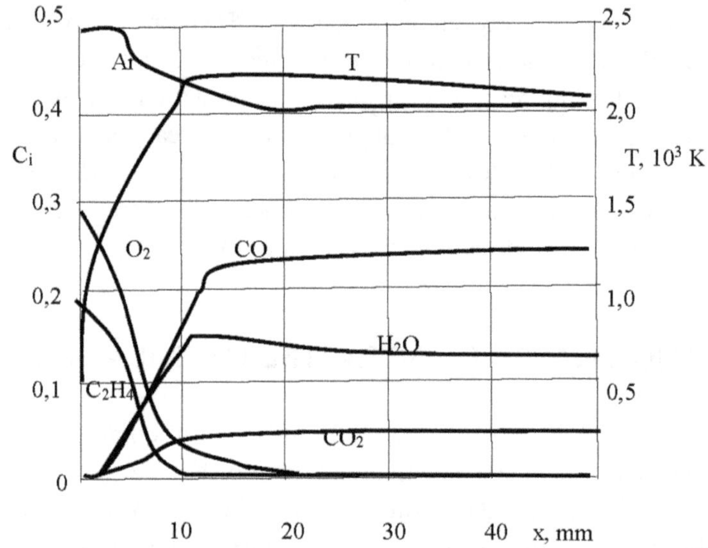

(University of Massachusetts Amherst, USA)

Figure 6 clearly shows that the change in the concentrations of the components occurs almost throughout the entire length of the combustion wave.

A detailed measurement of component concentrations and temperature was carried out in a laminar flame of a mixture of acylene, oxygen, argon at the University of Stuttgart (Warnatz et al., 1983). Figure 7 shows the results of these measurements.

Figure 7. Mole fractions (measured by mass spectrometry) and temperatures (determined by the sodium line inversion method) in a flat laminar flame of a premixed low pressure acylene-oxygen-argon mixture

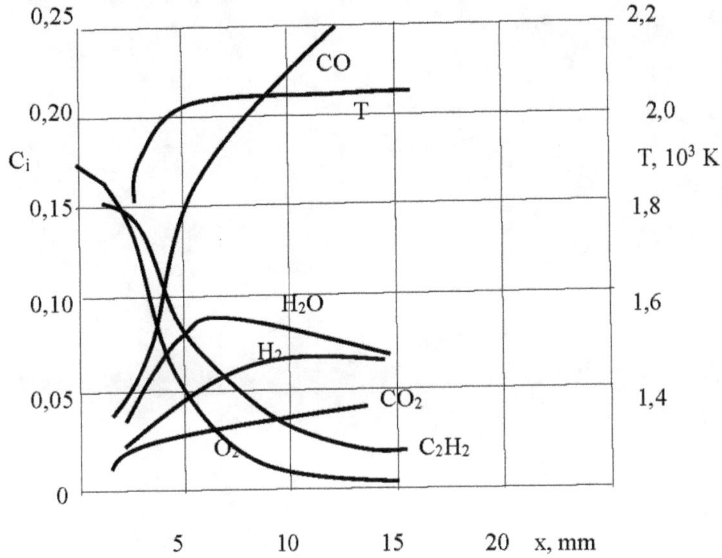

Figure 7 also convincingly proves that the change in the concentrations of constituents in the flame occurs throughout the flame.

CONCLUSION

The performed analysis of observational and experimental data indicates that many processes in excitable media in nature and technology propagate in a wave mode. Unfortunately, it should be recognized that many researchers, when analyzing the propagation of non-isothermal waves of chemical reactions (especially in Russia), were incorrectly oriented by the works of Academicians Ya.B. Zeldovich and D.A. Frank-Kamenetsky, who proposed the asymptotic method of a narrow reaction zone, assuming that the entire thickness of the flame is occupied by the heating zone, and the length of the chemical reaction zone is practically equal to zero. Experimental studies of the flame using modern methods (mass spectrometry, etc.) have shown that changes in the concentrations of the mixture components, even in highly activated reactions, occur almost throughout the entire thickness of the flame, and it is impossible to apply the asymptotic approximation of a narrow reaction zone.

In this regard, there is a need to develop non-asymptotic methods for analyzing waves in excitable media. This is the subject of the subsequent chapters of this work.

REFERENCES

Bagdasaryan, Kh. S. (1966). *Theory of radical polymerization*. Nauka.

Bei-Bienko, G. Ya. (1970). B.P. Uvarov (1889-1970) and his contribution to science and practice. *Entomological Review*, *49*(4), 915–922.

Butakov, A.A., & Zanin, A.M. (1978). Experimental study of the regularities of the flow of an exothermic reaction with a changing viscosity of the reagent in a tubular reactor. *Physics of combustion and explosion,* 14(5), 91-95.

Chauvin, R. (1963). *Les Socie'tes Animales. De L'Abeille av Gorille*. PLON.

Chechilo, N. M., & Enikolopyan, N. S. (1974). On the structure of the polymerization wave front and the mechanism of propagation of the polymerization reaction. *Reports AN USSR, 214*(5), 1131–1133.

Chechilo, N. M., & Enikolopyan, N. S. (1975). Influence of the concentration and nature of initiators on the propagation of the polymerization reaction. *Reports AN USSR, 221*(5), 1140–1142.

Chechilo, N. M., & Enikolopyan, N. S. (1976). On the role of pressure and initial temperature of the reaction mixture during the propagation of the polymerization reaction. *Reports AN USSR, 230*(1), 160–163.

Chechilo, N. M., Khvilivitsky, R. Ya., & Enikolopyan, N. S. (1972). On the phenomenon of propagation of the polymerization reaction. *Reports AN USSR, 204*(5), 1180–1181.

Chetverikov, S. S. (1905). Waves of life. From lepidopetrological observations for the summer of 1903. Diary of the Zoological Department of the Imperial Society of Lovers of Natural Science. *Anthropology and Ethnography*, *3*(6), 106–111.

Geidon, A. G., & Wolfgard, H. G. (1959). *Flame and its structure, radiation and temperature*. Metallurgizdat.

Malkin, A. Ya., & Zhirkov, P. V. (1984). Experimental investigation of flow in a tubular polymerization reactor. *Polym. Prok. Eng*, *2*(213), 207–218.

Michelson, V. A. (1930). *On the normal rate of ignition of explosive gas mixtures. Collected works. 1*. New Agronom.

Warnatz, J. (1983). The mechanism of high temperature combustion of propane and butane. *Combustion Science and Technology*, *34*(1-6), 177–200. DOI: 10.1080/00102208308923692

Warnatz, J., Maas, U., & Dibble, R. W. (2001). *Combustion. Physical and Chemical Fundamentals, Modeling, and Simulations, Experiments, Pollutant formation.* Springer.

Westmoreland, F. R. et al. (2006). Study of the flame structure by molecular beam spectrometry using electron impact ionization and photon ionization by synchrotron radiation. *Physics of combustion and explosion,* 42(6), 58-63.

Zaikin, A. I., & Zhabotinsky, A. M. (1977). Autowave processes in active concentration media. In Frank, G. M. (Ed.), *Bophysics of complex systems and radiation disturbances* (pp. 107–112). Science.

Zhizhin, G. V. (1995). Polymerization method. RF patent 2044746 (BI 27).

Zhizhin, G. V. (2011). Mathematical model of a solitary wave of a population of bark beetles. *Proceedings of the St. Petersburg State Forest Engineering Academy,* 196, 153-162.

Zhizhin, G. V. (2014). Mathematical model of the flight of gregarious locusts. *Biosphere*, *6*(2), 112–117.

Zhizhin, G. V., & Selikhovkin, A. V. (2012). *Mathematical modeling of the development and distribution of stem insect populations in the forests of Russia.* St. Petersburg State Forestry Technical University.

Chapter 3
Non-Asymptotic Methods for Describing Waves in Excitable Media

ABSTRACT

In the previous two chapters, it was shown that the narrow reaction zone method, which asymptotically assumes the length of the chemical reaction zone in the wave to be equal to zero, strictly speaking, contradicts the equations of conservation of energy and mass in the wave and modern experimental data proving that the concentrations of reagents change throughout the wave. In this chapter, non-asymptotic methods for calculating the structure and velocity of waves in excitable media are presented in a rather general form, taking into account the passage of kinetic processes in zones of infinite extent.

INTRODUCTION

Chemical reactions accompanied by the release of a large amount of heat are characterized by a non-linear dependence of the reaction rate on temperature. This leads to the possibility of propagation of waves of exothermic chemical reactions in space due to the transfer of heat by thermal conduction towards the cold reaction medium. To calculate the propagation velocity of such essentially non-isothermal chemical reactions, the narrow reaction zone method was proposed (Zeldovich, Frank-Kamenetsky, 1938), which assumed that asymptotically, due to the sharply nonlinear dependence of the reaction on temperature, the length of the chemical reaction zone in the wave can be considered equal to zero. However, modern methods for measuring the concentration of substances using molecular beam mass spectrometry (Warnatz, Maas, Dibble, 2001; Westmoreland et al., 2006) have

DOI: 10.4018/978-1-6684-7745-8.ch003

shown that in the flame of even highly activated chemical reactions, the change in the concentration of substances occurs over a significant length of the combustion wave, i. e. even for reactions with a large heat release, the asymptotic method of a narrow reaction zone is not applicable. In 1982, for the first time (Zhizhin, 1982), a new mathematical model of waves of exothermic chemical reactions was proposed, according to which the length of the chemical reaction zone, as well as the heating zone, is assumed to be semi-infinite. In general, the assumption of a semi-infinite reaction zone looks more natural than the assumption of a zero extension of the reaction zone, since it takes into account the stage of slow decay of a chemical reaction, which is important for any technological process. In further works of the author (Zhizhin, 1992, 1997, 2004) it was shown that such an assumption makes it possible to obtain an approximate solution for many problems of the propagation of waves of exothermic chemical reactions and that these solutions are in good agreement both with the results of experiments and with the results of computer calculations of non-stationary differential equations. The developed method is called the semi-infinite reaction zone method. The efficiency of the method was demonstrated by its application to the analysis of waves of exothermic reactions in various media: liquid, gaseous, dispersed, porous. In this chapter, the semi-infinite reaction zone method is described in a fairly general form for a condensed medium with an arbitrary heat source and a chemical reaction of the order of m, as well as for chemical exothermic reactions in a gas.

In excitable media, waves can occur that are not related to the release of heat in a chemical reaction. The reason for the formation of waves in these media is either the diffusion of molecules, or the mobility of living objects in the medium. The process of formation of these waves is described by non-stationary parabolic equations with a source. For the first time, this process was considered in general form in the work of A.N. Kolmogorov, I.G. Petrovsky, N.S. Piskunov (Kolmogorov, Petrovsky, Piskunov, 1937). As an application, this problem has been associated with the spread of dominant genes. In the work of Kolmogorov A. N., Petrovsky I.G., Piskunov N.S. under a number of assumptions about the initial distribution of the substance concentration over the spatial variable, as a result of solving the non-stationary equation, the value of the velocity of stationary wave propagation is found and it is shown that this value is the minimum of all possible velocity values. However, as was later shown (Samarskii et al. 1987), for other initial distributions of matter, the nonstationary diffusion equation can lead to other values of the wave velocity. Naturally, there is no reason to accept the value of the wave propagation velocity obtained by A.N. Kolmogorov and others under other conditions of the problem and, moreover, under other diffusion equations. The works of some authors who ignore this fundamental remark should be considered incorrect (Svirezhev, 1987; Ivanov, 1984; Kovalev, Vechernin, 1986; Berezovskaya, Karev, 1999). At the same time, the number of new mathematical models of wave propagation in excitable media is constantly growing. This includes the propagation of nerve impulses, attempts to explain the phenomenon of cardiac fibrillation, spin modes of wave propagation, etc. Research on the diffusion equation continues (for example, Volpert, 1987; Krinsky, 1968). However, there are no solutions for the velocity and structure of a wave with diffusion transfer in these works. It should be noted that, starting from the work of Kolmogorov, Petrovsky, Piskunov, many people are attracted by the problem of the solution of the non-stationary diffusion equation tending to a stationary solution as time tends to infinity. Here we are talking about the asymptotic output of the solution of a non-stationary equation to a stationary solution. In this case, the stationary solution itself is not considered. There are no methods for its consideration, except for numerical, given that kinetic processes occur in this solution over the entire wavelength. This does not allow one to obtain analytical expressions for the speed and structure of the wave in some approximation, and therefore does not allow one to study the dependence of these characteristics on the parameters of the problem. In this monograph, the asymptotic yield of the solution of the non-stationary diffusion equation to the stationary solution is considered for the waves chemical reactions of order m in condensed faces.

In 2001, for the first time, a method was proposed for an approximate determination of the speed and structure of a wave of a diffusion nature (Zhizhin, 2001). This method is called the method of infinite reaction zone, since it takes into account kinetic processes throughout the entire length of the wave. In this chapter, the model of an isothermal flame (a cold flame with first-order autocatalysis) is considered by the infinite reaction zone method. In addition, an analytical description of the wave system was obtained using the infinite reaction zone method, which was previously impossible.

3.1. COMBUSTION WAVES IN THE CONDENSED PHASE WITH AN ARBITRARY SOURCE

Let us consider the procedure for applying the semi-infinite reaction zone method to solve the problem of combustion wave propagation in a condensed phase with a rather arbitrary heat release function.

The system of non-stationary equations describing in a one-dimensional setting the change in temperature and the mass fraction of the reagent in the condensed phase (diffusion is neglected) has the form

$$c_p \rho \frac{\partial T}{\partial t} = \frac{\partial}{\partial x} \lambda \frac{\partial T}{\partial x} + \rho Q W, \tag{3.1}$$

$$\frac{\partial C}{\partial t} = -W, W = k_0\varphi'(C)\phi'(T). \tag{3.2}$$

There λ is the thermal conductivity; ρ is the density; c_p is the specific heat capacity of the medium (in the case of a condensed medium, these parameters can be considered constant); x is the spatial coordinate oriented along the propagation of the wave; Q is the specific thermal effect; t is the time; W is the reaction rate, which depends on the mass fraction of the reagent and temperature, represented as a product of functions of these variables $\varphi'(C)$and $\phi'(T)$; k_0 is the proportionality factor.

Introducing the wave coordinate $z=ut-x$, where u is the wave propagation velocity, we can pass from system (3.1), (3.2) to the system of ordinary differential equations

$$c_p\rho u\frac{dT}{dz} = \frac{d}{dz}\lambda\frac{dT}{dx} + \rho QW, \tag{3.3}$$

$$u\frac{dC}{dt} = -W. \tag{3.4}$$

It is assumed that $\varphi'(C)$ vanishes when the initial reagent is completely burned out (C=0). The desired wave solution must satisfy the boundary conditions, which consist in the tendency of the derivatives dT/dz, dC/dz to zero at $z\rightarrow\pm\infty$ (self-regulating waves are considered). We write equations (3.3), (3.4) and the boundary conditions for them in a dimensionless form, using the quantities accepted in the combustion theory (Chapter 1)

$$w^2\frac{d^2\eta}{d\zeta^2} - w^2\frac{d\eta}{d\zeta} + \frac{\varphi(n)\phi(\eta)}{C_0} = 0, \tag{3.5}$$

$$w^2\frac{dn}{d\zeta} + \frac{\varphi(n)\phi(\eta)}{C_0} = 0, \tag{3.6}$$

$$\eta = \frac{T-T_+}{T_+-T_0}, n = \frac{C}{C_0}, \zeta = \frac{z\rho u c_p}{\lambda}, -1 \le \eta \le 0, w = u\left(\frac{\rho c_p}{\lambda k_0}\right)^{1/2}, \varphi(n) = \varphi'(C_0 n),$$

$$\eta(-\infty) = -1, \eta(+\infty) = 0, \phi(\eta) = \phi'[\eta(T_+ - T_0)+T_0], C_0 = c_p\frac{T_+ - T_0}{Q},$$

$$0 \le n \le 1, n(-\infty) = 0, n(+\infty) = 0, \frac{d\eta}{d\zeta}(\pm\infty) = \frac{dn}{d\zeta}(\pm\infty) = 0, \varphi(0) = 0.$$

Here T_+ is the temperature of the medium at the end of the wave, T_0 is the initial temperature of the medium, C_0 is the initial mass fraction of the reagent.

Equation (3.5), taking into account equation (3.6), is integrated once

$$\frac{d\eta}{d\zeta} = \eta + n. \tag{3.7}$$

Solutions to the system of equations (3.6), (3.7) can be represented by trajectories in the phase plane (η,n) (Figure 1).

Figure 1. The phase plane of system (3.6), (3.7)

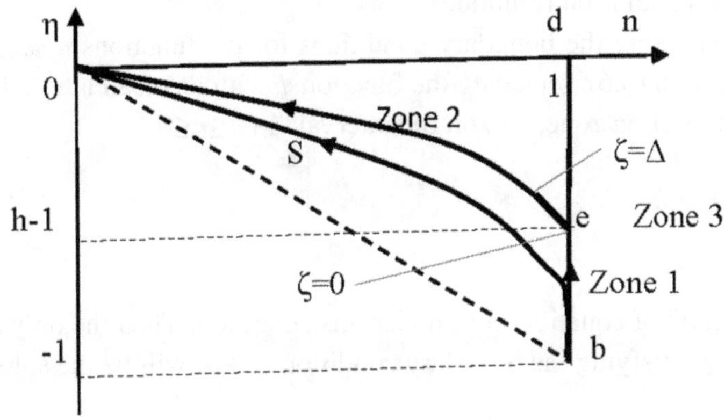

The dashed line in Figure 1 is the null isocline $\eta+n=0$, where $d\eta/d\zeta=0$. The point 0 with coordinates $\eta=0$, $n=0$ is a singular point (final equilibrium position), since the axis $n = 0$ is the zero isocline $dn/d\zeta=0$. Point b with coordinates $\eta=-1$, $n=1$ is not an equilibrium position, since, strictly speaking, the derivative $dn/d\zeta$ at it is not equal to zero. But at the same time, the system of equations (3.6), (3.7) does not have solutions satisfying the set boundary conditions necessary for the existence of the wave. At the same time, it is well known that exothermic chemical reactions propagate in waves. To resolve this contradiction, it is necessary to set the function $\phi(\eta)$ to zero at the point b. Then the desired wave solution is represented by the trajectory S (Figure 1), going from the initial equilibrium position b to the point o, i. e. the final equilibrium position. In the narrow reaction zone method, the trajectory S is modeled by a step bdo, i.e., in addition to the fact that the function is assumed to be equal to zero at point b, it is assumed that this function to be equal to zero on the segment bd passing through point b parallel to the axis η. This corresponds to the assumption that there is no chemical reaction in the initial temperature T_0 wave up to the maximum temperature T_+. A model closer to the wave solution S is the model proposed by the semi-infinite reaction zone method. In this method, the function $\phi(\eta)$ is set to zero not on the entire segment bd, but only on its part be. This part corresponds to the heating zone (zone 1, $-\infty\leq\zeta\leq0$), i. e. a change in temperature from the initial temperature $T_0(\zeta= -\infty)$ to a certain temperature of the reaction front $T_f<T_+$, which must be determined later . The value of the wave coordinate at the reaction front is assumed to be zero $\zeta= -\infty$. The relative length of the segment be

(source trim value) is $h = (T_f - T_0)/(T_+ - T_0)$. After point e, the process develops along the line eo, corresponding to the chemical reaction zone (zone 2, $0<\zeta<+\infty$). Both the temperature and the mass fraction of the reagent change along it. The patterns of these changes are determined by equations (3.6), (3.7). For given kinetic and initial parameters, the only values of the unknown parameters w and h will be found, hence the desired wave solution is unique.

Without violating the boundary conditions for the functions n,ϕ,η,φ, we will integrate equation (3.6), replacing the function ϕ with its mean integral value over the chemical reaction zone, i. e. on the interval $-1+h\leq\eta\leq 0$

Let

$$n = f(\zeta; h, w^2) \tag{3.9}$$

is the integral of equation (3.6) under this condition. Then the only solution of equation (3.7) satisfying the boundary condition at $+\infty$ will be the solution

$$\eta = \exp(\zeta) \int f(\zeta; h, w^2) \exp(-\zeta) d\zeta. \tag{3.10}$$

The fundamental point for obtaining an analytical solution in the semi-infinite reaction zone method is the introduction between the heating zone and the chemical reaction zone (Figure 3.1) of a certain buffer zone (zone 3, $0\leq\zeta\leq\Delta$) whose length Δ along the coordinate ζ is small and in the future after resolving the algebraic system of equations obtained as a result integration over zones tends to zero. Consider the integrals (3.9), (3.10) in this buffer region. Using equation (3.7), we write down the values of the derivative at the boundaries of the buffer zone

$$\frac{d\eta}{d\zeta}(\zeta = 0) = \eta(\zeta = 0) + n(\zeta = 0) = h - 1 + 1 = h,$$

$$\frac{d\eta}{d\zeta}(\zeta = \Delta) = \eta(\zeta = \Delta) + n(\zeta = \Delta).$$

Introducing the smallness of Δ functions $\eta(\zeta), n(\zeta)$ can be considered linear. Due to linearity, the derivatives $d\eta/d\zeta$ at the boundaries of the buffer zone must be equal, i.e., we have

$$h = \eta_\Delta + n_\Delta. \tag{3.11}$$

We write (3.10) in the Cauchy form. Then,

$$\eta_\Delta = -\exp(\Delta) \int_\Delta^\infty f(\tau; h, w^2) \exp(-\tau) d\tau, n_\Delta = f(\Delta; h, w^2). \tag{3.12}$$

From equations (3.5), (3.6), taking into account the linearity of the change in variables $\eta(\zeta),n(\zeta)$ in the buffer zone, we have

$$n_\Delta = -\frac{1}{C_0 w^2}\varphi\left(\frac{1+n_\Delta}{2}\right)\phi\left(\frac{\eta_\Delta + h - 1}{2}\right)\Delta + 1, \tag{3.13}$$

$$\eta_\Delta = \Delta h - 1 + h. \tag{3.14}$$

Eliminate η_Δ from (3.14) using (3.12)

$$\Delta h - 1 + h = -\exp(\Delta)\int_\Delta^\infty f(\tau; h, w^2)\exp(-\tau)\,d\tau, \tag{3.15}$$

We also eliminate η_Δ from the left side of (3.13) using (3.11), (3.14), canceling by Δ, which is not yet equal to zero,

$$h = \frac{1}{C_0 w^2}\varphi\left(\frac{1+n_\Delta}{2}\right)\phi\left(\frac{\eta_\Delta + h - 1}{2}\right). \tag{3.16}$$

Letting Δ tend to zero in (3.15), (3.16), we obtain

$$1 - h = \int_0^\infty f(\tau; h, w^2)\exp(-\tau)\,d\tau, \tag{3.17}$$

$$w^2 = (hC_0)^{-1}\varphi(1)\phi(h-1). \tag{3.18}$$

Equations (3.17), (3.18) define the desired values of the value of h (i. e, the temperature of the wave front) and the wave velocity. At the found values w^2,h, the integral (3.9) determines the profile of the mass fraction of the reagent in the reaction zone (in the heating zone $C = C_0$), and the integral (3.10) determines the temperature profile in the reaction zone. Assuming ϕ=0, using the boundary conditions, from (3.5) we find the temperature profile in the heating zone:

$$\eta = -1 + (h-1)\exp(\zeta). \tag{3.19}$$

It is easy to see that if we do not introduce a buffer zone, i.e. a priori set its length equal to zero, then as a result, by equating the derivatives $d\eta/d\zeta$ on both sides of the reaction front, we will have only one equation (3.17) for two unknown quantities w^2,h. Thus, the introduction of a buffer zone makes it possible to obtain a closed system of equations. The absence of a continuous transition at $\Delta\to 0$ indicates that the system of equations is singular with respect to the small parameter Δ. This situation can be explained using the concepts of non-standard analysis (Zhizhin, 1999). According to the concepts of non-standard analysis (Uspensky, 1987; Zvonkin, Shubin, 1984; Albeverio et al., 1990), applicable to various kinds of singular equations, the real number "zero" has a neighborhood of hyperreal numbers that do not vanish. Therefore, there is no continuous transition to the zero value of 6 in the region of real numbers.

Note that the semi-infinite reaction zone method essentially replaced the strong discontinuity at the reaction front in the narrow reaction zone method (discontinuity in the values of the relative mass fraction of the reagent n from 1 to 0) with a weak discontinuity - the discontinuity in the values of the derivative of the mass fraction of the reactant at the front.

3.2. THE WAVE OF A CHEMICAL REACTION OF ORDER *M* IN THE CONDENSED PHASE

Let a wave of a chemical reaction of order m propagate in a condensed medium, i. e. the functions $\varphi'(C)$and $\phi'(T)$ in (3.2) have the form

$$\varphi'(C) = C^m, \phi'(T) = \exp(-E/RT). \tag{3.20}$$

Here E is the activation energy, R is the universal gas constant.

Then

$$\varphi(n) = C_0 n^m, \phi(\eta) = \exp\left(-\frac{\beta}{\frac{\eta}{1+\sigma}+1}\right), f(\zeta; w^2, h) = \exp\left(-\frac{\langle\phi\rangle}{w^2}\zeta\right), \tag{3. 21}$$

$$\beta = \frac{E}{RT_+}, \sigma = \frac{T_0}{T_+ - T_0}.$$

The expression for $\langle\phi\rangle$ in accordance with (3.8) can be obtained using the expansion exp(-E/RT) (Frank-Kamenetsky, 1967)

$$\langle\phi\rangle = \frac{1+\sigma}{1-h}\frac{\exp(-\beta)}{\beta}\left[1 - \exp\left(\frac{h-1}{1+\sigma}\beta\right)\right]. \tag{3.22}$$

Substituting (3.21), (3.22) into (3.17), (3.18), after transformations, we obtain analytical solutions that are not reducible to each other for calculating the wave front temperature and wave velocity. If $m = 1$, then the analytical solutions have the form:

$$1 - \exp(\beta y - \beta) = \gamma^{-1}\exp\left(\beta\frac{y-1}{y}\right), \tag{3.23}$$

$$w^2 = M^2 \frac{\exp(-\beta/y)}{1+(y-1)y\beta}. \tag{3.24}$$

If $m > 1$, then the analytical solutions have the form:

$$1 - y = \beta^{-1}\gamma^{-1}\int_0^{\infty} \frac{\exp(-\tau)\,d\tau}{[(m-1)\exp(\beta/y)(1+y\beta\gamma-\beta\gamma)\tau+1]^{1/(m-1)}}, \tag{3.25}$$

$$w^2 = M^2 C_0^{m-1} \frac{\exp(-\beta/y)}{1+(y-1)y\beta}. \tag{3.26}$$

There $y = \frac{T_f}{T_+}, \gamma = \frac{R}{E}\frac{T_+^2}{T_+ - T_0}, M = \left(\frac{\lambda k_0}{\rho c_p}\right)^{1/2}.$

To verify the correctness of the semi-infinite reaction zone method, a numerical study of the system of non-stationary equations (3.1), (3.2), (3.20) was carried out under the following initial and boundary conditions: To verify the correctness of the semi-infinite reaction zone method, a numerical study of the system of non-stationary equations (3.1), (3.2), (3.20) was carried out under the following initial and boundary conditions:

$$t = 0, T = T_0, C = C_0; x = 0, T = T_+; x = +\infty, \partial T/\partial x = 0, C = C_0. \tag{3.27}$$

The numerical study was carried out until the stationary propagation of the chemical transformation wave was established (Zhizhin and Poritskaya, 1994). A purely implicit difference scheme was constructed using the control volume method. When calculating the nonlinear source term, an iteration cycle was organized with respect to temperature, and a non-uniform spatial grid adapted to the solution was used. Comparison of the results of the numerical solution of non-stationary equations and calculations by the method of the semi-infinite reaction zone was carried out according to the values of the wave velocity and its characteristic width. In the numerical solution of the nonstationary problem (3.1), (3.2), (3.20), (3.27), the characteristic wave width was determined by the formula

$$\bar{l} = l/L, l = \frac{T_+ - T_0}{(dT/dx)_{x=x_f}}, L = \left(\frac{\lambda}{\rho c_p k_0}\right)^{1/2}, \tag{3.28}$$

where the x_f coordinate corresponds to the inflection point in the temperature profile. When performing calculations using the method of a semi-infinite reaction zone, the wave width was determined by the relation

$$l = \frac{T_+ - T_0}{T_f - T_0}\frac{\lambda}{\rho u c_p}. \quad (3.29)$$

Figure 2 compares the results of numerical determination of the stationary wave velocity from the solution of a non-stationary problem (3.1), (3.2), (3.20), (3.27), with the results of calculation by the semi-infinite reaction zone method at m = 1 according to equations (3.23), (3.24).

Figure 2. Deviations of the analytical values of the wave velocity from her numerical values

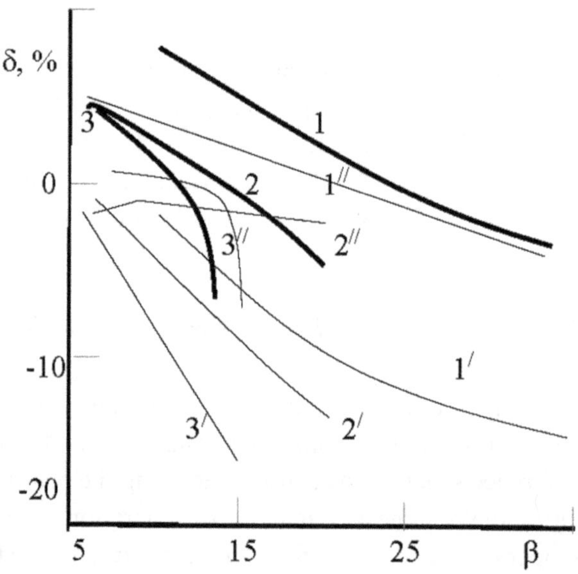

1 - 0.15, 2 - 0.175, 3 - 0.2; without a prime, the semi-infinite reaction zone method; one stroke, the narrow reaction zone method; two strokes, the method of matched asymptotic expansions.

The same figure shows comparative data on calculating the wave velocity $u' = M\gamma^{1/2}\exp(\beta/2)$ by the narrow reaction zone method (Novozhilov, 1961) and the wave velocity $u'' = u'\left(1 + \frac{\gamma\pi^2}{12} - \beta^{-1}\right)$ by the method of matched asymptotic expansions (Berman and Ryazantsev, 1972).

The quantity δ plotted along the ordinate axis was calculated as $\delta = \frac{u_a - u_n}{u_n}100$, where u_a is the velocity determined from the corresponding analytical solution, u_n is the wave velocity from the numerical solution of non-stationary equations. It can be seen from Figure 3.2 that the semi-infinite reaction zone method gives, on average, the closest wave speed values to the speed values obtained by solving non-stationary equations, compared to other analytical methods. Velocity calculations by the methods of matched asymptotic expansions and semi-infinites of the reaction zone differ slightly from each other. The larger γ, the stronger the coincidence of these methods, although the deviation from the numerical calculation increases with increasing β. The narrow reaction zone method for β > 10 over the entire range of γ gives approximately on 7,5% lower wave velocities compared to the methods of the semi-infinite reaction zone and matched asymptotic expansions, as well as the largest deviations from the numerical calculation (16%), underestimating the velocity values. Parameter values γ and β, for which the results of calculating the velocity in Figure 3.2 are shown, belong to the region of existence of a stable wave solution of the system (3.1), (3.2), (3.20), (3.27), (region I, Figure 3).

Figure 3. Region of existence of a stable nondegenerate quasi-stationary reaction wave

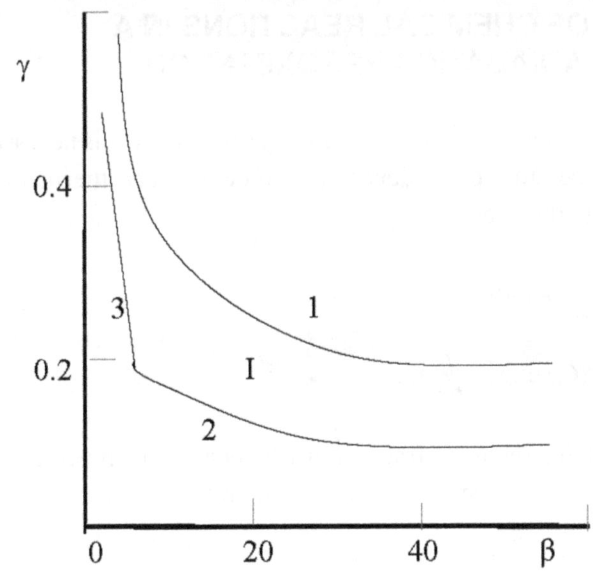

Area borders I are curves 1, 2, 3. Curve 1 corresponds to the boundary of the absence of degenerate regimes. The equation of such a curve for a reaction in the gas phase in the case of similar concentration and temperature fields has the form (Aldushin, Lugovoi, Merzhanov, 1978)

$$\gamma - 1/\beta = \Gamma, \Gamma = 0.1 \div 0.3. \quad (3.30)$$

In this paper, for the reaction in the condensed phase, it was found that condition (3.30) can be used by assuming $\Gamma = 0.13$. The other boundary $9.1\gamma - 2.5/\beta = 1$ (curve 2) corresponds to the transition to self-oscillating regimes ($3 \geq m \geq 1$) obtained at $m = 1$ (Shkadinsky, Khaikin, Merzhanov, 1971). Curve 3, $\gamma\beta - 1 = 0$, The curve that appears at small β limits the physical region corresponding to positive values of the absolute temperature.

Thus, the study showed that the semi-infinite reaction zone method correctly describes stationary waves of chemical reactions of order m in the condensed phase in the region of existence of stationary solutions. This is proved by comparing the results of calculating the speed and structure of these waves using the semi-infinite reaction zone method and the results of calculating the non-stationary equations for the propagation of waves of chemical reactions.

3.3. WAVES OF CHEMICAL REACTIONS IN A GAS IN THE ADIABATIC APPROXIMATION

When considering waves of exothermic gas reactions, it is necessary to take into account the temperature dependences of the density and thermal conductivity of the gas. We write the heat

$$c_p \rho u \frac{dT}{dz} = \frac{d}{dz} \lambda \frac{dT}{dx} + Q\Phi(T), \tag{3.31}$$

where $\Phi(T) = k_0 C\rho \exp\left(-\frac{E}{RT}\right)$,

it is believed that the mass fraction of C depends on temperature T.

This dependence is determined by the integral conservation law

$$(T - T_0) c_p / Q = C_0 - C. \tag{3.32}$$

Since the reactant burns out completely at the end of the wave, (3.32) can be written in another form

$$(T_+ - T) c_p / Q = C(T). \tag{3.33}$$

Relationship (3.33) is strictly satisfied only for the initial point, i.e. when $T = T_0$, $C = C_0$. The assumption that relation (3.33) is satisfied and for any point in the wave coincides in its meaning with the idea of the adiabatic approximation (Hacken, 1980), according to which, in this case, it is considered that the value of the mass fraction of the reagent C quickly follows temperature changes, so that the integral conservation law is approximately satisfied for any point in the wave.

It can be assumed that the thermal conductivity of the gas is proportional to the root of the temperature

Let us introduce the dimensionless wave coordinate $\zeta = z\left(\frac{c_p\rho_+ k_0}{\lambda_+}\right)^{1/2}$ and wave velocity $w = u\rho\left(\frac{c_p}{\lambda_+\rho_+ k_0}\right)^{1/2}$, where the index "+" indicates that the values of these variables are taken at the temperature T_+ at the end of the wave. Equation (3.31) taking into account (3.33) and relations $\rho/\rho_+ = T_+/T$ have the form

$$\frac{d}{d\zeta}\frac{\lambda}{\lambda_+}\frac{d\theta}{d\zeta} - w\frac{d\theta}{d\zeta} + (1-\theta)\frac{e^{-\beta/\theta}}{\theta} = 0, \tag{3.35}$$

$$\theta = T/T_+, \beta = E/(RT_+).$$

Using (3.34), we reduce equation (3.35) to a system of two first-order equations

$$\frac{d\Theta}{d\zeta} = w\Theta\theta^{-1.2} - (1-\theta)\frac{e^{-\beta/\theta}}{\theta}, \tag{3.36}$$

$$\frac{d\theta}{d\zeta} = \Theta\theta^{-1.2}. \tag{3.37}$$

System (3.36), (3.37) has two zero isoclines in the phase plane: the Θ- axis, on which $d\theta/d\zeta=0$, and the curve

$$\Theta = (1-\theta)w^{-1}\theta^{1/2}\exp(-\beta/\theta),$$

on which $d\Theta/d\zeta=0$ (see the dotted line in Figure 4).

Figure 4. Phase diagram of a wave of an exothermic chemical reaction in the gas phase

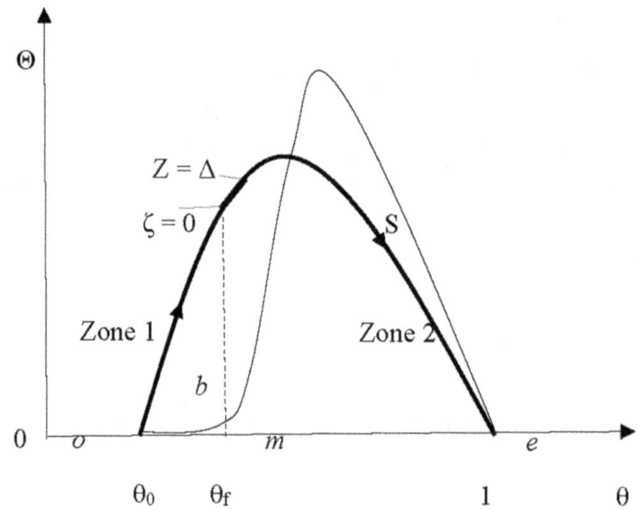

System (3.36), (3.37) has two equilibrium positions - two singular points: the origin o (Θ=0,θ=0) and point e(Θ=0,θ=1). Thus, strictly speaking, the wave solution can only leave the point o and enter the point e. However, such a solution does not have a physical meaning, since the value θ=0 corresponds to the zero value of the absolute temperature. Therefore, it is quite justified, given that the value exp(-β/θ) tends to zero as it decreases more than any degree of θ, to approximate the zero isocline (3.38) with a composite line *ombe*, consisting of a segment of the *om* axis θ, adjacent to the origin, and part of the zero isocline (3.38) *be*, connected by a vertical segment *mb* (Figure 4). Such an approximation leads to the fact that each point of the segment *om* is an equilibrium position, and from each equilibrium position of this segment there is a single trajectory and the angle of inclination for all these trajectories is the same as at point o. It is equal to w. From the discriminant of the system (3.36), (3.37) in a neighborhood of the point e we find the eigenvalues and eigendirections at this singular point:

$$\lambda_{1,2} = 0.5w \pm \sqrt{0.25w^2 + \exp(-\beta)} = (d\Theta/d\theta)_{1,2},$$

i.e. the equilibrium position e is of saddle type.

In the phase plane (Θ,θ), the desired wave solution (solid line in Figure 4) comes from some point of the segment *om* axis θ, for example, at a value of $\theta=\theta_0$, and enters the point *e* along the separatrix *S* of the saddle. Part of this wave solution from θ_0 to the value θ at the point $m(\theta=\theta_f)$, obviously, corresponds to the heating zone (zone 1,$(-\infty\leq\zeta\leq0)$ since the accepted approximation of the zero isocline (3.38) just corresponds to the cutoff of the source $\Phi(T)$ of equation (3.31) at the $T=T_f$ temperature of the reaction front ($\zeta=0$). The rest of the wave solution corresponds to the chemical reaction zone (zone 2, $0\leq\zeta\leq+\infty$).

Following the method of the semi-infinite reaction zone, we replace the value $A=\theta^{-1/2}\exp(-\beta/\theta)$ in (3.36) with its mean integral value over the reaction zone

$$\langle A\rangle = \frac{\int_0^1 \theta^{-1/2}\exp(-\beta/\theta)\,d\theta}{1-y}, y = T_f/T_+. \tag{3.39}$$

System (3.36), (3.37), taking into account (3.39), has two solutions

$$\theta^{1/2}\frac{d\theta}{d\zeta} = -\left(\frac{w}{2}+\sqrt{\frac{w^2}{4}+\langle A\rangle}\right)(1-\theta)$$

, (3.40)

$$\theta^{1/2}\frac{d\theta}{d\zeta} = \left(-\frac{w}{2}+\sqrt{\frac{w^2}{4}+\langle A\rangle}\right)(1-\theta). \tag{3.41}$$

Solution (3.40) has no physical meaning, since it corresponds to a decrease in temperature along the wave, and therefore is discarded. Equation (3.41) can be coupled again. Thus, in the reaction zone we have

$$\frac{1-\theta^{1/2}}{1+\theta^{1/2}}e^{2\theta^{1/2}} = K_1 e^{-\left(-\frac{w}{2}+\sqrt{\frac{w^2}{4}+\langle A\rangle}\right)\zeta}, K_1$$

is constant of integration. (3.42)

In the heating zone from the system (3.36), (3.37) we get

$$d\Theta/d\theta = w. \tag{3.43}$$

Since for $\Theta=0$, if $\theta = \theta_0$, then from (3.43) we have

$$\theta^{1/2}\frac{d\theta}{d\zeta} = w(\theta-\theta_0). \tag{3.44}$$

Integrating (3.44), we find the relationship between the spatial coordinate and the temperature in the heating zone

$$\left(\frac{\theta^{1/2} - \theta_0^{1/2}}{\theta^{1/2} + \theta_0^{1/2}}\right)^{\theta_0} e^{2\theta^{1/2}} = K_2 e^{w\zeta}, K_2$$

is constant of integration. (3.45)

To obtain an analytical solution, as before, we introduce a buffer zone 3 between the heating and reaction zones, the length of which will tend to zero in the future. In the buffer zone, we consider θ to be a linear function of the spatial coordinate, i. e. in equation (3.36) it is necessary to set to zero the second derivative of θ

$$\frac{d\theta}{d\zeta} = w\theta^{1/2} \pm \sqrt{w^2\theta - 2(1-\theta)\theta^{-1/2}\exp(-\beta/\theta)}. \quad (3.46)$$

Let us equate the first derivatives of temperature at ζ=0, (θ = y). Using (3.44), (3.46) we obtain

$$\frac{w(y-\theta)}{y^{1/2}} = wy^{1/2} \pm \sqrt{w^2 y - 2(1-y)y^{-1/2}\exp(-\beta/y)}. \quad (3.47)$$

From (3.47) it follows that only the sign "-" in front of the root has a physical meaning. Let's solve this equation for speed

$$w = \sqrt{2\frac{1-y}{y^2-\theta_0}y^{1/2}\exp(-\beta/y)}. \quad (3.48)$$

Let us equate the first derivatives of temperature at $\Delta \geq 0$. Using (3.41), (3.46) we get

$$\left(-\frac{w}{2} + \sqrt{\frac{w^2}{4} + \langle A\rangle}\right)\frac{1-y}{y^{1/2}} = wy^{1/2} - \sqrt{w^2 y - 2(1-y)y^{-1/2}\exp(-\beta/y)}$$

. (3.49)

Let us calculate the mean integral value A in (3.39) by the methods of perturbation theory (Naife, 1984), assuming β to be a large value

$$\int_y^1 \frac{e^{-\beta/t}}{t^{1/2}}dt \approx \frac{e^{-\beta}}{\beta} - y^{3/2}\frac{e^{-\beta/y}}{\beta}, \text{ i.e.}$$

$$\langle A\rangle = \frac{e^{\beta} - y^{3/2}e^{-\beta/y}}{\beta(1-y)}. \quad (3.50)$$

Substituting (3.50) into (3.49), after transformations, we obtain an equation for calculating y

$$\frac{e^{-\beta+\beta/y}}{y^{1/2}} = y + 2\beta\frac{1-\theta_0}{y+\theta_0}. \tag{3.51}$$

Since, according to (3.32), taking into account the equality C_+, = 0, T_+ is determined by the values T_0, C_0, equation (3.51) allows us to calculate the reaction front T_f from the initial data and, therefore, from (3.48) the wave speed. Equations (3.45), (3.42) can be used to calculate the temperature profiles in the heating zone and the reaction zone by determining the constants from the boundary condition $\theta = y, \zeta = 0$

$$K_2 = \left(\frac{y^{1/2}-\theta_0^{1/2}}{y^{1/2}+\theta_0^{1/2}}\right)^{\theta_0}\exp(2y^{1/2}),$$

$$K_1 = \left(\frac{1-y^{1/2}}{1+y^{1/2}}\right)\exp(2y^{1/2}).$$

Note that the introduction of a buffer zone led to equations (3.47), (2.44), equalities of the first derivatives of temperature at the boundaries of zone 3 instead of one equation (3.49) in the absence of this zone. This, as before, made it possible to close the system of equations for calculating the structure and velocity of an exothermic reaction wave in a gas using the semi-infinite reaction zone method.

3.4. WAVES IN EXCITABLE MEDIA, AT EACH POINT OF WHICH KINETIC PROCESSES OCCUR. INFINITE REACTION ZONE METHOD

In contrast to the waves of exothermic chemical reactions, waves can arise in excitable media, at each point of which kinetic processes occur. Examples are waves of biological populations and waves of isothermal chemical reactions. The semi-infinite reaction zone method is not suitable for their analytical description. Here it is required to create an infinite reaction zone method, i.e. a method that takes into account the flow of kinetic processes throughout the wave from minus infinity to plus infinity along the wave coordinate. Consider two simple examples of such waves.

Waves of Logistic Populations

The nonstationary equation for the change in the concentration N of individuals in the logistic population in a one - dimensional range, taking into account the chaotic mobility μ of individuals, has the form (Svirezhev, 1987)

$$\frac{\partial N}{\partial t} = \mu \frac{\partial^2 N}{\partial x^2} + F(N). \tag{3.52}$$

Here $F(N)$ is the function describing the local law of the population growth, and

$$F(N) = (B - D)N, \tag{3.53}$$

where B and D are the fertility and mortality functions.

Equation (3.53) describes the fact that the local law of the population growth is determined by two processes — birth and death. In the model of the logistic population it is assumed, that mortality D is a linearly increasing function of the concentration of individuals

$$D(N) = \alpha_1 + \alpha_2 N, \alpha_1 > 0, \alpha_2 > 0.$$

Here α_1 is natural mortality. Increase in mortality with increasing concentration N is due to increased competition with limited resources (food, space, etc.). In addition, it is assumed that the fertility function B it is determined only by the physiological limits of fertility and is independent of N, i. e. $B = m =$ constant, where m is the so - called natural fertility or fecundity. Given these assumptions and equation (3.53), equation (3.52) takes the form

$$\frac{\partial N}{\partial t} = \mu \frac{\partial^2 N}{\partial x^2} + \alpha N(1 - N/N_+). \tag{3.54}$$

Here $\alpha = m - \alpha_1$ is Malthusian parameter, and $N_+ = (m - \alpha_1)/\alpha_2$ is the capacity of the medium, i.e. maximum possible concentration of individuals in the environment.

If the competition between individuals of the population is completely absent ($\alpha_2 = 0$), $N_+ = \infty$, then the law of the local population growth takes a simple form $F(N) = \alpha N$. In this case, the population density increases indefinitely (the Malthus model of exponential growth) and the wave solution, as a transition from one equilibrium position to another, does not exist ($\alpha > 0$) .

To find the stationary wave solution of equation (3.54), one can introduce the wave coordinate $z = ut - x$, here u is the wave velocity. One also introduces dimensionless variables

$$n = N/N_+, \varsigma = z(\alpha/\mu)^{1/2}, w = u(\alpha\mu)^{-1/2}.$$

Then equation (3.54) can be written as a system

$$\frac{dp}{d\varsigma} = wp - F(n), F(n) = (1-n)n, \tag{3.55}$$

$$\frac{dn}{d\varsigma} = p. \tag{3.56}$$

Cold Flame With First - Order Auto - Catalysis

Cold (isothermal) flames can occur in various mixtures. It was observed in mixtures of hydrocarbons with air (Neiman, 1939), carbon disulfide with oxygen (Voronkov and Semyonov, 1939), fluorine with hydrogen (Fedotov and Chaikina, 1972), and nitrogen trichloride with air (Azatyan et al., 1976). Chemical oxidation reactions in a cold flame occur at relatively low temperatures with little thermal effect. The detailed mechanism of these reactions is very complex. It is known that these are chain reactions. The foundations of the theoretical description of a cold flame were laid on the example of the oxidation of carbon disulfide (Voronkov and Semyonov, 1939). A separate chapter of the monograph will be devoted to the analysis of this description. For now, let us emphasize the main features of a cold flame, which must be taken into account in its mathematical description. Firstly, the oxidation reaction in a cold flame is autocatalytic in nature, i. e. the reaction rate has a factor or factors that increase as the concentration of some intermediate reactant decreases. Secondly, due to the low temperature of a cold flame, when describing it, it is necessary to take into account the influence of a chemical reaction throughout the flame, considering it as a wave, i. e. from $-\infty$ to $+\infty$ space variable.

The non-stationary equation for the development of a cold flame in its simplest form with a first-order autocatalytic reaction has the form

$$\frac{\partial N}{\partial t} = D\frac{\partial^2 N}{\partial x^2} + k_0 N(N_+ - N). \tag{3.57}$$

Here D is the diffusion coefficient, k_0 is the rate constant of a chemical reaction, N_+ is the concentration of the substance at the end of the wave, t is time, x is spatial coordinate.

By introducing the wave coordinate $z=ut-x$, where u is the wave velocity and dimensionless quantities

$$n = N/N_+, \zeta = z(k_0 N_+/D)^{1/2}, w = u(k_0 N_+/D)^{-1/2},$$

the equation (3.57) can be written as a system (3.55), (3.56).

Approximate Analytical Solution of the Stationary Wave Equation in an Excitable Medium With Kinetic Processes Over the Entire Length of the Wave

The solutions of the autonomous system (3.55), (3.56) can be represented by trajectories in the phase plane (p,n) . Their locations are determined by the zero isocline $p=n(1-n)/w$, on which $dp/d\varsigma=0$, the zero isocline $p=0$, on which $dn/d\varsigma=0$, as well as by special points $a(0,0)$, $b(0,1)$. The eigenvalues in the special point a (the initial equilibrium position) are equal to the eigendirections at this point

$$\lambda_{1,2} = w/2 \pm \sqrt{w^2/4 - 1} = (dp/dn)_{1,2}$$

, (3.58)

and the eigenvalues at the special point b (the final equilibrium position) are equal to the eigendirections at the special point b

$$\lambda_{3,4} = w/2 \pm \sqrt{w^2/4 + 1} = (dp/dn)_{3,4}. \tag{3.59}$$

Let us consider in more detail than in the well - known work of A.N. Kolmogorov, I.G. Petrovsky, N.S. Piskunov (Kolmogorov, Petrovsky, Piskunov, 1937), the behavior of trajectories in the phase plane, omitting the already known details. From (3.59) it follows that the final equilibrium position is of a saddle type, since $\lambda_1>0$, $\lambda_2<0$.

From (3.58) it follows that the initial equilibrium position can be of the type of an unstable node with $w\geq2$ or a focus with $w<2$. Since at $w<2$ trajectories will fall into the region of negative concentrations, this area of parameters is not of interest. One confine ourselves to the values of speed $w\geq2$. Then the point a has the type of an unstable node, since $\lambda_{1,2}>0$ and has two proper directions, and both are located in the quadrant $p>0$, $n>0$. The wave solution leaves the point a and enters point b, i.e. boundary conditions must be satisfied $p=n=0$ at $\varsigma\to-\infty$ and $p=0$, $n=1$ at $\varsigma\to+\infty$. Obviously, for any value of speed, there is a trajectory emanating from point a, and entering point b (Figure 5).

Figure 5. Phase diagram of waves in the logistic population

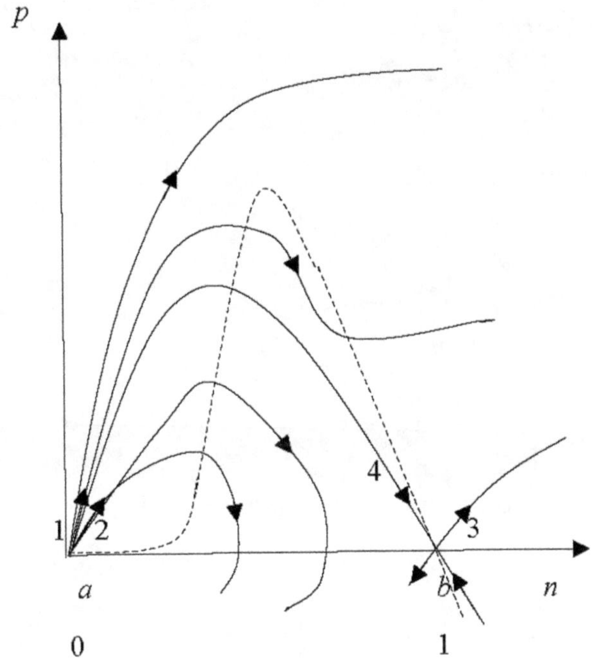

However, with which eigenvalues λ_1 or λ_2 the trajectory leaves point *a,* if it enters point *b*, is not known in advance, since both options can be. In the first case, the trajectories emanating from point *a* along its own direction 2 intersect the zero isocline $p=n\ (1-n)/w$ and, without entering point *b*, intersect the axis $p = 0$. In this case, one of the trajectories emanating from point *a* along its own direction 1 enters point *b*. In the second case, the trajectories emanating from point *a* in its own direction 1, not reaching point *b*, leave in their own direction3 points *b*. For $w=2$ the eigenvalues of the point *a* coincide; their own directions coincide, and the point *a* has the type of a degenerate unstable node. Figure 1 shows the case when the trajectory emanating from point *a* in its own direction 1 enters the special point *b* along the own direction 4. The corresponding integral curve is shown in Figure 6.

Figure 6. Wave profile of conquering the logistic population: n_f is the concentration of individuals at the wave front (ς=0), n_Δ is the concentration of individuals at the bound of zone 2 ($\varsigma=\Delta$)

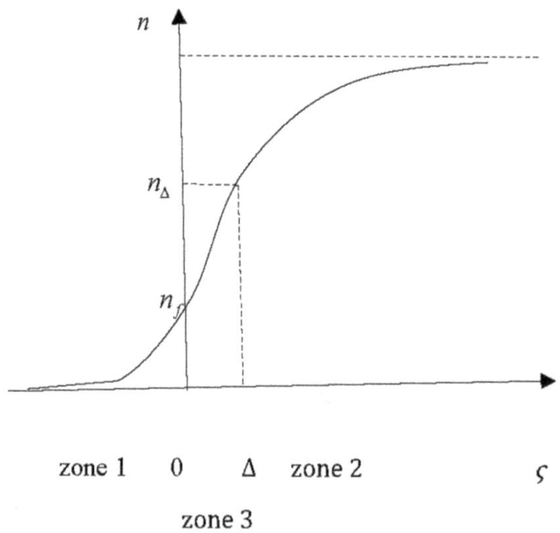

It represents the conquest wave profile since the wave, moving along the spatial coordinate x, with a speed u, fills the space with the maximum concentration of individuals N_+ (according to the introduced wave coordinate, the directions x and z are opposite). We divide the wave into three zones (Figure 3.6). In zone 1 ($-\infty<\varsigma<0$), we consider the solution of the system linearized in the vicinity of the initial equilibrium state a to be fair

$$n = n_f \exp(\lambda_{1,2}\varsigma). \tag{3.60}$$

Here n_f is the concentration of individuals at the wave front $\varsigma=0$, $\lambda_{1,2}$ is the eigenvalue at the point a (which of them will be established later).

In zone 2 ($\Delta\leq\varsigma<+\infty$), we consider the solution of the system linearized in the vicinity of the final equilibrium state b to be valid

$$n = 1-(1-n_\Delta)\exp(\lambda_4(\varsigma-\Delta)). \tag{3.61}$$

The conditions in zone 1 and zone 2 ensure that the approximate solution of the system emerges from the initial equilibrium position and enters the final equilibrium position with almost the same eigenvalues and proper directions as the exact solution of system (3.55), (3.56).

In zone 3, we linearize the right - hand side of equation (3.55) in the vicinity of the maximum of the trajectory in the phase plane (expand the right - hand side of (3.56) into the Taylor series in the vicinity of the maximum and limit the first nonzero member of the series)

$$\frac{dp}{d\varsigma} = (2n_m - 1)(n - n_m). \tag{3.62}$$

Here n_m is concentration at maximum of trajectory.
We integrate (3.62) in view of (3.56)

$$(n - n_m)(1 - 2n_m)^{1/2}/p_m = \sin((1 - 2n_m)^{1/2}\varsigma + A). \tag{3.63}$$

Here

$$A = \arcsin((n_f - n_m)(1 - 2n_m)^{1/2}/p_m), \tag{3.64}$$

$$p_m = n_m(1 - n_m)/w$$

is sense of p at maximum of trajectory. (3.65)

Put in (3.63) $\varsigma=\Delta$, then we have

$$(n_\Delta - n_m)(1 - 2n_m)^{1/2}/p_m = \sin((1 - 2n_m)^{1/2}\Delta + A). \tag{3.66}$$

Here n_Δ is sense of n at $\varsigma=\Delta$.
Differentiating (3.60), (3.61) once, we equate the first derivatives on the border of zones 1 - 3 and 2 – 3

$$n_f^2\lambda_{1,2}^2 = p_m^2 - (1 - 2n_m)(n_f - n_m)^2, \tag{3.67}$$

$$\lambda_4^2(1 - n_\Delta)^2 = p_m^2 - (1 - 2n_m)(n_\Delta - n_m)^2. \tag{3.68}$$

Differentiating (3.60), (3.61) twice, we equate the second derivatives on the border of zones 1 - 3 and 2 – 3

$$n_f\lambda_{1,2}^2 = (2n_m - 1)(n_f - n_m), \tag{3.69}$$

$$\lambda_4^2(1 - n_\Delta) = (1 - 2n_m)(n_\Delta - n_m). \tag{3.70}$$

So, we have a system of nine algebraic equations (3.58), (3.59), (3.64) - (3.70) with respect to nine unknowns $n_f, n_m, n_\Delta, p_m, A, \lambda_{1,2}, \lambda_4, w, \Delta$, i. e. system is closed. Solving it, can get the equation to determine the speed of the wave

$$\lambda_4^2(4w^2(-K_{1,2} \pm T_{1,2} - 1)^{2} - (-K_{1,2} \pm T_{1,2})^{-1}) = 1. \tag{3.71}$$

Here

$$K_{1,2} = 0.5(\lambda_4^2 + \lambda_{1,2}^2 - w^2\lambda_{1,2}^2 + w\lambda_4^2), T_{1,2} = (K_{1,2}^2 + (w^2 - 1)\lambda_{1,2}^2\lambda_4^2).$$

After finding the velocity in (3.71), the remaining parameters are determined by the equations already obtained.

The results of the calculation of the left side of equation (3.71) (*Y)* as functions of speed are shown in Figure 7.

Figure 7. Decision of equation (3.71)

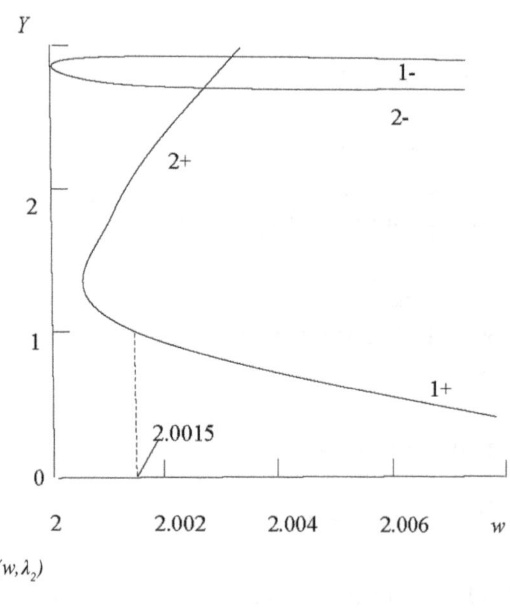

1 – Y(w,λ_1), 2 – Y (w,λ_2)

The branches of the function Y (w), corresponding to the values of λ_1, are designated 1. The branches of the function Y (w), corresponding to the values of λ_2, are designated 2. The signs "+" and "-"at 1 and 2 correspond to the signs $T_{1,2}$. From Figure 3.7 it follows that equation (3.71) satisfies only the branch Y (w), corresponding to the eigenvalue λ_1 at the sign "+" at T_1. The value of speed satisfying equation (3.71) is 2.0015. It is very close to the speed value of 2, found by A.N.

Kolmogorov, I.G. Petrovsky, & N.S. Piskunov as a result of a complex analysis of the nonstationary equation (3.54).

The main attention in this work was paid to determining the shape and velocity of a population wave in a stationary mode. With the help of a mathematical study of the nonstationary equation (3.52), they showed that over time the initial perturbation of the concentration of a population in the form of a step transforms and the population wave takes a stationary form of a hollowed-out step. For it the wave propagation velocity tends to a constant value — the minimum of all possibility the values of the diffusion coefficient and the coefficient of proportionality in the law of reproduction of the population. Unfortunately, the possibility of applying the results of A.N. Kolmogorov, I.G. Petrovsky, N. S. Piskunov is limited by the framework of a very simple model of biological population and accepted form of the initial perturbation of the concentration. In the monograph by A.A. Samara et al. (Samara, Galaktionov, Kurdyumov, & Mikhailov, 1987) showed that if the initial concentration perturbation is different from the step, then the propagation velocity of a stationary wave may differ from the minimum. This agrees with the numerical study of non-stationary equations of waves of exothermic chemical reactions, carried out in this chapter of the monograph. They prove the dependence of solutions of non-stationary partial differential equations on possible initial conditions of the original problem. Therefore, it should be considered incorrect the assumptions of some authors (for example, Svirezhev, 1987; Ivanov, 1984), who consider more modern models of the population, that the wave propagation velocity has a minimum value of 2. In each case, the wave propagation velocity of the biological population and in the study of equations of mathematical physics (Polozhiy, 1964), it must be determined on the basis of this particular system of equations, as an eigenvalue of the problem. Moreover, this eigenvalue must correspond to the solution of the problem of the profile of the population wave, as an eigenvalue function.

Many authors generally do not set the task of determining the wave velocity of a biological population (see, for example, Keller, & Segel, 1971; Ivanitsky, Medvedinsky, & Tsyganov, 1991), considering it to be an unknown parameter, and classify possible solutions (Berezovsky, & Karev, 1999) depending on wave velocity values. At the same time, it is as if they forget that the speed of wave propagation is a function of the conditions in which the population is located and the characteristics of the population itself. All these conditions and characteristics should be included in the equations of the population under consideration, and thus these equations themselves should determine the value of the wave velocity. Only those profiles can be considered in the wave, which correspond to the found speed values. The reason for this situation is the lack of methods for determining the speed of stationary propagation of a wave of a biological population.

3.5. WAVE SYSTEMS

The system of waves is a set of trajectories of the corresponding phase space, each of which is a wave solution going from one equilibrium position to another. Previously, the question of the existence of a system of waves in one-component models of excitable media described by nonlinear parabolic equations was considered (Volpert, 1987). The infinite reaction zone method allows one to proceed to the analytical description of the wave system, which was previously impossible. The non-stationary equation for the change in the concentration H of a certain substance in an excitable medium for a sufficiently arbitrary form of the kinetic function (source function) in a one-component one-dimensional approximation has the form

$$\frac{\partial N}{\partial t} = D\frac{\partial^2 N}{\partial x^2} + f(N). \tag{3.72}$$

Let the function $f(N)$ goes to zero at some values of concentration $N_i(i=1\div k)$ and we are interested in the transition from the state with $N=N_\nu$ to the state with $N=N_\mu(\nu,\mu=1\div k)$. Let's introduce the wave coordinate $z=ut-x$, u is the speed of the transition wave. Then (3.72) can be written as

$$dp/dz = pu/D - f(N)/D$$

, (3.73)

$$dN/dz = p. \tag{3.74}$$

We linearize the right side of (3.73) in a neighborhood of the point $\nu(p=0, N=N_\nu)$

$$dp/dz = pu/D - (N - N_\nu)f_N(N_\nu)/D, \tag{3.75}$$

and in a neighborhood of the point $\mu(p=0, N=N_\mu)$

$$dp/dz = pu/D - (N - N_\mu)f_N(N_\mu)/D. \tag{3.76}$$

We will assume that the points ν and μ, representing the initial and final equilibrium positions, respectively, are simple singular points in the phase plane (p, N), i. e. $f_N(N_\nu) \neq 0, f_N(N_\mu) \neq 0.$. The eigenvalues at these points are found from the discriminants of systems (3.74), (3.75) and (3.74), (3.76), respectively. They coincide with their own directions at these points

$$\lambda_{\nu(1,2)} = (dp/dN)_{\nu(1,2)} = u/(2D) \pm \sqrt{u^2/(2D)^2 - f_N(N_\nu)/D}, \tag{3.77}$$

$$\lambda_{\mu(1,2)} = (dp/dN)_{\mu(1,2)} = u/(2D) \pm \sqrt{u^2/(2D)^2 - f_N(N_\mu)/D}. \tag{3.78}$$

To find a solution starting from the point ν, it is necessary to choose a positive one from two eigenvalues (3.77), if they are real. If $f_N(N_\nu) < 0$, then $\lambda_{\nu 1}$ satisfies this condition and the point ν has the type of a saddle. If $f_N(N_\nu) > 0$, then two options are possible. If, in addition, $u^2/(4D) > f_N(N_\nu)$, then both eigenvalues (3.77) are real and positive. In this case, it is necessary to determine which of the two eigendirections the desired wave solution leaves the point ν. If $u^2/(4D) < f_N(N_\nu)$then the point ν has a focus type and both eigenvalues must be used to describe the wave solution in the neighborhood of this point.

To find a solution that enters the point μ, it is necessary to choose $\lambda_{\mu 2}$ from (3.78). In this case, the inequality $f_N(N_\mu) < 0$must be satisfied, i. e. the point μ can only be a saddle. We restrict ourselves to the case when both points ν and μ are of saddle type. Then the solutions in the neighborhood of these points have, respectively, the form

$$N = N_\nu + (N_f - N_\nu)\exp(\lambda_{\nu 1} z), \tag{3.79}$$

$$N = N_\mu + (N_\Delta - N_\mu)\exp(\lambda_{\mu 2}(z - \Delta)). \tag{3.80}$$

Here N_f is the value of the concentration at the boundary of the vicinity of the point ν at $z = 0$ (wave front); N_Δ is the value of the concentration at the boundary of the vicinity of the point μ at $z = \Delta$. We divide the length of the wave into three zones: the first zone is the neighborhood of the point $\nu(-\infty<z<0)$; the second zone is a neighborhood of the point μ; the third zone is the space between the first two zones. The third zone contains the extremum of the trajectory leaving ν and entering the point μ of the phase plane (p, N). If we assume that the extremum in this zone is unique, then either it is a maximum if the wave solution is located in the positive half-plane ($p>0$, $\nu<\mu$), or it is a minimum if the wave solution is located in the negative half-plane ($p<0$, $\nu>\mu$). We linearize the right side of (3.73) in the vicinity of the trajectory extremum ($p=p_m$, $N=N_m$) and divide the resulting equation by (3.74)

$$dp/dN = -(N - N_m) f_N(N_m)/(Dp). \tag{3.81}$$

We integrate equation (3.81) two times, taking into account (3.74)

$$(N - N_m)(f_N(N_m)/D)^{1/2} = p_m \sin((f_N(N_m)/D)^{1/2} z + E, \tag{3.82}$$

$$E = \arcsin((N_f - N_m) p_m^{-1} (f_N(N_m)/D)^{1/2}), \tag{3.83}$$

$$p_m = -f(N_m)/u. \tag{3.84}$$

On the boundary of zone 3 at z = Δ from (3.84) we have

$$(N_\Delta - N_m)(f_N(N_m)/D)^{1/2} = p_m \sin((f_N(N_m)/D)^{1/2}\Delta + E. \tag{3.85}$$

Differentiating (3.84), (3.80), (3.82) once, we equate the derivatives *dN/dz* at the boundary of zones 1 - 3, 2 – 3

$$(N_f - N_\nu)^2 \lambda_{\nu 1}^2 = p_m^2 - (N_f - N_m)^2 f_N(N_m)/D, \tag{3.86}$$

$$(N_\Delta - N_\mu)^2 \lambda_{\mu 2}^2 = p_m^2 - (N_\Delta - N_m)^2 f_N(N_m)/D. \tag{3.87}$$

Differentiating (3.84), (3.80), (3.82) two time, we equate the derivatives d^2N/dz^2 at the boundary of zones 1 - 3, 2 – 3

$$(N_f - N_\nu)\lambda_{\nu 1}^2 = -(N_f - N_m)f_N(N_m)/D, \tag{3.88}$$

$$(N_\Delta - N_\mu)\lambda_{\mu 2}^2 = -(N_\Delta - N_m)f_N(N_m)/D. \tag{3.89}$$

Resolving the resulting algebraic system of equations, we reduce it to a system of two equations with respect to *u* and N_m

$$u^2(N_\nu - N_m)^2 f_N(N_m)/D = f(N_m)\lambda_{\nu 1}^2(\lambda_{\nu 1}^2 + f_N(N_m)/D)^{-1}, \tag{3.90}$$

$$u^2(N_\mu - N_m)^2 f_N(N_m)/D = f(N_m)\lambda_{\mu 2}^2(\lambda_{\mu 2}^2 + f_N(N_m)/D)^{-1}. \tag{3.91}$$

After finding *u* and N_m by (3.90), (3.91). (3.77), (3.78) the values N_f, N_Δ are determined from relations (3.88), (3.89). The width of zone 3 is found from (3.85), (3.86). Concentration profiles in zones 1, 2, 3 are calculated according to (3.79), (3.80), (3.82) respectively.

For a numerical study, it is necessary to specify a specific type of function *f*(*N*). We define it as a fifth-order polynomial

$$f(N) = k(N_1 - N)(N_2 - N)(N_3 - N)(N_4 - N)(N_5 - N),$$
$$N_5 > N_4 > N_3 > N_2 > N_1 \geq 0, k > 0.$$

Let us introduce dimensionless quantities

$$n = N/N_5, \zeta = zN_5^2(k/D)^{1/2}, w = u(N^4 kD)^{-1/2}.$$

Then equations (3.73), (3.74) take the form of system (3.55), (3.56), where we have

$$F(n) = (n - n_1)(n - n_2)(n - n_3)(n - n_4)(1 - n). \quad (3.92)$$

This system has five equilibrium positions in the phase plane

(p,n): $a(p = 0, n = n_1), b(p = 0, n = n_2),$
$c(p = 0, n = n_3), d(p = 0, n = n_4), e(p = 0, n = 1)$

.

Without changing the qualitative features of the system, let's match the point a with the origin, i. e. let's put $n_1 = 0$. The eigenvalues at the points a, c, e are equal:

$$\lambda_{a(1,2)} = (dp/dn)_{a(1,2)} = w/2 \pm \sqrt{w^2/4 + n_2 n_3 n_4}, \quad (3.93)$$

$$\lambda_{c(1,2)} = (dp/dn)_{c(1,2)} = w/2 \pm \sqrt{w^2/4 + n_3(n_3 - n_2)(n_4 - n_3)(1 - n_3)}, \quad (3.94)$$

$$\lambda_{e(1,2)} = (dp/dn)_{e(1,2)} = w/2 \pm \sqrt{w^2/4 + (1 - n_2)(1 - n_3)(1 - n_4)}. \quad (3.95)$$

Since the second terms under the root in (3.93) - (3.95) are positive, the points a, c, e are of saddle type. Between these points, connections are possible by trajectories corresponding to wave solutions (Figure 8).

Figure 8. The waves solutions of system (3.55), (3.56), (3.92)

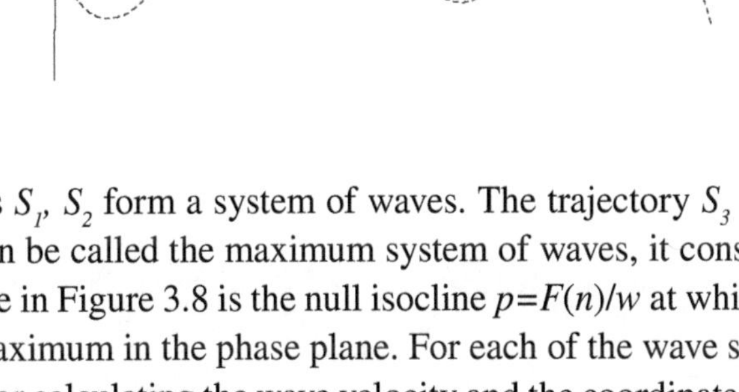

Trajectories S_1, S_2 form a system of waves. The trajectory S_3 (according to the values of p) can be called the maximum system of waves, it consists of one wave. The dashed line in Figure 3.8 is the null isocline $p=F(n)/w$ at which the wave solutions have a maximum in the phase plane. For each of the wave solutions S_1, S_2, S_3 the equations for calculating the wave velocity and the coordinates of the trajectory maximum have the following form

$$S_1 : (F_m/w^2) = F'_{nm} n_m^2 \lambda_{a1}^2 (\lambda_{a1}^2 + F'_{nm})^{-1} = F'_{nm} \lambda_{c2}^2 (n_3 - n_m)^2 (\lambda_{a1}^2 + F'_{nm})^{-1}, \quad (3.96)$$

$$S_2 : (F_m/w^2) = F'_{nm} (n_3 - n_m)^2 \lambda_{c1}^2 (\lambda_{c1}^2 + F'_{nm})^{-1} = F'_{nm} \lambda_{e2}^2 (1 - n_m)^2 (\lambda_{e1}^2 + F'_{nm})^{-1}, \quad (3.97)$$

$$S_3 : (F_m/w^2) = F'_{nm} n_m^2 \lambda_{a1}^2 (\lambda_{a1}^2 + F'_{nm})^{-1} = F'_{nm} \lambda_{e2}^2 (1 - n_m)^2 (\lambda_{e2}^2 + F'_{nm})^{-1}. \quad (3.98)$$

The wave solution leaves the point a in its own direction $(dp/dn)_{a1}$. From (3.93) it follows that wave S_3 must have a speed greater than wave S_1, since the final concentration in wave S_3 is greater than the final concentration in wave S_1 $(1 > n_3)$. The wave solution enters the point e in its own direction $(dp/dn)_{e2}$. It follows from (3.95) that the wave solution S_2 must have a greater velocity than the wave solution S_3, since $|(dp/dn)_{e2}|$ decreases with increasing w. Wave S_1, S_2, S_3 calculations were carried out using the obtained analytical solutions, as well as directly using the system of equations *(3.55), (3.56), (3.92)*. The results of calculations based on analytical solutions and differential equations are in good agreement with each other (the discrepancy between them does not exceed 10%). Differential equations were calculated by the continuous Runge-Kutta method of the 8th order. The exit from the initial equilibrium position was set along the own direction of the singular point. The value of the velocity was varied until the trajectory entered the specified final equilibrium position with high accuracy (shooting method). As a result of calculations, the regions of existence of waves in the space of parameters n_1, n_2, n_3 and the dependence of the wave speed on these parameters are determined. Figure 9 shows the domains of existence of solution S_1 (indicated by arrows) in the plane (n_4, n_3) for various values of n_2.

Figure 9. The domains of existence of wave S_1

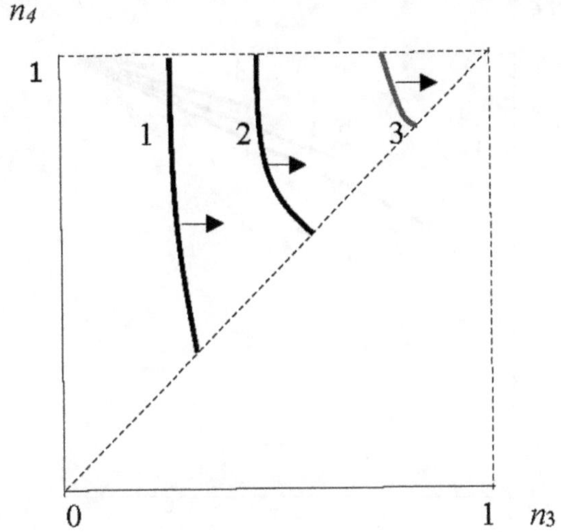

$1 - n_2 = 0.1$, $1 - n_2 = 0.2$, $1 - n_2 = 0.3$

The domains of existence are limited by straight lines $n_4 = n_3$, $n_4 = 1$ and lines of zero velocity at $n_2 = \text{const}$. The equation of this line is found by integrating the system (3.55), (3.56), (3.92) at $w = 0$ and the boundary conditions of wave S_1

$$n_4 = n_3(n_3(3 - 2n_3) + n_2(3n_3 - 5))(n_3(5 - 3n_3) + n_2(5n_3 - 10))^{-1}. \qquad (3.99)$$

Figure 9 shows that as n_2 increases, the area of existence of wave S_1 sharply decreases and completely disappears at $n_2 > 0.4$.

Figure 10 shows the domains of existence of solution S_2 in the plane (n_4, n_3) for various values of n_2.

Figure 10. The areas of existence of the wave S_2

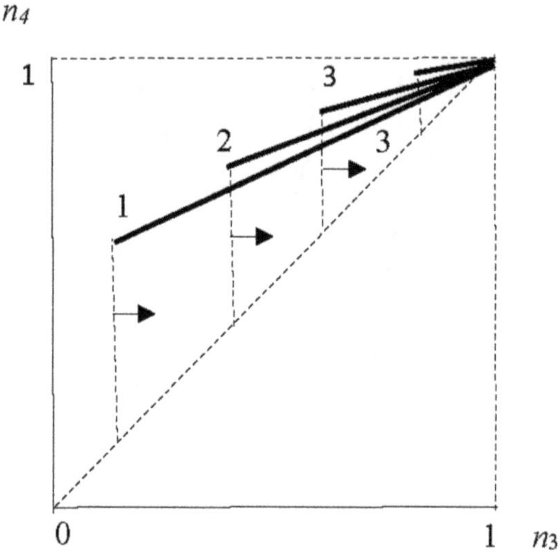

$1 - n_2 = 0.2$, $1 - n_2 = 0.4$, $1 - n_2 = 0.6$, $1 - n_2 = 0.8$

The regions of existence of this wave (indicated by arrows near the dotted vertical lines) are limited by straight lines $n_4 = n_3$, $n_2 = n_3$ and lines of zero velocity at n_2 =const. The equations of this line are found by integrating the system (3.55), (3.56), (3.92) with w =0 and the boundary conditions of wave S_2

$$n_4 = \frac{n_3^5(3 - 2n_3)+2 - 3n_3 + n_2(n_3^4(3n_3 - 5)+5n_3 - 3)}{n_3^4(-3n_3 + 5)-5n_3 + 3 + n_2(n_3^3(3n_3 - 10)+10n_3 - 5)}. \qquad (3.100)$$

Figure 10 shows that the domain of existence also decreases as n_2 increases. However, this region is also preserved for large n_2, disappearing only when equality $n_2 = 1$ is reached.

Figure 11 shows the domains of existence of solution S_3 (indicated by arrows) in the plane (n_4, n_3) for various values of n_2.

Figure 11. The domains of existence of wave S_3

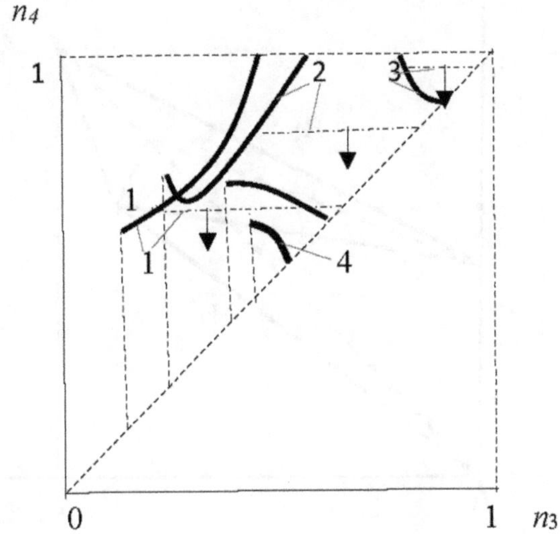

$1 - n_2 = 0.1$, $1 - n_2 = 0.2$, $1 - n_2 = 0.3$, $1 - n_2 = 0.4$

The domains of existence of solution S_3 are limited by straight lines $n_4 = n_3$, $n_3 = n_2$ and lines of zero velocity at n_2 = const. The equation of this line is found by integrating the system (3.55), (3.56), (3.92) at $w = 0$ and the boundary conditions of wave S_3

$$n_4 = (n_3 3 - 2 + n_2(3 - 5n_3))(n_3 5 - 3 + n_2(5 - 10n_3)). \tag{3.101}$$

In addition, the areas of existence of wave S_3 are limited by the line of wave S_3 break, i.e. an abrupt transition of wave S_3 into wave S_1, which occurs at a finite speed and its decrease. These lines (dashed and dotted) in the parameter space are determined numerically, and they significantly limit the region of existence of the S_3 wave. The line of zero wave velocity S_3 (3.101) differs significantly from the lines of zero wave velocity S_1 and S_2 in that at n_2 >0.2764 it consists of two parts that do not have common points. Up to n_2 = 0.2764, the line of zero velocity is continuous and turns into two perpendiculars (parallel to the axes) resting on the bisector $n_4 = n_3$ and at n_2 =0.2764 (in order not to clutter up the figure, the perpendiculars are not shown on it). Wave S_3 does not exist at $n_2 > 0.5$.

Based on the analysis carried out, it can be said that the area of existence of the system of waves S_1 and S_2 (the simultaneous existence of these waves) represents a certain pyramid in the parameter space (n_1, n_2, n_3) (Figure 12), consisting of two pyramids – H_1 and H_2.

Figure 12. The region of existence of the system of waves S,and S,

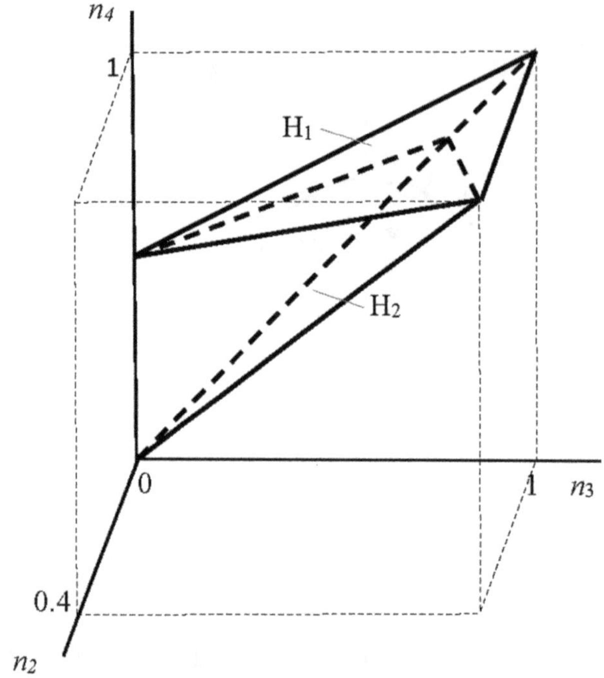

H_1 is the region of parameters where the speed of wave S_2 is less than the speed of wave S_1; H2 is the range of parameters where the speed of wave S_2 is greater than the speed of wave S_1

The area of the pyramid H1 (large n) combines the modes when the speed of the wave S_2 is less than the speed of the wave S_1. Wave S_2 does not catch up with wave S_1 and wave C3 is not formed: in the range of parameters of the pyramid H1, wave C3 does not exist. The area of the H2 pyramid (small n) unites the modes when the speed of the wave S_2 is greater than the speed of the wave S_1. Wave S_2 can catch up with wave S_1 and form wave S_3: in the area of values of pyramid parameters H2 there are waves S_1, S_2, and S_3.

Calculations have shown that there is a large range of parameter values (to the left of the lines n_2 = constant of equation (3.99) in the plane (n_4, n_3). Figure 9, when wave S_1 does not exist, but waves S_2 and S_1 exist. Thus, the statement of A.I. Volpert (Voltert, 1987) that a necessary and sufficient condition for the existence of the S_3 wave is the condition for the existence of the S_1 and S_2 waves when the speed of the S_2 wave is greater than the speed of the S_1 wave, should be recognized as incorrect.

To illustrate the satisfactory agreement between the numerical solutions of differential equations and calculations based on analytical solutions obtained by the infinite reaction zone method, Figure 13 shows the results of calculating the dependences of the S_1 and S_2 wave velocities on the parameter n_4, n_2 =0.1, n_3 =0.5. It can be seen from the diagram that the speed of the S_1 wave increases with increasing n_4, while the speed of the S_2 wave decreases.

Figure 13. Dependences of the S_2 and S_1 wave velocities on the parameter n_4 (n_2= 0.1, n_3=0.5)

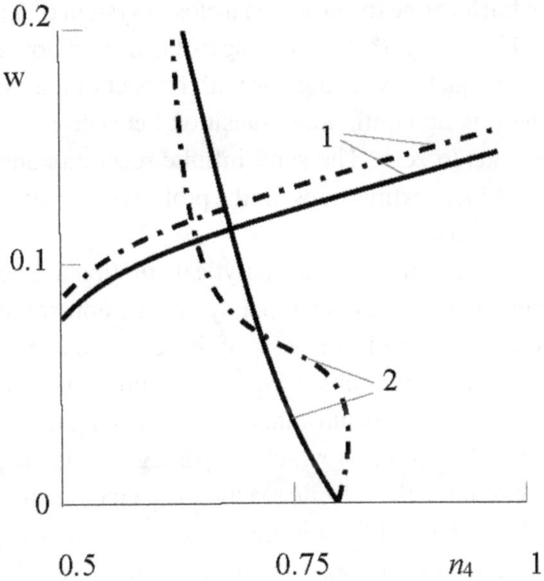

The solid line is the numerical solution of differential equations, the dash-dotted line is the calculation by the infinite reaction zone method.

CONCLUSION

Non-asymptotic methods for describing waves in excitable media are developed, taking into account the available experimental data on these waves, described in the previous chapter, and the criticism of the theoretical assumptions of asymptotic approaches, presented in the first chapter. For waves of exothermic chemical reactions, the main non-asymptotic method is the semi-infinite reaction zone method, in which the wave length is divided into two semi-infinite zones: the heat conduction zone (heating zone) and the chemical reaction zone. Thus, chemical reactions according to this model occur at every point of the semi-infinite zone, in contrast to the narrow reaction zone method, in which chemical reactions occur in a zone of zero extent. The fundamental point for obtaining an analytical solution in the semi-infinite reaction zone method is the introduction of a buffer zone between the heating zone and the chemical reaction zone, the length of which is small and, after resolving the system of algebraic equations obtained as a result of integration over the zones, tends to zero. The introduction of the buffer zone made it possible to obtain a closed system of equations and eventually solve this system. If we a priori set the length of the buffer zone to zero, then a closed system of algebraic equations cannot be obtained. This proves that, according to the ideas of non-standard analysis, the original system of equations is singular with respect to the extent of the buffer zone. Therefore, there is no continuous transition between systems as the length of the buffer zone tends to zero. The semi-infinite reaction zone method made it possible to obtain analytical expressions for the profiles of parameter changes along the wave and wave velocity.

The results of the calculation by the analytical solution are compared with the results of a numerical study of the original system of non-stationary differential equations carried out within the framework of this chapter. A comparison of these results and the results of calculations using other approximate analytical methods proves the correctness of the semi-infinite reaction zone method.

For waves in excitable media, for which the passage of kinetic processes at each point of the wave is essential, the infinite reaction zone method has been developed. These waves include isothermal flames and waves of biological populations. The reason for the propagation of such waves is the mobility of the elements of the medium through which these waves run. Often these waves are called diffusion waves, although in the case of biological populations such a name is not entirely correct. In the infinite reaction zone method, the wavelength is divided into two semi-infinite zones, in each of which kinetic processes occur. In these zones, the solutions are assumed to be well-posed in the vicinity of the initial and, accordingly, the final equilibrium positions. A buffer zone of finite length is introduced between the zones. The conditions at the boundaries of the zones are determined from the solvability of the algebraic system of equations obtained as a result of integration over the zones. Thus, it is possible to obtain a closed system of algebraic equations and describe solutions for the most diverse examples of waves in excitable media, at each point of which kinetic processes occur. The results of calculations using the infinite reaction zone method are compared with the results of a numerical study of the initial non-stationary differential equations carried out within the framework of this chapter. Comparisons of these comparisons testify to the correctness of the infinite reaction zone method. The infinite reaction zone method made it possible to analytically describe a system of waves consisting of several wave solutions passing into each other. The results of calculations based on analytical solutions are also compared in this case with the results of a numerical study of the corresponding systems of differential equations. The correctness of the infinite reaction zone method is also proved in these cases. The erroneousness of the statements of A.I. Volpert on the condition for the existence of a wave solution that combines two wave solutions.

REFERENCES

Albeverio, S., & Fensteid, J., Heeg-Kran, R., Lindström, T. (1990). *Non-standard methods in stochastic analysis and mathematical physics*. Moscow: Mir.

Aldushin, A. P., Lugovoy, V. D., & Merzhanov, A. G. (1978). Conditions for the degeneracy of a stationary combustion wave. *Reports AN USSR*, *243*(6), 1434–1442.

Azatyan, V. V., ... (1976). Isothermal propagation of a rarefied nitrogen trifluoride flame. *Proceedings of the Academy of Sciences of the USSR. Chemical series*. 7, 1459 - 1461.

Berezovskaya, F. S., & Karev, G. P. (1999). Traveling wave bifurcations in population models with taxis. *Успехи физических наук*, *169*(9), 1011–1024. DOI: 10.3367/UFNr.0169.199909d.1011

Berman, V. S., & Ryazantsev, Yu. S. (1972). Application of the theory of matched asymptotic expansions to the calculation of stationary thermal propagation of an exothermic reaction front in a condensed medium. *Journal of Applied Mechanics and Technical Physics*, *5*, 106–112.

Fedotov, V. G., & Chaikina, A. M. (1972). Investigation of the cold flame of the reaction of fluorine with hydrogen in a flow. *Reports of the Academy of Sciences of the USSR*, *203*(2), 406–408.

Frank-Kamenetsky, D. A. (1967). *Diffusion and heat transfer in chemical chemistry*. Nauka.

Ivanitsky, G. R., Medvedinsky, A. B., & Tsyganov, M. A. (1991). From disorder to ordering – on the example of the movement of microorganisms. *Успехи физических наук*, *161*(3), 13–71. DOI: 10.3367/UFNr.0161.199104b.0013

Ivanov, N. N. (1984). Solitary waves in population theory. *Proceedings of the Academy of Sciences of the USSR. Biological series*. 5, 656-663.

Keller, E. F., & Segel, L. A. (1971). Traveling bands of chemotaxis bacteria: A theoretical analysis. *Journal of Theoretical Biology*, *42*(3), 397–429.

Kolmogorov, A. N., Petrovsky, I. G., & Piskunov, N. S. (1937). The study of the diffusion equation, combined with the increase in the amount of matter and its application to one biological problem. *Bulletin of Moscow State University. Series A.Mathematics and Mechanics*, *1*(6), 1–26.

Kovalev, O. N., & Vechernin, V. V. (1986). Description of a new wave process in populations on the example of the introduction and dispersal of the ragweed leaf beetle. *Entomological Review*, *65*(1), 21–38.

Krinsky, V. I. (1968). Fibrillation in excitable media. *Problems of Cybernetics*, *20*, 59–80.

Naife, A.Kh. (1984). *An Introduction to Perturbation Methods*. Moscow: Mir.

Novozhilov, B. V. (1961). The rate of propagation of the front of an exothermic reaction in the condensed phase. *Reports of the Academy of Sciences of the USSR*, *441*(1), 151–153.

Polozhiy, G. N. (1964). *Equations of mathematical physics*. High Schol.

Samarsky, A. A., Galaktionov, V. A., Kurdyumov, S. P., & Mikhailov, A. P. (1987). *Blow-up Regimes in Problems for Quasilinear Parabolic Equations*. Nauka.

Shkadinsky, G.V., & Khaikin, B.I., Merzhanov, A.G. (1971). Propagation of a pulsating front of an exothermic reaction in a condensed phase. *Physics of combustion and explosion*. 7(1), 19 - 28.

Svirezhev, Yu. M. (1987). *Nonlinear waves, dissipative structures and catastrophes in ecology*. Nauka.

Uspensky, V. A. (1987). *What is non-standard analysis?* Nauka.

Volpert, A. I. (1987). On the propagation of waves described by nonlinear parabolic equations. In Petrovsky, I. G. (Ed.), *Selected works. Differential equations. Probability theory*. Nauka.

Voronkov, V. G., & Semenov, N. N. (1939). Cold flame propagation in combustible mixtures containing 0.03% carbon disulfide. *Journal of Physical Chemistry*, *13*(12), 1695–1727.

Warnatz, J., Maas, U., & Dibble, R. W. (2001). *Combustion. Physical and Chemical Fundamentals, Modeling, and Simulations, Experiments, Pollutant formation*. Springer.

Westmoreland, F. R. et al. (2006). Study of the flame structure by molecular beam spectrometry using electron impact ionization and photon ionization by synchrotron radiation. *Physics of combustion and explosion,* 42(6), 58-63.

Zeldovich, Ya. B., & Frank-Kamenetsky, D. A. (1938). Theory of thermal propagation of the flame. *Journal of Physical Chemistry*, *12*(1), 100–105.

Zhizhin, G. V. (1982). Wave structure of frontal polymerization. *Reports of the Academy of Sciences of the USSR*, *263*(6), 1399–1402.

Zhizhin, G. V. (1992). *Macrokinetics in frontal polymerization reactors*. Polytechnic.

Zhizhin, G. V. (1997). *Mathematical models of waves of radical polymerization.*

Zhizhin, G. V. (1999). Semi-infinite reaction zone method in the theory of self-regulating waves of exothermic reactions and non-standard analysis. Mathematical methods in engineering and technology. Proceedings of the 12th International Conference. Veliky Novgorod, 14 – 17.

Zhizhin, G. V. (2001). *Anthropogenic impact on the development of biological populations in the environment. The impact of anthropogenic impact on public health and environmental quality. Materials of the regional conference*. St. Petersburg State University.

Zhizhin, G. V. (2004). *Self-regulating waves of chemical reactions and biological populations*. Science.

Zhizhin, G.V., & Poritskaya, I.Ya. (1994). Self-regulating waves of exothermic reactions of the n-th order in condensed media. *Physics of combustion and explosion*. 6, 61 - 68.

Zvonkov, A. K., & Shubin, M. A. (1984). Non-standard analysis and singular perturbations of ordinary differential equations. *Advances in Mathematical Sciences*, *39*(2), 77–127.

Chapter 4
Waves of Radical Polymerization

ABSTRACT

The semi-infinite reaction zone method is used to describe stationary waves of radical polymerization of flat, cylindrical and spherical shapes. For waves of radical polymerization of the above geometry, analytical expressions are obtained for calculating the profiles of changes in the mass fractions of the mixture components along the spatial coordinate. According to these expressions, fundamentally new designs of continuous frontal polymerization reactors were developed. Such reactors were manufactured at machine-building plants and tested at chemical plants. Continuous frontal polymerization reactors with axial and central symmetry have a specific productivity several times higher than the specific productivity of standard polymerization reactors.

INTRODUCTION

In 1982, the idea of the semi-infinite reaction zone method was proposed specifically for the analysis of the structure of radical polymerization waves (Zhizhin, 1982). At this time, there was a need to create a small-barite continuous polymerization reactor with a standing wave reaction instead of huge batch polymerization reactors. At the same time, experimental studies of polymerization waves (Chapter 2) showed that the temperature profile in the polymerization wave differs significantly from the temperature profile in the combustion wave in the approximation of the narrow reaction zone method. It is characterized by a stretched "tail" at the end of the wave, where the reaction slowly decays.

In this regard, taking into account the practical problem of creating a reactor, it was necessary to build a mathematical model of the wave, free from the limitations of the narrow reaction zone method. Such a model has been created. Algebraic ex-

DOI: 10.4018/978-1-6684-7745-8.ch004

pressions are obtained for calculating a reactor with a flat reaction front. However, considering the hydrodynamics in this reactor (the polymerization wave in this reactor must be stationary, and the medium moves through the wave), the issues of adhesion of the viscous polymer formed during the reaction to the reactor walls turned out to be very significant. If the reactor has the shape of a pipe, then the polymer sticks to the walls of the pipe, and the low-viscosity monomer, moving along the axis of the pipe, outstrips the polymer, and a breakdown of the reactor is formed. Therefore, the creators of the reactor faced the question of eliminating the influence of the reactor walls. This can be done if the reaction medium moves in space without walls, i.e. or from the center of the sphere, perpendicular to the outer walls of the reactor, or from the axis of the pipe, perpendicular to the walls of the pipe (RF Patent, Zhizhin, 1995). In this chapter, the semi-infinite reaction zone method is used to calculate a flat, cylindrical, and spherical polymerization front. Based on the analytical expressions obtained for the cylindrical and spherical reaction front, the corresponding reactor designs were calculated. These reactors were created and tested on different monomers at different plants.

WAVE OF IRREVERSIBLE RADICAL POLYMERIZATION REACTION

Consider a wave of an irreversible radical polymerization reaction, assuming that the rate of the depolymerization reaction is low (either the temperature is low or the depolymerization reaction is retarded by high pressure). Let us take into account the influence of the gel effect in a rather simple form (Zhizhin, 1990). The system of differential equations of radical polymerization in coordinates associated with the wave, under the accepted conditions, has the form

$$\lambda \frac{d^2 T}{dz^2} - \rho u c_p \frac{dT}{dz} + Qf\rho C_1^r C_2^n k_{02} \exp\left(-\frac{E_2 + \Delta VP}{RT}\right) = 0, \tag{4.1}$$

$$u\frac{dC_1}{dz} + k_{01} C_1 C_2^l exp\left(-\frac{E_1}{RT}\right) = 0, \tag{4.2}$$

$$u\frac{dC_2}{dz} + f\rho C_1^r C_2^n k_{02} \exp\left(-\frac{E_2 + \Delta VP}{RT}\right) = 0. \tag{4.3}$$

Here C_1 is the mass fraction of the initiator in solution; C_2 is the mass fraction of the monomer; T is temperature; P is pressure (at large values it affects the kinetics of the polymerization reaction); E_1 is the activation energy of the initiation reaction; E_2 is the activation energy of the polymerization reaction; Q is the specific thermal effect of the polymerization reaction (it is believed that the specific thermal effect of the initiation reaction is immeasurably small); ΔV is activation volume, n and l are the orders of polymerization and initiation reactions with respect to the monomer, r is the order of polymerization reactions with respect to the initiator; R is the universal gas constant; k_{01} is the pre-exponent of the initiation reaction; k_{02} is the pre-exponent of the polymerization reaction; the thermophysical characteristics of the solution (density ρ, specific heat capacity c_p, thermal conductivity λ) are assumed to be constant, since it is believed that the mass fraction of the initiator in the solution is very small, and the thermophysical characteristics of other components of the solution (solvent, monomer, polymer) are close.

In equations (4.1) - (4.3), the values of l=1, r=0.5, n=1.5 correspond to the radical polymerization of ethylene, the values l=0, r=0.5, n=1 correspond to the radical polymerization of methyl methacrylate and styrene (Bagdasaryan, 1966). Function f takes into account the influence of the gel - effect on the reaction rate. Taking into account that the influence of the gel effect significantly affects at large depths of polymerization, and also that the polymerization reaction in a wave proceeds rather quickly, we will take this form of the function f (Zhizhin, 1990)

$$f = e^{a(1-C_{2+}/C_{20})}, a = 2.6. \quad (4.4)$$

Here, the index "+" corresponds to the state at the end of the wave (z=+∞), and the index "0" corresponds to the state at the beginning of the wave (z=-∞), a is a constant (in the general case, it depends on pressure). Let's accept fair for the majority of initiators equality r=0.5. Substituting equation (4.3) into equation (4.1), using the substitution ($C = C_1^{1/2}$), we transform the system (4.1) - (4.3) to the form

$$\lambda\frac{d^2T}{dz^2} - \rho u c_p \frac{dT}{dz} - Q\rho u\frac{dC_2}{dz} = 0, \quad (4.5)$$

$$\frac{dC}{dz} = -\frac{k_{01}}{2u}C C_2^l exp\left(-\frac{E_1}{RT}\right), \quad (4.6)$$

$$\frac{dC_2}{dz} = -\frac{k_{02}}{u}fC C_2^n \exp\left(-\frac{E_2 + \Delta VP}{RT}\right). \quad (4.7)$$

Integrating equation (4.5), determining the integration constant from the condition $T=T_+$, $C_2=C_{2+}$, $dT/dz=0$, $z=+\infty$, we obtain the equation

$$\frac{dT}{dz} = \frac{\rho u c_p}{\lambda}(T - T_+) + \frac{Q\rho u}{\lambda}(C_2 - C_{2+}). \quad (4.8)$$

The system of equations (4.6) - (4.8) is autonomous. Its solutions can be represented by trajectories in three-dimensional phase space (T, C, C_2). The set of trajectories of the phase space at $l > 0$ and $l = 0$ have certain differences.

Consider the case $l > 0$ that corresponds, for example, to the radical polymerization of ethylene. In the open part of the positive octant of the phase space (T, C, C_2), which unites solutions having a physical meaning, there is a single zero surface - the plane F

$$c_p(T - T_+) + Q(C_2 - C_{2+}) = 0.$$

At its points $dT/dz=0$. The remaining zero surfaces are coordinate planes $T=0$, $C=0$, $C_2=0$, at the points of which $dC/dz=dC_2/dz=0$. Direct intersections of the plane F with the coordinate planes are a set of singular points of the system (4.6) – (4.8)

$$a)\ T = 0, C_2 = T_+ c_p/Q + C_{2+}\ ;$$
$$b)\ (T - T_+)c_p/Q + C_2 - C_{2+} = 0\ ;$$
$$c)\ T = T_+ + C_{2+}Q/c_p, C_2 = 0.$$

The trajectories corresponding to the wave can only be among the trajectories emanating from the points of the line a, entering the points of the lines δ or b (Figure 1).

Figure 1. Phase diagram of waves of radical polymerization at $l > 0$

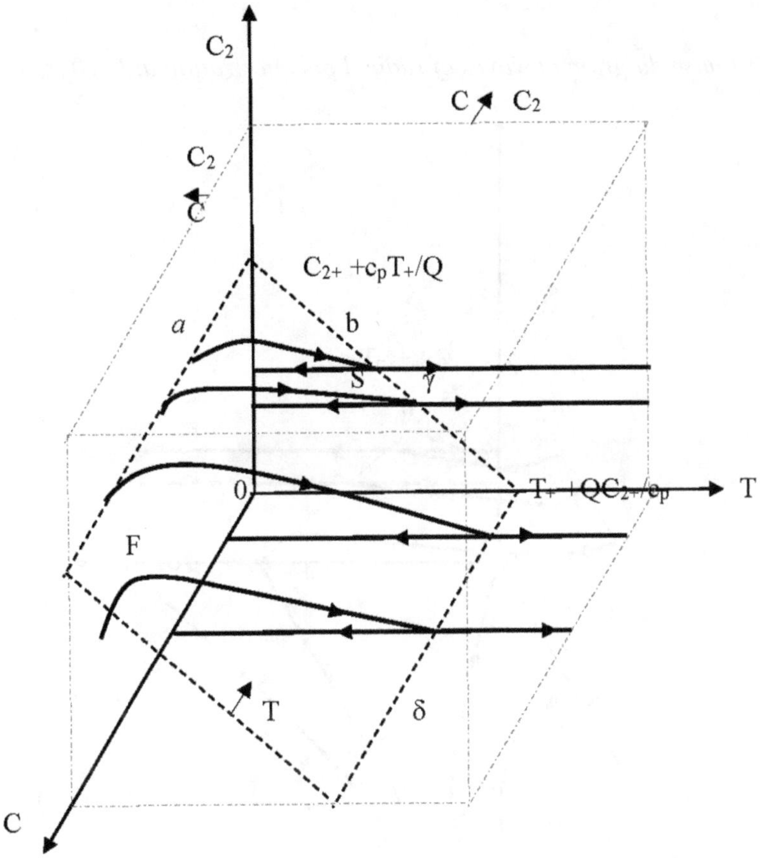

The equilibrium positions δ correspond to the complete consumption of the monomer in the polymerization reaction ($C_{2+} = 0$), and the equilibrium positions *b* correspond to the complete consumption of the initiator. Of practical interest are the trajectories that enter equilibrium positions *b*, since in industrial polymerization processes the initial concentration of the initiator is many times lower than the initial concentration of the monomer, and the initiator rapidly disappears during the reaction. In addition, under the boundary conditions set, only the trajectories entering the singular point γ of the line *b* are of interest.

When $l = 0$, which corresponds, for example, to the polymerization of methyl methacrylate, the phase space has no direct positions of equilibrium *b*, since in this case the plane $C_2=0$ is not zero for the derivative dC/dz in this case. Instead, on the plane C_2=0 there is a trajectory parallel to the C axis, ending at a point C_2=0, T=T_+, C_+=0 (Figure 2), which has no physical meaning.

Figure 2. Phase diagram of waves of radical polymerization at l = 0

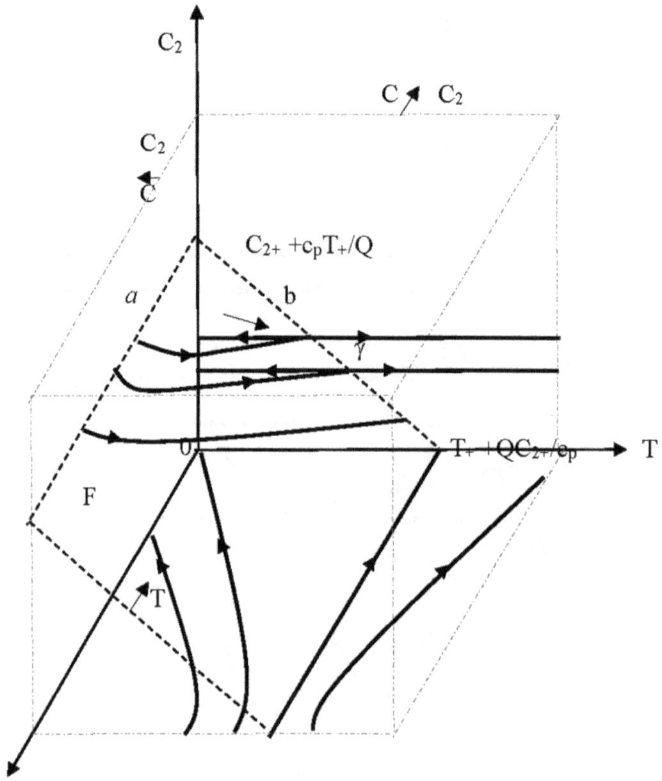

Thus, if at $l > 0$ wave solutions are fundamentally possible with simultaneous vanishing of the initiator and monomer concentrations, then at $l = 0$ (polymerization of methyl methacrylate), all trajectories end at a finite monomer concentration.

In order to exclude the zero value of the absolute temperature from the initial equilibrium positions, we assume that the temperature in the initial equilibrium positions is greater than zero, i. e. we set the boundary condition C_2=C_{20}, T=T_0, dT/dz=0, z= -∞. Then from (4.8) we obtain the integral energy conservation law

$$c_p(T_+ - T_0) = Q(C_{20} - C_{2+}). \tag{4.9}$$

In accordance with the semi-infinite reaction zone method, we distinguish two semi-infinite zones in the structure of the radical polymerization wave: the heating zone (1) and the reaction zone (2). Between them, we temporarily introduce a buffer zone (3) with a linear change $T, C_1^{1/2}, C_2$ along the coordinate z, the thickness of which, after obtaining and resolving the algebraic system of equations, tends to zero. The values of the exponents in (4.6), (4.7) will be calculated as average integral values over the reaction zone. We will also calculate the values of the degree of the mass fraction of the monomer in these equations. Then equations (4.6), (4.7) take the form

$$\frac{dC}{dz} = -\frac{k_{01}}{2u} C\langle C_2^l\rangle\langle\exp\left(-\frac{E_1}{RT}\right)\rangle, \tag{4.10}$$

$$\frac{dC_2}{dz} = -\frac{k_{02}}{u} fC\langle C_2^n\rangle\langle\exp\left(-\frac{E_2+\Delta VP}{RT}\right)\rangle. \tag{4.11}$$

The eigenvalues at the final singular point $\gamma(0, C_{2+}, T_+)$ of the system (4.8), (4.10), (4.11) are found from the discriminant

$$\begin{vmatrix} -\frac{k_{01}}{2u}\langle C_2^l\rangle\left\langle\exp\left(-\frac{E_1}{RT}\right)\right\rangle - \xi, & 0, & 0 \\ -\frac{k_{02}}{2u} f\langle C_2^n\rangle\left\langle\exp\left(-\frac{E_2+\Delta VP}{RT}\right)\right\rangle, & -\xi, & 0 \\ 0, & \frac{Q\rho u}{c_p}, & \frac{\rho u c_p}{\lambda} - \xi \end{vmatrix} = 0 \tag{4.12}$$

$$\xi_1 = \frac{\rho u c_p}{\lambda}, \xi_2 = -\frac{k_{01}}{2u}\langle C_2^l\rangle\langle\exp\left(-\frac{E_1}{RT}\right)\rangle, \xi_3 = 0. \tag{4.13}$$

Since the eigenvalues $\xi_{1,2}$ are real and have different signs, $\xi_3=0$, then the special point $(0, C_{2+}, T_+)$ has the saddle type and the particular solution corresponding to the only trajectory entering this point has the form

$$\begin{aligned} T - T_+ &= K_1\exp(\xi_2 z), \\ C_2 - C_{2+} &= K_2\exp(\xi_2 z), \\ C &= K_3\exp(\xi_2 z). \end{aligned} \tag{4.14}$$

Revealing the uncertainty at the point γ, we find our own direction at this point

$$\left(\frac{dT}{dC_2}\right)_\gamma = \frac{\xi_1}{\xi_2 - \xi_1}\frac{Q}{c_p}, \tag{4.15}$$

$$\left(\frac{dC}{dC_2}\right)_\gamma = \frac{k_{01}}{2k_{02}f}\frac{\langle C_2^l\rangle\langle\exp\left(-\frac{E_1}{RT}\right)\rangle}{\langle C_2^n\rangle\langle\exp\left(-\frac{E_2+\Delta VP}{RT}\right)\rangle}. \tag{4.16}$$

From equations (4.14) - (4.16) we find

$$K_1 = \frac{K_2 \xi_1}{\xi_2 - \xi_1}\frac{Q}{c_p}; K_3 = K_2 \frac{k_{01}}{2k_{02}f}\frac{\langle C_2^l \rangle \langle \exp\left(-\frac{E_1}{RT}\right)\rangle}{\langle C_2^n \rangle \langle \exp\left(-\frac{E_2 + \Delta VP}{RT}\right)\rangle}. \tag{4.17}$$

The value of the constant K_2 is determined from the condition of equality of derivatives dT/d z on the right and left boundaries of zones (1) and (3), respectively. Taking into account that practically no reaction occurs in zone (1), from (4.1) we obtain

$$T = (T_f - T_0)\exp(\xi_1 z) + T_0. \tag{4.18}$$

Here T_f is the temperature value on the right boundary (1), on which z = 0 and

$$dT/dz = \xi_1 (T_f - T_0). \tag{4.19}$$

Differentiating the first of equalities (4.14) with z = Δ (Δ is the width of the buffer zone), we have

$$dT/dz = K_1 \xi_2 \exp(\xi_2 \Delta). \tag{4.20}$$

Comparing (4.19) and (4.20), we get

$$K_2 = \frac{\xi_2 - \xi_1}{\xi_2}\frac{c_p}{Q}\left(T_f - T_0\right)\exp\left(-\Delta \xi_2\right). \tag{4.21}$$

Using (4.17) and (4.21), we eliminate the integration constants at (4.14)

$$\begin{aligned} T &= T_+ + (T_f - T_0)\xi_1 \xi_2^{-1}\exp(\xi_2 z - \xi_2 \Delta), \\ C_2 &= C_{2+} - (T_f - T_0)(\xi_1 - \xi_2)c_p Q^{-1}\xi_2^{-1}\exp(\xi_2 z - \xi_2 \Delta), \\ C &= \frac{\xi_2 - \xi_1}{\xi_2}\frac{c_p}{Q}(T_f - T_0)\exp(-\Delta \xi_2)\frac{k_{01}}{2k_{02}f}\frac{\exp(\xi_2 z)\langle C_2^l \rangle \langle \exp\left(-\frac{E_1}{RT}\right)\rangle}{\langle C_2^n \rangle \langle \exp\left(-\frac{E_2 + \Delta VP}{RT}\right)\rangle} \end{aligned} \tag{4.22}$$

At the boundary of zones 2 and 3 at z = 0, based on (4.22), the values of the variables are

$$T_\Delta = T_+ + (T_f - T_0)\xi_1 \xi_2^{-1}, \tag{4.23}$$

$$C_2 = C_{2+} - (T_f - T_0)(\xi_1 - \xi_2)c_p Q^{-1}\xi_2^{-1}, \tag{4.24}$$

$$C_\Delta = \frac{\xi_2 - \xi_1}{\xi_2}\frac{c_p}{Q}(T_f - T_0)\frac{k_{01}}{2k_{02}f}\frac{\langle C_2^l\rangle\langle\exp\left(-\frac{E_1}{RT}\right)\rangle}{\langle C_2^n\rangle\langle\exp\left(-\frac{E_2 + \Delta VP}{RT}\right)\rangle}. \tag{4.25}$$

Taking into account the linearity of the change in variables C, C_2, T over zone 3, integrating (4.6), (4.7), (4.19) over this zone, we obtain

$$C_\Delta = -\frac{k_{01}}{2u}\frac{C_0 + C_\Delta}{2}\left(\frac{C_{20} + C_{2\Delta}}{2}\right)^l \Delta \exp\left(-\frac{E_1}{R}\frac{2}{T_f + T_\Delta}\right) + C_0, \tag{4.26}$$

$$C_{2\Delta} = -\frac{k_{02}}{2u}\frac{C_0 + C_\Delta}{2}f\left(\frac{C_{20} + C_{2\Delta}}{2}\right)^n \Delta \exp\left(-\frac{E_2 + \Delta V}{R}\frac{2}{T_f + T_\Delta}\right) + C_{20}, \tag{4.27}$$

$$T_\Delta = \xi_1(T_f - T_0)\Delta + T_0. \tag{4.28}$$

From (4.13), (4.23) we find the formula for the wave velocity

$$u = \left[\frac{\lambda k_{01}}{2\rho c_p}\frac{T_+ - T_\Delta}{T_f - T_0}\left\langle C_2^l\right\rangle\left[\left\langle\exp\left(-\frac{E_1}{RT}\right)\right\rangle\right]\right]^{1/2}, \tag{4.29}$$

and from (4.26) we find a formula for calculating the initial mass fraction of the initiator

$$C_0 = C_\Delta\frac{1 + \frac{k_{01}}{4u}\left(\frac{C_{20} + C_{2\Delta}}{2}\right)^l \Delta \exp\left(-\frac{E_1}{R}\frac{2}{T_f + T_\Delta}\right)}{1 - \frac{k_{01}}{4u}\left(\frac{C_{20} + C_{2\Delta}}{2}\right)^l \Delta \exp\left(-\frac{E_1}{R}\frac{2}{T_f + T_\Delta}\right)}. \tag{4.30}$$

Here $C_{2\Delta}$ is given by equation (4.24), and

$$\Delta = \frac{T_\Delta - T_f}{T_f - T_0}\frac{\lambda}{\rho u c_p}. \tag{4.31}$$

Substitute (4.30) into (4.27)

$$C_{2\Delta} = -\frac{k_{02}f}{u}\frac{C_\Delta\left(\frac{C_{20} + C_{2\Delta}}{2}\right)^n \Delta \exp\left(-\frac{E_2 + \Delta V}{R}\frac{2}{T_f + T_\Delta}\right)}{1 - \frac{k_{01}}{4u}\left(\frac{C_{20} + C_{2\Delta}}{2}\right)^l \Delta \exp\left(-\frac{E_1}{R}\frac{2}{T_f + T_\Delta}\right)} + C_{20}. \tag{4.32}$$

Substituting equations (4.24), (4.26) into (4.32) and using (4.31), (4.23), (4.9), we obtain

$$\frac{2^n\langle C_2^n\rangle}{\xi_1-\xi_2}\langle\exp\left(-\frac{E_2+\Delta VP}{RT}\right)\rangle =$$

$$=\frac{\left[C_{20}+C_{2+}-\frac{c_p}{Q}(T_f-T_0)\frac{\xi_1-\xi_2}{\xi_1}\right]^n\xi_1^{-1}\exp\left(-\frac{E_2+\Delta VP}{R}\frac{2}{T_f+T_\Delta}\right)}{1-\frac{k_{01}}{4u2^l}\left[C_{20}+C_{2+}-\frac{c_p}{Q}(T_f-T_0)\frac{\xi_1-\xi_2}{\xi_1}\right]^2\Delta\exp\left(-\frac{E_1}{R}\frac{2}{T_f+T_\Delta}\right)}. \tag{4.33}$$

Let's take Δ to zero. Then from (4.29), (4.25), respectively, we have

$$u=\left[\frac{\lambda k_{01}}{2\rho c_p}\frac{T_+-T_f}{T_f-T_0}\langle C_2^l\rangle\left[\left\langle\exp\left(-\frac{E_1}{RT}\right)\right\rangle\right]\right]^{1/2}, \tag{4.34}$$

$$C_0=\frac{c_p}{Q}(T_+-T_0)\frac{k_{01}}{2k_{02}f}\frac{\langle C_2^l\rangle\langle\exp\left(-\frac{E_1}{RT}\right)\rangle}{\langle C_2^n\rangle\langle\exp\left(-\frac{E_2+\Delta VP}{RT}\right)\rangle}. \tag{4.35}$$

The equation for determining the temperature of the reaction front is found from (4.33) at $C_{2\Delta}=C_{20}$, $\Delta=0$

$$\langle C_2^n\rangle\langle\exp\left(-\frac{E_2+\Delta VP}{RT}\right)\rangle\frac{C_{20}^{-n}\xi_1}{\xi_1-\xi_2}=\exp\left(-\frac{E_2+\Delta VP}{RT_f}\right). \tag{4.36}$$

Using (4.23), from (4.36) we have

$$\langle C_2^n\rangle\langle\exp\left(-\frac{E_2+\Delta VP}{RT}\right)\rangle\frac{C_{20}^{-n}(T_+-T_f)}{T_+-T_0}=\exp\left(-\frac{E_2+\Delta VP}{RT_f}\right). \tag{4.37}$$

Let us now calculate the average integral values of the degrees of mass fractions and exponentials in the reaction zone:

$$\langle C_2^l\rangle=\frac{\int_{C_{20}}^{C_{2+}}C_2^l\,dC_2}{C_{2+}-C_{20}}=\frac{1}{1+l}\frac{C_{20}^{1+l}-C_{2+}^{1+l}}{C_{20}-C_{2+}}, \tag{4.38}$$

$$\langle C_2^n\rangle=\frac{\int_{C_{20}}^{C_{2+}}C_2^n\,dC_2}{C_{2+}-C_{20}}=\frac{1}{1+n}\frac{C_{20}^{1+n}-C_{2+}^{1+n}}{C_{20}-C_{2+}}, \tag{4.39}$$

$$\langle e^{-\frac{E_1}{RT}} \rangle = \frac{\int_{T_f}^{T_+} \exp\left(-\frac{E_1}{RT}\right) dT}{T_+ - T_f} = \frac{e^{-\beta_1} - e^{y\beta_1 - 2\beta_1}}{\beta_1 (1-y)}, \tag{4.40}$$

$$\langle e^{-\frac{E_2 + \Delta VP}{RT}} \rangle = \frac{\int_{T_f}^{T_+} \exp\left(-\frac{E_2 + \Delta VP}{RT}\right) dT}{T_+ - T_f} = \frac{e^{-\beta_2} - e^{y\beta_2 - 2\beta_2}}{\beta_2 (1-y)}.$$

Here

$$\beta_1 = E_1 / RT_+, \beta_2 = (E_2 + \Delta VP)/RT_+, y = T_f / T_+.$$

When integrating in (4.40), (4.41), D.A. Frank-Kamenetsky. It can be shown that the numerical result of integrating (4.40), (4.41) with the help of this expansion practically does not differ from the exact integration of these equations.

Substituting (4.38) - (4.41) into (4.34), (4.35), (4.37), we obtain the equations necessary for the calculation. For the polymerization of styrene and methyl methacrylate (l=0, n =1, C_{20} =1), they have the following form

$$u^2 = \frac{M^2}{2\beta_1} \frac{e^{-\beta_1} - e^{y\beta_1 - 2\beta_1}}{y - \varsigma/(1+\varsigma)}, \tag{4.42}$$

$$\left(1 - \frac{i}{2\varsigma}\right) \frac{1+\varsigma}{\beta_2} = \frac{e^{\beta_2 (y-1)/y}}{1 - e^{\beta_2 (y-1)}}, \tag{4.43}$$

$$\frac{C_0 f k_{02}}{k_{01}} = \frac{\beta_2 e^{\beta_2 - \beta_1}}{\beta_1 (2\varsigma / i - 1)} \frac{1 - e^{\beta_1 (y-1)}}{1 - e^{\beta_2 (y-1)}}. \tag{4.44}$$

Here

$$M = \left(\frac{\lambda k_{01}}{\rho c_p}\right)^{1/2}, i = \frac{\varsigma c_p}{Q}(T_+ - T_0), \varsigma = \frac{T_0}{T_+ - T_0}.$$

Figure 3 shows the results of calculating the wave velocity of the radical polymerization of methyl methacrylate (Zhizhin, 1992) according to the analytical solution obtained by the semi-infinite reaction zone method, depending on the initial concentration of the CPC initiator (dicyclohexylperoxydicarbonate) at pressure P = 3.5 10^8 Pa, T_0 =295 K (line 1).

Figure 3. Dependence of the wave velocity of the irreversible reaction of radical polymerization on the initial concentration of the initiator

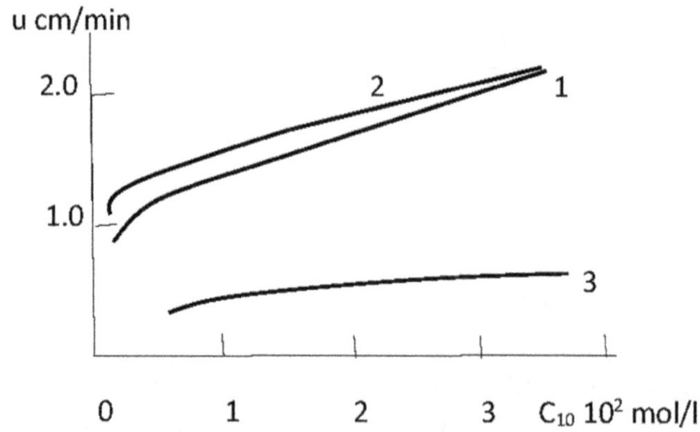

The same figure shows the experimental data From Figure 3 it can be seen that the values of the wave speed obtained by the semi-infinite reaction zone method are in good agreement with the experimental data. It should be noted that when calculating the wave velocity of radical polymerization by the method of a narrow reaction zone, without any justification, an almost 10 times increased value of the thermal conductivity coefficient of the solution was used in comparison with the value of this coefficient characteristic of most organic liquids (Davtyan et al., 1977). This led to the agreement between the calculated and experimental data in this work.

WAVE OF REVERSIBLE RADICAL POLYMERIZATION REACTION

At sufficiently high temperatures and in the absence of high pressure, the chain of monomer molecules connected to each other begins to detach molecules, i.e. a depolymerization reaction develops and the polymerization reaction becomes reversible.

Studies of the waves of the reversible reaction of radical polymerization (Zhizhin, 1997) showed that the depolymerization reaction naturally limits the degree of polymerization (conversion value) in the wave process. However, the wave velocity in this case may turn out to be higher than the wave velocity of irreversible radical polymerization at the same initial temperature and mass fractions of the monomer and initiator.

The reason for this increase in the speed of the polymerization wave lies in the reversible radical control effect of the initiator. If in the wave of irreversible radical polymerization, an increase in the concentration of the initiator inevitably leads to an increase in temperature at the end of the wave and, as a result, to an increase in the wave speed (and at the same time to a deterioration in the quality of the polymer). then in the wave of reversible radical polymerization, the temperature increase is limited by the reverse reaction. But the rate of chemical processes nevertheless increases with an increase in the concentration of the initiator, which leads to the possibility of increasing the wave speed without increasing the temperature at the end of the wave. This, apparently, can explain the high productivity values of frontal polymerization reactors, which are often observed in experiments (Zhizhin, 1990), in comparison with the calculated values determined without taking into account the depolymerization reaction.

In the coordinates associated with the wave, the equation for temperature (4.5) in the case of a reversible reaction of radical polymerization retains its form, and the equations for the mass fractions of the monomer and initiator after the introduction of the variable $C = C_1^{1/2} k_{02}/k_{01}, l = 0, n = 1$ are transformed (Zhizhin, 1997 a, b)

$$\frac{dC}{dz} = -\frac{k_{01}}{2u} C \exp\left(-\frac{E_1}{TR}\right), \tag{4.45}$$

$$\frac{dC_2}{dz} = -\frac{k_{01}}{2u} C \exp\left(-\frac{E_2}{TR}\right)\left[C_2 - \frac{k_{03}}{k_{02}} \exp\left(-\frac{Q}{RT}\right)\right]. \tag{4.46}$$

Here

$$k_{01} = \varphi k_{i0}, k_{02} = k_{p0}\left(2\frac{k_{i0}\varphi f}{k_{t0}}\right)^{1/2}, k_{03} = k_{d0}\left(\frac{2k_{i0}\varphi f^2}{k_{t0}}\right)^{1/2}, E_2 = E_p + \frac{1}{2}E_1 + \frac{1}{2}E_t,$$

E_1, E_p, E_t are activation energies, respectively, of the reactions of the decomposition of the initiator, the reactions of chain growth and chain termination; $Q=E_d-E_p$ there is a thermal effect of the reaction; E_d is the activation energy of the depolymerization reaction; T is the temperature; R is the universal gas constant; φ is the efficiency of the initiator; k_{i0} is the pre-exponential of the decay rate constant of the initiator; k_{p0} is the pre-exponent of the chain growth rate constant; k_{t0} is the pre-exponent of the chain termination rate constant; k_{d0} is the pre-exponent of the rate constant of the depolymerization reaction.

Considering the wave solutions of the system (4.5), (4.45), (4.46), it is necessary to look for functions that satisfy these equations under the boundary conditions

$$z = \pm\infty, dT/dz = dC/dz = dC_2/dz = 0.$$

The integral of equation (4.5) under the condition on $z=\pm\infty$ still has the form (4.8).

The system of equations (4.5), (4.45), (4.46) is autonomous. In the phase space (C_2,T,C), it has two zero surfaces in the open part: the plane F –

$$T - T_+ + (C_2 - C_{2+})\,Q/c_p = 0, \tag{4.47}$$

at the points of which $dT/dz=0$, and the cylindrical surface L (Figure 4),

$$C_2 = \frac{k_{03}}{k_{02}}\exp\left(-\frac{Q}{RT}\right), \tag{4.48}$$

at the points of which $dC_2=0$.

Figure 4. Phase diagram of waves of the reversible reaction of radical polymerization at l = 0

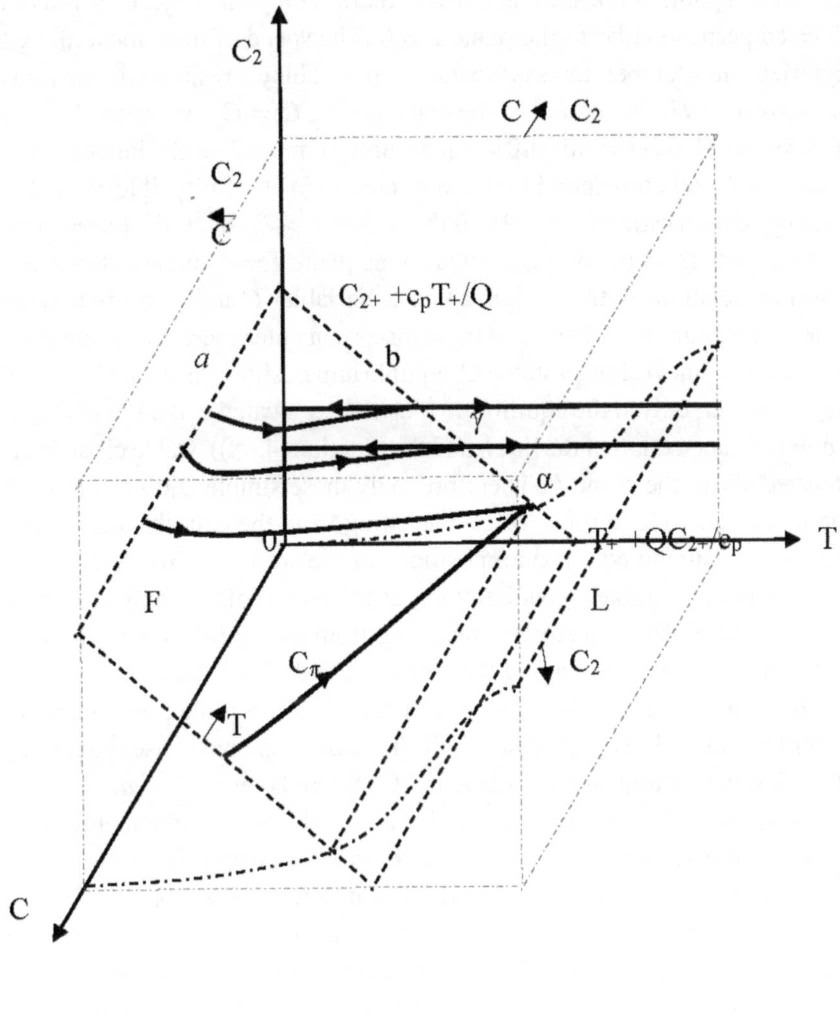

In addition, zero surfaces are coordinate planes $T = 0$ and $C = 0$ on which $dC/dz=dC_2/dz=0$.

Zero surfaces divide the phase space into regions with constant signs of the derivatives of the phase coordinates of the independent variable. Areas of positive derivatives are indicated by arrows at the corresponding zero surfaces. The derivatives dC/dz, dC_2/dz tend to zero very quickly as they approach the $T = 0$ plane due to a sharp decrease in values $e^{-E/RT}$ at $T \to 0$. So long before strict equality $T = 0$, these derivatives are practically equal to zero.

The equilibrium positions of the system (4.5), (4.45), (4.46) form two direct intersections of the plane F with the coordinate planes $T = 0$ (line a) and $C = 0$ (line b). From each point of the straight a line a, there is only one trajectory. All of them are located perpendicular to the plane $T = 0$. The speed of movement along these trajectories at low temperatures is practically zero. This corresponds to the vanishing of the derivative dT/dz. Let us assume that a $T = T_0$, $C_2 = C_{20}$ the value dT/dz can be considered equal to zero with sufficient accuracy ("cutoff" of the kinetic function). This condition will correspond to the wave condition at infinity; it leads to the integral energy conservation law (4.9). In the region $T > T_0$, the trajectories gradually begin to deviate from the perpendicular to the plane $T = 0$ and are attracted to the equilibrium positions of the straight a line b. Variables C and C_2 begin to decrease, and the temperature increases. The trajectories can enter the equilibrium positions of the straight a line b along a stable C_2 equilibrium positions is located on the $C = 0$ plane). However, only at the equilibrium positions γ located to the left of the point α (the point of intersection of the line b and the function (4.48)), stable eigendirections are located above the plane F. Therefore, only these simple equilibrium positions of the line b can include trajectories emanating from the equilibrium positions of the line a. For simple equilibrium positions of the straight a line b located to the right of the point α, stable eigendirections, as follows from the analysis, are located under the plane F. The trajectories emanating from the equilibrium positions of the straight line a cannot, judging by the signs of the derivatives, fall into the region under the plane F. The equilibrium position α itself is a complex singular point (the trajectories in its vicinity are located in the three-dimensional phase space), capable of attracting the set trajectories emanating from points on the line a.

They are characterized by large initial values of the moss fraction of the initiator. In their initial part, these trajectories are located far enough from the line b and, approaching the surface L, they cannot cross it. The intersection of the surface L is accompanied by a change in the sign of the derivative dC_2/sz from negative to positive. However, it is impossible to intersect the surface L, i. e. to be under this surface so that the mass fraction of C_2 increases. Therefore, the trajectories turn around the surface L and rush to the only equilibrium position on this surface - to the point α. We can single out some boundary initial value of the variable C such (C_π) that the trajectories with initial values less than C_π enter the equilibrium positions of the straight a line b of type γ. Moreover, each trajectory of this set to its unique position of equilibrium of the line b. If the initial value C is greater than C_π, then all trajectories that satisfy this condition, coming from the straight a line a, enter the same singular point - the equilibrium position γ. From a physical point of view, this situation can be understood as follows. At small values of the initial mass fraction of the initiator, an increase in the initial mass fraction of the initiator leads to an increase in the final temperature in the wave and to a decrease in the mass fraction of the residual monomer. When the initial mass fraction of the initiator exceeds the threshold value, an increase in the initial mass fraction of the initiator cannot lead to an increase in temperature at the end of the wave and to a decrease in the mass fraction of the residual monomer due to a significant limitation of the polymerization process by the depolymerization reaction.

Recall that values T_+, C_{2+} are parameters of a system of equations (4.5), (4.45), (4.46) since they are included in one of the equations of this system. And they also determine the final state of the medium in the wave. These parameters define the line b and characterize the end point of the trajectory located on this line. The question arises: on what part of the straight a line *b* is there a point with coordinates T_+, C_{2+} to the left of the point α, to the right of the point α, or coincides with it? These cases are determined by relation (4.48). If the given value C_{2+} is less than $\frac{k_{03}}{k_{02}}\exp\left(-\frac{Q}{RT}\right)$, then the point ($T_+$, C_{2+})is located to the right of point α. Under this condition, there is no wave solution. If $C_{2+} > \frac{k_{03}}{k_{02}}\exp\left(-\frac{Q}{RT}\right)$, then the point ($T_+$, C_{2+}) is located to the left of the point α. In this case, an increase in T_+ is possible with an increase in the initial mass fraction of the initiator. If $C_{2+} = \frac{k_{03}}{k_{02}}\exp\left(-\frac{Q}{RT}\right)$, the point ($T_+$, C_{2+}) coincides with point α. In this case, an increase in the initial mass fraction of the initiator does not lead to an increase in the final temperature T_+ in the wave. The described qualitative pattern significantly distinguishes the wave of reversible radical polymerization from the wave of irreversible radical polymerization, in which T_+ and C_{2+} can be set quite arbitrarily (if only the conservation laws are fulfilled) and in all cases $0 < C_{2+} < 1$ there is a wave solution.

THE WAVES OF TYPE γ

The wave solutions tending to equilibrium positions γ (we will call them waves of the γ type) are structurally close to the wave solutions of irreversible radical polymerization. To obtain an analytical solution, using the semi-infinite reaction zone method, we divide the γ-type wave into two semi-infinite zones: the heating zone and the reaction zone. We temporarily introduce a buffer zone between them. In the buffer zone, a linear change of variables along the spatial coordinate is assumed, and its extent after the solution of the system of algebraic equations tends to zero. As a result (omitting the transformations similar to those given for the wave of the irreversible reaction of radical polymerization), we obtain algebraic relationships between the parameters of the system (Zhizhin, 1997)

$$C_{20} - \frac{k_{03}}{k_{02}}e^{-\beta_3/y} = \left[\frac{C_{20} + C_{2+}}{2} - \frac{k_{03}}{k_{02}}\frac{e^{-\beta_3} - e^{-\beta_3(y-2)}}{\beta_3(1-y)}\right]\frac{e^{-\beta_2} - e^{-\beta_2(y-2)}}{e^{-\beta_2/y}}\frac{1+\varsigma}{\beta_2}, \tag{4.49}$$

$$C_0 = \frac{i\beta_2}{2\varsigma\beta_1}\frac{e^{-\beta_1} - e^{-\beta_1(y-2)}}{e^{-\beta_2} - e^{-\beta_2(y-2)}}\left[\frac{C_{20} + C_{2+}}{2} - \frac{k_{03}}{k_{02}}\frac{e^{-\beta_3} - e^{-\beta_3(y-2)}}{\beta_3(1-y)}\right]^{-1}, \beta_3 = \frac{Q}{RT_+}. \tag{4.50}$$

The velocity of a wave of type γ is determined by relation (4.42).

THE WAVES OF TYPE α

We will call wave solutions tending to equilibrium positions α-type waves. As follows from a qualitative analysis of the initial system of equations, trajectories with a large initial mass fraction of the initiator ($C_0 > C_\pi$), upon reaching a temperature close to equilibrium, unfold almost perpendicular to the temperature axis. As a result, in the future, the temperature along the trajectory remains almost constant, and the mass fractions of the reagents sharply decrease to their limiting values $C_0=0$, $C_2=C_{2+e}$. Thus, for waves of this type, the approximation of a narrow reaction zone is applicable, i.e. in the reaction zone, the temperature is constant and is found from the solution of the system (4.49), (4.9), if at (4.49) $T=T_+$, $C_2=C_{2+}$. Let us denote this temperature value as T_{+e}. In this case, in the heating zone, we have the solution

$$T = (T_{+e} - T_0)e^{\xi_1 z} + T_0, C = C_0, C_2 = C_{20}. \tag{4.51}$$

We will assume that at the boundary of the heating and reaction zones at z = 0, the second derivative of the temperature with respect to z from the side of the heating zone is equal to zero. Then from (4.5), taking into account (4.46), we have

$$\left.\frac{dT}{dz}\right|_{z=0} = -\frac{QC_0}{uc_p}k_{01}\left(C_{20} - \frac{k_{03}}{k_{02}}e^{-\frac{Q}{RT_{+e}}}\right)e^{-\frac{E_2}{RT_{+e}}}. \tag{4.52}$$

Differentiating (4.51), comparing the result with (4.52), taking into account (4.48), (4.9), we find an equation for calculating the speed of a wave of type α

$$u^2 = M^2 C_0 e^{\beta_{2e}}. \tag{4.53}$$

DECISION ANALYSIS.

Both from a theoretical and practical point of view, when calculating waves of radical polymerization, the question of the influence on the characteristics of waves (its speed, maximum temperature, length) of the main control parameters, a given kinetic system - the initial mass fraction of the initiator and the initial temperature of the mixture, is of interest.

However, due to the complexity of the equations, even for the model of irreversible radical polymerization, such an analysis was not carried out for a long time. This was due to the fact that analytical solutions make it relatively easy to calculate the characteristics of the wave at a given final temperature in the wave, rather than a given initial mass fraction of the initiator.

Nevertheless, the calculation, as it were, in the reverse order is possible and implemented (Zhizhin, 1997) for waves of reversible and irreversible radical polymerization. The calculation was carried out for a mixture of methyl methacrylate with benzoyl peroxide. The initial mass fraction of the monomer in all calculations was taken equal to unity (C_{20} =1). The kinetic and thermophysical values of the parameters were taken as follows: E_1 =124.24 kJ/mol, E_2 = 81.7 kJ/mol, k_{01} =3 10^{13} s, c_p = 2.053 kJ/mk, λ = 0.13 W/mk, ρ = 0.81 g/cm^3, μ = 100g \ mol is the molecular weight of methyl methacrylate. The value of the pre-exponential of the depolymerization reaction (more precisely, in the accepted notation, the ratio of the pre-exponential $\frac{k_{03}}{k_{02}}$) was found using the thermodynamic relation in the equilibrium state (Ham, 1972; Oudian, 1974), which in this case can be written as

$$RT_{+e} \ln M_{+e} = \Delta H^0 - T_{+e} \Delta s^0. \tag{4.54}$$

Here Δs^0=-117.32 J/(mol K), ΔH^0= - 54470 J/mol = - Q are change in entropy and enthalpy of methyl methacrylate at . Let's accept T_{+e}= 473 K. Then from (4.54) we have M_{+e}= 13 mol/l or C_{2+e}=0.16, and from (4.48) we find $\frac{k_{03}}{k_{02}}$ = $1.64 \cdot 10^5$. The calculation results are presented in Figure 5, Figure 6, Figure 7.

Figure 5. Dependence of the final temperature of the wave of the reversible reaction of radical polymerization on the control parameters

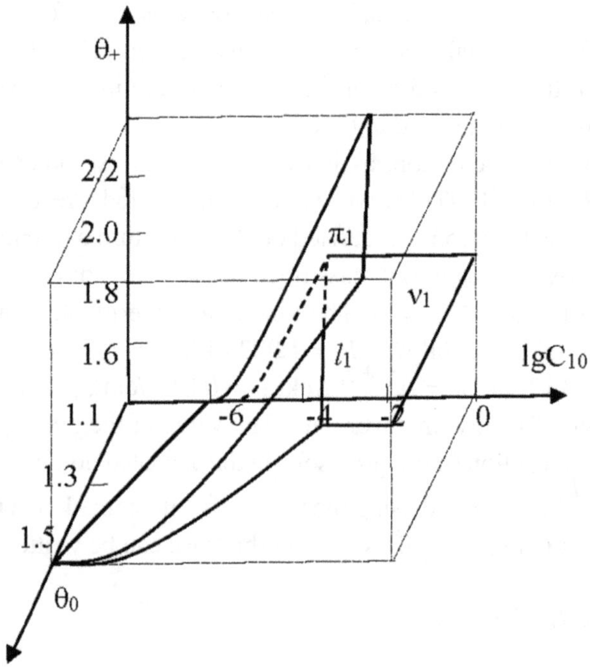

Figure 6. Dependence of the velocity of the wave of the reversible reaction of radical polymerization on the control parameters at l=0

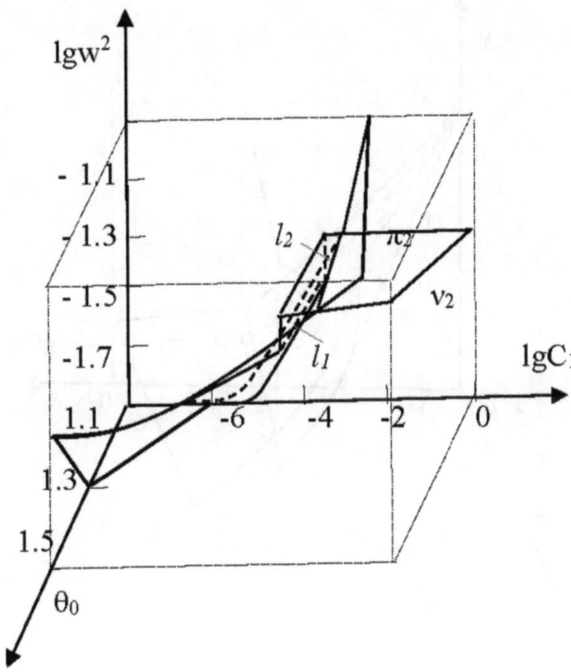

Figure 7. Dependence of the temperature of the front wave of the reversible reaction of radical polymerization on the control parameters at l=0

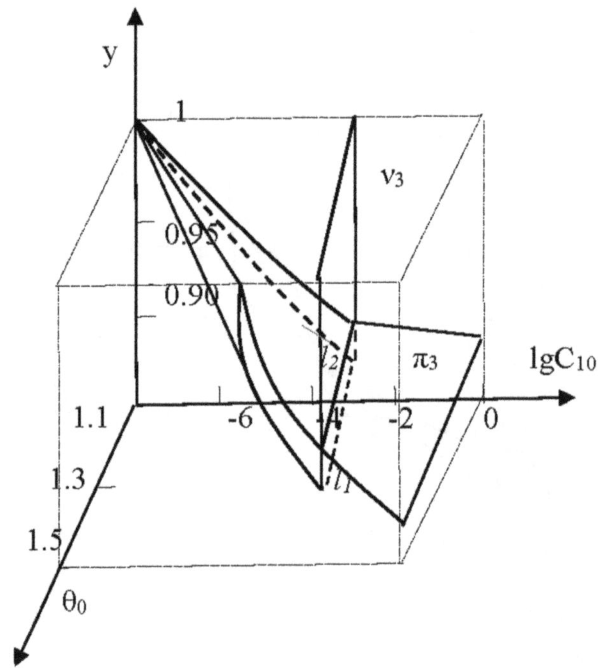

The surfaces π_1, π_2, π_3 in the parameter spaces in these figures correspond to a wave of irreversible radical polymerization.

Figure 5 shows the dependences of the dimensionless temperature $\theta_+ = T_+ c_p / Q$ at the end of the wave on the control parameters of the system $\theta_0 = T_0 c_p / Q, C_{10}$. It follows from the calculations that an increase in both the initial mass fraction of the initiator and the initial temperature of the mixture leads to an increase in the temperature at the end of the wave (both "reversible" and "irreversible"). In this case, the effect of changing the initial mass fraction of the initiator is more significant at low temperatures, and the effect of changing the initial temperature is more significant at a small initial fraction of the initiator. The solutions for "reversible" waves lie below (in θ_+) the solutions for irreversible waves, and in the region where waves of type γ exist, this difference is small and decreases with decreasing C_{10}. In the region of existence of α-type waves (to the right of the line of limiting states l_1), the value of θ_+ does not depend on C_{10} and slightly increases with increasing θ_0. Figure 6 shows the dependence of the wave speed $w^2 = u^2 \rho c_p / \lambda k_{01}$ on the control parameters C_{10}, θ_0. The wave speed, as well as the temperature at the end of the wave (by the same qualitative features) increases with an increase in the initial values of the mass fraction of the initiator and temperature. The speed of "reversible" waves of type γ is less than the speed of "irreversible" waves. However, after the line of limiting states l_1 is reached and the transition to waves of type α, an abrupt increase in the speed of "reversible" waves is observed, accompanied by a predominance of the speed of "reversible" waves over the speed of "irreversible" waves in the vicinity of the jump on the right.

In the region of existence of solutions α, the wave velocity increases with increasing initial mass length of the initiator $\approx C_{10}^{1/4}$. For some values C_{10}, θ_0 corresponding to the line l_2 (Figure 6), the velocities of "reversible" and "irreversible" waves are compared, and to the right of this line the speed of the "reversible" wave becomes less than the speed of the "irreversible" wave.

Figure 7 shows the dependence of the relative temperature of the wave front $y=T_f/T_+$ on the control parameters of the system. It follows from the figure that with an increase in the initial values of the mass fraction of the initiator and temperature, the difference between the temperature of the wave front and the temperature at the end of the wave increases for both "irreversible" and "reversible" waves of the γ type. Moreover, this difference is more significant for a "reversible" wave. Upon reaching the limit states (i.e., the line l_1), the "reversible" wave of type γ passes into a wave of type α, which is accompanied by an abrupt increase in the relative temperature of the wave front to unity.

EXPERIMENTAL STUDIES OF CONTINUOUS RADICAL POLYMERIZATION REACTORS

Prototypes of continuous radical polymerization reactors were developed and manufactured at LenNIIkhimmash (St. Petersburg) under the direction of Zhizhin G.V. They had two modifications with axial symmetry (cylindrical frontal polymerization reactor) and with central symmetry (spherical frontal polymerization reactor). The flow of the reaction medium in both modifications of these reactors was directed perpendicular to the reactor walls in order to avoid the possible effect of reactor breakdown by a monomer jet, which is typical for tubular reactors. The reactor was started up as follows. The internal volume of the reactor with the outlet valve closed was filled with the initial reaction mixture of monomer and initiator. Then the electric heating coil was switched on and the reaction mixture in the reactor was heated in the convective mixing mode. After the induction time had elapsed, the polymerization reaction began in the reactor, which proceeded in the volumetric thermal explosion mode and quickly ended. After that, the outlet valve was opened, and the metering pump was switched on at the minimum flow rate. In this case, the initial reaction mixture entered the reactor and partially displaced the final polymer product from the reaction volume. After passing the transient processes, a stationary temperature distribution was established in the reactor, and it began to operate in the continuous mode. During the entire experiment, the thermocouple readings were continuously recorded by recording devices. The thermocouple readings in the stationary mode confirmed the existence of a stable stationary cylindrical polymerization wave. Experiments confirmed the absence of a breakdown mode in these reactors. Several copies of each type of reactor were made, which were tested in various Russian organizations at existing production facilities. The geometric and regime characteristics of the reactors were determined by the method of a semi-infinite reaction zone, taking into account their design concept. The results of mathematical research, calculations and experiments are published in various scientific publications and service documents of LenNIIkhimmash (Zhizhin,1982, 1984, 1985, 1986, 1990, 1992, 1995, 1997 a, b, 1999, 2001, 2004; Zhizhin, Segal, 1983, 1985, 1988; Zhizhin, Obukhova, 1997; Zhizhin, Poritskay, 1994). Due to their special significance, these materials are not presented in this monograph.

The main thing is that the possibility of stable operation of continuous frontal polymerization reactors of a special design with high specific productivity has been proven. To obtain high-quality polymers under industrial conditions, additional technological studies are required that reflect the specifics of polymers and initiators. Unfortunately, these works in Russia were suspended due to the instability of the political system in Russia.

CONCLUSION

The semi-infinite reaction zone method in this chapter is applied for an approximate analytical solution of the system of differential equations describing stationary waves of radical polymerization. The reaction medium in this case contains the monomer, the initiator and the polymer formed during the reaction. The effect of the gel effect on the reaction rate is taken into account. In addition, the possibility of a depolymerization reaction associated with chain termination is taken into account. The semi-infinite reaction zone method in this complex system made it possible to obtain an analytical solution and completely describe the dependences of the structure of the polymerization wave and its rate on the control parameters of the system: the initial concentration of the initiator and the initial temperature of the mixture. This made it possible to calculate the concentration profiles of the components along the wave and the temperature distribution along the wave. To do this, it was necessary to solve the complex nonlinear algebraic system of equations formed as a result of integration, which contains many unknown variables and parameters, which is of independent interest. The effectiveness of the semi-infinite reaction zone method was successfully demonstrated, which divides the wave of length into two semi-infinite zones (heating and reaction) with the introduction of a buffer zone between them, the length of which, after resolving the algebraic system of equations, tends to zero. The study of systems of differential equations of the reversible reaction of radical polymerization in three-dimensional phase space revealed the existence of two types of waves tending to qualitatively different equilibrium positions in the phase space. At low initial concentrations of the initiator, an increase in the mass fraction of the initiator leads to an increase in the temperature at the end of the wave and an increase in the wave speed. However, the excess of the initial mass fraction of the initiator of a certain limit value leads to a sharp change in the nature of the wave solution. In this case, the temperature at the end of the wave is already independent of the initial concentration of the initiator and remains constant. The wave velocity of the reversible reaction of radical polymerization under these conditions can exceed the wave velocity of the irreversible reaction of radical polymerization at close values of the control parameters. The solution of systems of equations for radical polymerization waves was used in the calculation and design of experimental frontal polymerization reactors with axial and central symmetry. Experiments on these reactors confirmed the correctness of the semi-infinite reaction zone method.

REFERENCES

Bagdasaryan, Kh. S. (1966). *Theory of radical polymerization*. Nauka.

Frank-Kamenetsky, D. A. (1967). *Diffusion and heat transfer in chemical chemistry*. Nauka.

Ham, D. (Ed.). (1972). *Polymerization of vinyl monomers*. Chemistry.

Oudian, J. (1974). *Fundamentals of polymer chemistry*. Moscow: Mir.

Zhizhin, G. V. (1982). Wave structure of frontal polymerization. *Reports of the Academy of Sciences of the USSR*, *263*(6), 1399–1402.

Zhizhin, G. V. (1984). On the calculation of the wave structure of frontal polymerization. *Kinetics and Catalysis*, *2*, 292–298.

Zhizhin, G. V. (1985). Determination of the coordinates of the location of the front of radical polymerization. *Kinetics and Catalysis*, *4*, 1002–1005.

Zhizhin, G.V. (1990). *Macrokinetics in frontal polymerization reactors*: Abstract of Doc. dissertations. St. Petersburg: LTI named after Lensoviet.

Zhizhin, G. V. (1992). *Macrokinetics in frontal polymerization reactors*. Polytechnic.

Zhizhin, G. V. (1995). *Polymerization method*. RF patent 2044746 (BI 27).

Zhizhin, G. V. (1997 a). *Mathematical models of waves of radical polymerization*. Northwestern Correspondence Polytechnic Institute.

Zhizhin, G. V. (1997b). Stationary waves of reversible radical polymerization. *Chemical Physics*, *16*(3), 114–123.

Zhizhin, G. V. (1999). Semi-infinite reaction zone method in the theory of self-regulating waves of exothermic reactions and non-standard analysis. Mathematical methods in engineering and technology. *Proceedings of the 12th International Conference*. Veliky Novgorod, 14 - 17.

Zhizhin, G. V. (2001). *Anthropogenic impact on the development of biological populations in the environment. The impact of anthropogenic impact on public health and environmental quality. Materials of the regional conference*. St. Petersburg State University.

Zhizhin, G. V. (2004). *Self-regulating waves of chemical reactions and biological populations*. Science.

Zhizhin, G. V.. (1986). On stationary states of a spherical reaction front in a fluid with variable viscosity. *Kinetics and Catalysis*, *6*, 1310–1314.

Zhizhin, G. V., & Obukhova, I. A. (1997). Gel - effect at radical polymerization. *Mathematical Modelling*, *9*(11), 3–13.

Zhizhin, G.V., & Poritskaya, I.Ya. (1994). Self-regulating waves of exothermic reactions of the n-th order in condensed media. *Physics of combustion and explosion*. 6, 61 - 68.

Zhizhin, G. V., & Segal, A. S. (1983). Hydrodynamic stability of a cylindrical reaction front accompanied by a strong increase in viscosity. *Journal of Applied Mechanics and Technical Physics*, *2*, 62–71.

Zhizhin, G. V., & Segal, A. S. (1985). Stability and multiplicity of stationary states of the spherical and cylindrical reaction front in the condensed phase. *Abstracts of the All-Union Conference Dynamics of Processes and Apparatuses of Chemical Technology*. Cherkasy, 26 - 27.

Zhizhin, G. V., & Segal, A. S. (1988). Hydrodynamic stability of a spherical reaction front accompanied by a strong increase in viscosity. *Proceedings of the Academy of Sciences of the USSR. Mechanics of liquid and gas*. 3, 46-53.

Chapter 5
Waves of Could Flame

ABSTRACT

The infinite reaction zone method is used to solve mathematical problems about the propagation of the carbon disulfide oxidation reaction in the form of a cold flame. It is shown that this method makes it possible to obtain approximate analytical solutions for both one-component and two-component cold flame models. The correctness of N.N. Semyonov and V.G. Voronkov on the existence of a soliton solution in a two-component model of a cold flame. This solution was found, which N.N. Semenov and V.G. Voronkov. It was found that the approximate analytical solution obtained by the infinite reaction zone method for a one-component model of a cold flame practically coincides with the exact solution of D.A. Frank-Kamenetsky, which proves the correctness of the infinite reaction zone method. For a two-component model of a cold flame, calculations were carried out using the obtained solution, and their results were compared with the available experimental data and calculations for a single-component model of a cold flame.

INTRODUCTION

In the first chapter, the problem of flame propagation during the interaction of carbon disulfide with oxygen was considered in the simplest setting as a problem of a cold flame with first-order autocatalysis. In this formulation, the problem coincides with the problem of the propagation of a wave of a biological population in its simplest form (Kolmogorov, Petrovsky, Piskunov, 1937). For the solution, the infinite reaction zone method was used to obtain analytical expressions for the distributions of concentrations along the wave and the wave velocity. Cold flames have been observed in mixtures of hydrocarbons with air (Neiman, 1938), fluorine with hydrogen (Fedotov and Chaikina, 1972), nitrogen trichloride with air (Azatyan et al., 1976), and other mixtures. In all these flames, the thermal effect of the chemical reaction is small, and the chemical reaction occurs throughout the wave.

DOI: 10.4018/978-1-6684-7745-8.ch005

Here, for the analysis it is necessary to use the infinite reaction zone method. This chapter considers the carbon disulfide oxidation model (Semyonov, Voronkov, 1939), a one-component cold flame model with second-order autocatalysis with a maximum concentration of an intermediate reactant at the end of the wave and with a zero concentration of the intermediate reactant at the end of the wave, as well as a two-component cold flame model with second-order autocatalysis. In the first chapter, the problem of flame propagation during the interaction of carbon disulfide with oxygen was considered in the simplest setting as a problem of a cold flame with first-order autocatalysis.

5.1. COLD FLAME MODEL N.N. SEMYONOV AND V.G. VORONKOV

Let us now turn to a discussion of the cold flame model proposed (Semyonov and Voronkov, 1939) for the analysis of a specific carbon disulphide oxidation reaction $CS_2+O_2\rightarrow\ldots$. It was assumed that the reaction between the initial substance A (fuel CS_2) and another initial substance B (oxidizer O_2) proceeds in a chain way through two active centers N_1 and N_2. Moreover, in one of the reactions, for example, the center N_1), not only the center N_2 is formed, but also an intermediate substance ξ

$$N_1 + B \xrightarrow{k_1} N_2 + \xi,$$

$$N_2 + A \xrightarrow{k_2} N_1 + \ldots$$

Here k_i is the constant of the rate of reaction i. The intermediate substance disappears in the future in two ways:

1) by reaction with the starting materials:

$$\xi + A + B \xrightarrow{k_4} \text{final product},$$

2) by a reaction between two molecules ξ:

$$\xi + \xi \xrightarrow{k_3} N_1 + \text{final product}.$$

Moreover, in the last reaction, along with the final product, an active center N_1 is formed, i.e. branching of the chain occurs (quadratic branching, or mutual assistance of chains).

Assuming that the lifetime of the intermediate product is relatively long, the mixture is poor in fuel, so that the concentration of the oxidizer can be considered approximately constant, using a quasi-stationary approximation with respect to active centers, a system of equations for the stationary propagation of a cold flame was obtained, which can be written as

$$D_\xi \frac{d^2 n}{dz^2} - u\frac{dn}{dz} - k_4 b(1-\eta)n + f(1-\eta)n^2 = 0, \tag{5.1}$$

$$D_A \frac{d^2 \eta}{dz^2} - u\frac{d\eta}{dz} + f(1-\eta)n^2 = 0. \tag{5.2}$$

Here

$$n = [\xi]/[CS_2]_0, \eta = \Delta[CS_2]/[CS_2]_0, f = [CS_2]_0 k_1 k_3/(k_5+k_6), \Delta[CS_2]$$

is the amount of fuel that has disappeared to a given coordinate z, $[CS_2]_0$ is the initial concentration of the fuel in the initial mixture, $b = [B]/[CS_2]_0$ the dimensionless concentration of the oxidizing agent, k_5 and k_6 are the rate constants of the chain termination reaction in the volume and on the surface of the tube through which the flame propagates, respectively.

The system of equations (5.1), (5.2) is a two-component cold flame model. The solution of this system would have to determine, in the opinion of the authors, the dependences of ξ, η at z in the cold flame wave, and also to find the unknown value of the wave velocity under the boundary conditions $\xi=\eta=0$, $z=-\infty$ and $\xi=0$, $\eta=1$, $z=+\infty$. And in the middle of the interval ξ has a maximum. After the derivation of system (5.1), (5.2), N.N. Semyonov and V.G. Voronkov wrote: "We failed to solve the system (5.1, (5.2) and had to make a new assumption, which is that the speed of the flame front is determined mainly by the first stage of the reaction, where the main mass of the fuel is converted into intermediate products. In this case, we can get approximate solution, setting n = η in equation (5.1) and integrating this equation alone".

Equation (5.1) then takes the form

$$D\frac{d^2 n}{dz^2} - u\frac{dn}{dz} - k_4 b(1-n)n + f(1-n)n^2 = 0, \tag{5.3}$$

This equation is a one-component cold flame model.

Further, the authors write (Semenov, Voronkov, 1939): Unfortunately, we could not prove mathematically that the solution obtained in this case can be considered as the first approximation of the system (5.1), (5.2). However, the hypothesis physically accepted by us seems probable." Why did N.N. Semenov and V.G. Voronkov fail to mathematically prove the validity of their model? And here's why. that equation (5.3) has an exact solution, from which follows the expression for the wave velocity.

According to (5.4), the wave velocity is equal to zero at $f = 2bk_4$. For $f < 2bk_4$, there is no stationary wave solution, i. e. Disruption of the stationary regime of flame propagation occurs at zero velocity. In experiments, it was observed that the disruption of the stationary flame propagation occurs at a finite wave velocity. This contradiction was the reason for abandoning the one-component cold flame model. Moreover, the two-component cold flame model proposed by N.N. Semyonov and V.G. Voronkov, expressed by equations (5.1), (5.2), was also essentially rejected, since later the two-component model of a cold flame was studied in a different form (Novozhilov, Posvyansky, 1973). It was assumed that the intermediate substance interacts in this reaction only with the oxidizing agent, and this leads to a change in the form of the equations describing the stationary propagation of a cold flame.

In this chapter we will show that a qualitative analysis of equation (5.3) reveals the existence of wave solutions in this equation in the form of waves of conquest and solitary waves. therefore, the rejection of this equation in the description of the cold flame model with second-order autocatalysis was not justified. The existence of one exact solution does not exclude the existence of other solutions.

5.2. ONE-COMPONENT MODEL OF A COLD FLAME WITH SECOND-ORDER AUTOCATALYSIS (THE CONCENTRATION OF THE INTERMEDIATE REACTANT AT THE END OF THE WAVE IS MAXIMUM)

Let us qualitatively study equation (5.3), trying to find wave solutions satisfying this equation. Approximate analytical solutions will be sought using the infinite reaction zone method.

By introducing dimensionless quantities

$$n = [\xi]/[A]_0, w = u[A]_0^{-1}k(k_2k_3D)^{-1/2},$$

$$\beta = [B]kk_4(k_2k_3[A])^{-1}, \varsigma = z[A]_0\left(\frac{k_2k_3}{Dk}\right)^{1/2},$$

equation (5.3) is reduced to the system of equations (3.55), (3.56), where

$$F(n) = n(1-n)(n-\beta). \qquad (5.5)$$

The absolute value F, which is the reaction rate, has two factors that increase with decreasing n. This corresponds to second-order autocatalysis. The zero isocline in the phase plane (p, n), on which $dp/d\varsigma=0$, is given by the equality $p=w^{-1}(1-n)(n-\beta)$. This null isocline (dashed line in Figure 1) intersects the null isocline p = 0, at which $dn/d\varsigma=0$, at three points $a(0,0)$, $b(0,1)$, $c(0,\beta)$. They represent the equilibrium positions of the system. From the discriminant of the system (3.55), (3.56), taking into account (5.5), in the vicinity of the point a, we find the eigenvalues equal to the eigendirections at this point

$$\lambda_{5,6} = w/2 \pm \sqrt{w^2/4 + \beta} = (dp/dn)_{5,6}, \tag{5.6}$$

those the singular point a in this case has the type of a saddle, since $\lambda_5>0$, $\lambda_6<0$.

From the discriminant of the system (3.55), (3.56), taking into account (5.5), in the vicinity of the point b, we find the eigenvalues equal to the eigendirections at this point

$$\lambda_{7,8} = w/2 \pm \sqrt{w^2/4 + 1 - \beta} = (dp/dn)_{7,8} \tag{5.7}$$

Since $\beta < 1$, it follows from (5.7) that the point b has the saddle type $\lambda_7>0$, $\lambda_8<0$. We are interested in the possibility of the existence of a solution connecting points a and b, i. e. this solution is a separatrix going from saddle a to saddle b (trajectory S_1 in Figure 1). It will correspond to the integral curve in Figure 2 (wave of conquest).

Figure 1. Phase plane trajectory of a one-component cold flame model with second-order autocatalysis

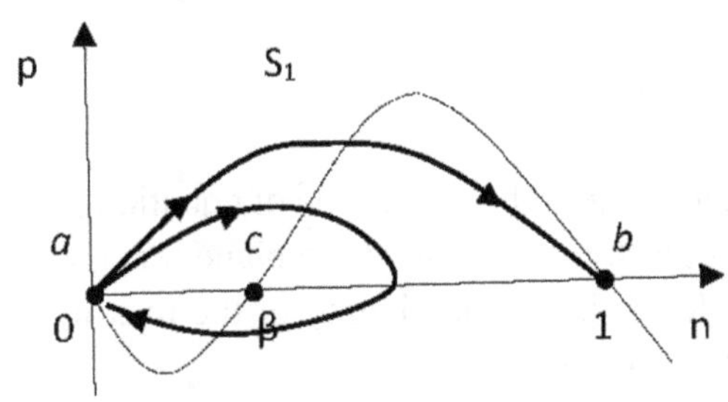

S_1 - solution with the maximum concentration of the intermediate reagent at the end of the wave; - solution with zero concentration of the intermediate reagent at the end of the wave

Figure 2. Phase plane trajectory of a one-component cold flame model with second-order autocatalysis

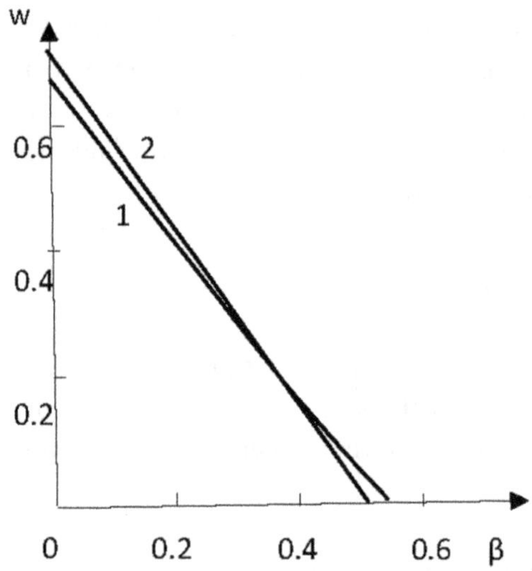

S_2 - solution with the maximum concentration of the intermediate reagent at the end of the wave; - solution with zero concentration of the intermediate reagent at the end of the wave

We divide the wave into three zones (Figure 2). In zone 1 ($-\infty<\varsigma<0$), we consider the solution of the system linearized in the vicinity of the equilibrium state *a* to be fair

$$n = n_f \exp(\lambda_5 \varsigma). \tag{5.8}$$

Here n_f is the concentration at the wave front $\varsigma=0$.

In zone 2 ($\Delta \leq \varsigma < +\infty$, we consider the solution of the system linearized in the vicinity of the final equilibrium state *b* to be valid

$$n = 1 - (1 - n_\Delta)\exp(\lambda_8(\varsigma - \Delta)). \tag{5.9}$$

In zone 3, we linearize the right - hand side of equation (3.55) in the vicinity of the maximum of the trajectory S_1 in the phase plane (expand the right - hand side of (3.56) into the Taylor series in the vicinity of the maximum and limit the first nonzero member of the series)

$$\frac{dp}{dn} = -F'_{nm}(n - n_m)p^{-1}. \tag{5.10}$$

Here n_m is concentration at maximum of trajectory, F'_{nm} is derivative F on n in point of maximum of trajectory S_1.

We integrate (5.10) in view of (3.56)

$$(n - n_m)(F'_{nm})^{1/2}/p_m = \sin((F'_{nm})^{1/2}\varsigma + E). \tag{5.11}$$

Here

$$E = \arcsin((n_f - n_m)F'^{1/2}_{nm}/p_m), \tag{5.12}$$

$$p_m = -F_m/w. \tag{5.13}$$

Put in (5.11) $\varsigma=\Delta$, then we have

$$(n_\Delta - n_m)(F'_{nm})^{1/2}/p_m = \sin((F'_{nm})^{1/2}\Delta + E). \tag{5.14}$$

Here n_Δ is sense of n at $\varsigma=\Delta$.

Differentiating (5.8), (5.9), (5.11) once, one equates the first derivatives on the border of zones 1 - 3 and 2 – 3

$$n_f^2\lambda_5^2 = p_m^2 - F'_{nm}(n_f - n_m)^2, \tag{5.15}$$

$$\lambda_8^2(1 - n_\Delta)^2 = p_m^2 - F'_{nm}(1 - 2n_m)(n_\Delta - n_m)^2. \tag{5.16}$$

Differentiating (5.8), (5.9) twice, we equate the second derivatives on the border of zones 1 - 3 and 2 – 3

$$n_f\lambda_5^2 = -F'_{nm}(n_f - n_m), \tag{5.17}$$

$$\lambda_8^2(1 - n_\Delta) = F'_{nm}(n_\Delta - n_m). \tag{5.18}$$

Resolving the resulting algebraic system of equations, we reduce it to a system of two equations with respect to w and n_m

$$(1 - n_m)(n_m - \beta) = \lambda_5 w(1 + \lambda_5^2/\alpha_1)^{-1/2}, \tag{5.19}$$

$$n_m(n_m - \beta) = -\lambda_8 w(1 + \lambda_8^2/\alpha_1)^{-1/2}. \tag{5.20}$$

Here

$$\alpha_1 = 3(n_1 - n_m)(n_m - n_2), n_{1,2} = (\beta + 1 \pm \sqrt{\beta^2 - \beta + 1})/3.$$

After finding the velocity in (3.71), the remaining parameters are determined by the equations already obtained.

System (5.19), (5.20) determines, in particular, the dependence of the velocity w at the parameter β. Figure 3 shows the results of calculating this dependence (line 1). As a comparison, here is also the dependence of the wave velocity on the parameter β (line 2), which is described in a dimensionless form by the relations (see (5.4))

$$w = (1 - 2\beta)2^{-1/2}, p = n(1 - n)2^{-1/2}.$$

Figure 3. Dependence of the velocity in the one-component cold flame model on the parameter at the maximum concentration of the intermediate reagent at the end of the wave

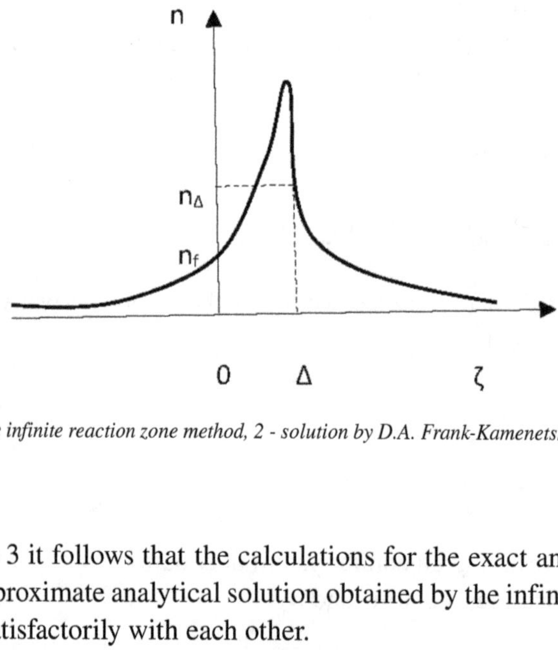

1 - solution by the infinite reaction zone method, 2 - solution by D.A. Frank-Kamenetsky (Frank-Kamenetsky, 1967).

From Figure 3 it follows that the calculations for the exact analytical solution (5.4) and the approximate analytical solution obtained by the infinite reaction zone method agree satisfactorily with each other.

Since the solution with boundary conditions $n=0$, $\varsigma=-\infty$; $n=1$, $\varsigma=+\infty$, obtained in this section leads to a zero value of the flame velocity at the limit of the existence of its steady-state regime (and this contradicts experiments), we proceed to analyze the possibility of describing soliton-like solutions with respect to the concentration of the intermediate reagent using a one-component cold flame model.

5.3. ONE-COMPONENT MODEL OF A COLD FLAME WITH SECOND-ORDER AUTOCATALYSIS (THE CONCENTRATION OF THE INTERMEDIATE REACTANT AT THE END OF THE WAVE IS ZERO)

In this section, it will be shown that the physical intuition of N.N. Semyonov and V.G. Voronkov was justified, and the reaction-diffusion equation in a one-component approximation can be the basis of solitary cold flame waves, despite the fact that this equation does not have exact solutions of the soliton type (Loskutov, Mikhailov, 1990). The infinite reaction zone method makes it possible to approximately find such solutions based on the one-component reaction-diffusion equation.

A qualitative study of equation (5.3) showed that in the phase plane there can be a trajectory in the form of a loop of the separatrix S_2, outgoing and entering the equilibrium position *a*. Let us analytically consider the possibility of the existence of a trajectory emanating from the equilibrium position *a* with an eigenvalue (Figure 1) and entering the same equilibrium position with an eigenvalue, i. e. trajectory S_2. Let's divide the length of the desired wave solution (Figure 4) into three zones. In zone 1 and zone 2, we consider the solution of the system (3.55), (3.56), (5.5) linearized in a neighborhood of the point *a* to be valid. They have the form (5.8) and

$$n = n_{\Delta} \exp(\lambda_6(\varsigma - \Delta)). \tag{5.21}$$

Modeling of solutions in zones 1, 2 takes into account the fundamental feature of exact solutions in the vicinity of the point *a* - the output and input of solutions in eigendirections (5.6). In this case, both kinetic processes and transfer processes are taken into account. In zone 3 ($0<\varsigma<\Delta$) containing the maximum of the soliton, we expand the right side of equation (3.56) in the vicinity of the maximum in a Taylor series, restricting ourselves to the first non-zero term of the series

$$\frac{dn}{d\varsigma} = -F_m(\varsigma - \varsigma_m). \tag{5.22}$$

Differentiating (5.22) we get that in zone 3 instead of (3.55) the equation $dp/d\varsigma = -F_m$ is used. Comparing this equation with (3.55), we can conclude that, from a physical point of view, linearization (5.22) corresponds to neglecting the convective motion in zone 3 and averaging kinetic processes over the volume of the zone. Such assumptions seem to be quite acceptable, since kinetic processes are most active in this zone and diffusion transfer is highly developed due to large concentration gradients.

Integrating (5.22) we get

$$n = n_m - 0.5F_m(\varsigma - \varsigma_m)^{2}. \tag{5.23}$$

On the boundaries of zone 3 from (5.23) we have

$$n_f = n_m - 0.5F_m\varsigma_m^2, \tag{5.24}$$

$$n_\Delta = n_m - 0.5F_m(\Delta - \varsigma_m)^{2}. \tag{5.25}$$

Differentiating (5.8), (5.21) once, we equate the first derivatives at the boundaries of zones 1 - 3, 2 - 3, taking into account (5.22)

$$n_f\lambda_5 = \varsigma_m F_m, \tag{5.26}$$

$$n_\Delta\lambda_6 = -(\Delta - \varsigma_m)F_m. \tag{5.27}$$

Let us set the derivative $d^2n/d\varsigma^2$, at the boundary of zones 1 - 3, 2 - 3 equal to zero and substitute in the obtained expressions the value of the first derivative $dn/d\varsigma$ from the algebraic system of equations

$$n_f\lambda_5 = F_f/w, \tag{5.28}$$

$$n_\Delta\lambda_6 = F_\Delta/w, \tag{5.29}$$

where the meaning of the indices for F is the same as for n.

Omitting transformations with respect to the resolution of the algebraic system of equations, we present the result. The system can be reduced to one equation with respect to z_1

$$1 = n_m\Psi z_2 + \frac{z_1(1-n_m)(\beta_0-1)-1}{(1-\beta_0\Psi z_2)(1-n_m)z_1(\beta_0-1)}, \tag{5.30}$$

$$n_m = \frac{\beta_0\Psi - z_1(\beta_0-1)}{\Psi(\beta_0\Psi-1)-z_1(\beta_0-1)}, \tag{5.31}$$

$$z_2 = z_1(1-n_m)(\beta_0-1)\Omega, \tag{5.32}$$

$$\Omega = \frac{1-z_1(1-n_m)(\beta_0-1)+\sqrt{1+2z_1}+\sqrt{z_1}\sqrt{(1-n_m)(\beta_0-1)+2}}{z_1(1-n_m)(\beta_0-1)-1+\sqrt{1+2z_1}+\sqrt{z_1}\sqrt{(1-n_m)(\beta_0-1)+2}}, \tag{5.33}$$

$$\Psi = (-1+\sqrt{1+2z_1})/z_1, \tag{5.34}$$

$$z_1 = \nu_1^2(1-n_m)^{-1}(\beta_0-1)^{-1}, \tag{5.35}$$

$$\nu_1 = \lambda_5 \beta^{-1/2}, \tag{5.36}$$

$$\beta_0 = n_m / \beta. \tag{5.37}$$

The calculation algorithm is as follows. By setting the value of β_0, we find from (5.30) - (5.34) z_1. Then, using the values β_0, z_1 from (5.31), we find n_m. Using the obtained values β_0, z_1, n_m we determine ν_1 according to (5.35). Since in accordance with (5.6), (5.36)

$$\nu_{1,2} \lambda_{5,6} / \beta^{1/2} = 0.5\nu \pm \sqrt{0.25\nu^2 + 1}, \tag{5.38}$$

$$\nu = w / \beta^{1/2}, \tag{5.39}$$

then from (5.38) we have

$$\nu = \nu_{1,2} - \nu_{1,2}^{-1}. \tag{5.40}$$

Consequently, after finding ν_1, one can find ν by (5.40). From the values β_0, n_m from (5.37) we find β, after which from (5.39) we calculate w, and from (5.38) we find the values of $\lambda_{5,6}$. The coordinate of the soliton maximum is found from the equation

$$\varsigma_m = \lambda_5^{-1}(-1 + \sqrt{1 + 2z_1}). \tag{5.41}$$

The width of zone 3 is calculated from the ratio

$$\Delta = \lambda_6^{-1} + \varsigma_m + \sqrt{(\lambda_6^{-1} + \varsigma_m)^2 + 2\varsigma_m(\lambda_5^{-1} - \lambda_6^{-1})}. \tag{5.42}$$

and the values of concentrations at the boundaries of the zones are determined by (5.26), (5.27). The concentration profiles for each zone are calculated according to (5.8), (5.21), (5.23).

Using the infinite reaction zone method, we will show that a one-component model of a cold flame leads to the existence of a peninsula of cold flame propagation, which is characteristic of a non-thermal flame (Azatyan et al., 1976; Azatyan, 2002; Ayvazyan, Sinelnikova, 1994). The results of calculating the dependence of the cold flame wave velocity on the cutoff parameter using the infinite reaction zone method are shown in Figure 4.

Figure 4. Dependence of the velocity in a single-component model of a cold flame wave on the parameter b for second-order autocatalysis and zero concentration of the intermediate reagent at the end of the wave

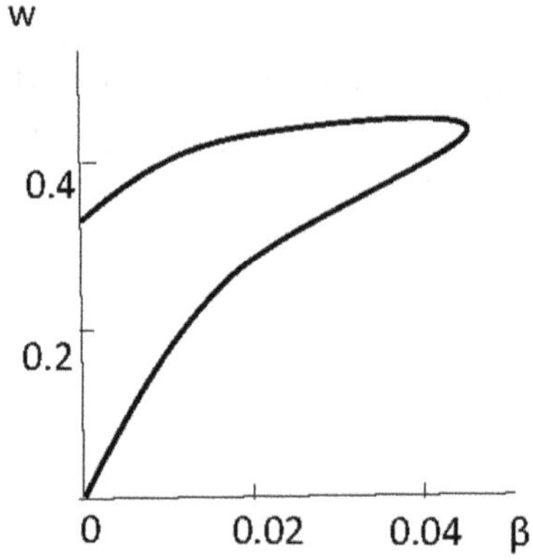

From Figure 4 it follows that the critical value of the breakage parameter β, corresponding to the disruption of a stationary flame, is 0.0465. The velocity value corresponding to the breakdown of a stationary flame differs from zero and is equal to 0.42. The dependence w(β) in Figure 4 is represented by two branches corresponding to fast and slow waves. The lower branch with a lower speed, as is observed in similar flame propagation problems (Zel'dovich et al., 1980), is apparently unstable. Using the experimental dependences of the kinetic parameters of the carbon disulfide oxidation reaction on pressure and temperature (Semenov, Voronkov, 1939), we can find an expression for the parameter β as a function of pressure P and temperature T

$$\beta = g(CP^2 + B)(AP)^{-1}, 2gBA^{-1} = K_1 \exp(\alpha_1/T), A(2gC)^{-1} = K_2 \exp(-\alpha_2/T).$$

Considering that the value of the parameter β at the limiting point of the existence of a stationary regime of a cold flame is 0.0465, the last expression can be reduced to the following dependence of pressure on temperature

$$P = \beta_m K_2 \exp(-\alpha_2/T)\left(1 \pm \sqrt{1 - K_1 K_2^{-1} \beta_m^{-2}\left(\exp\frac{\alpha_1 + \alpha_2}{T}\right)}\right).$$

Here K_1=0.2; α_1=700; α_2=1150; K_2=4.5•$10^4\gamma^{-1}$; T is temperature (in K); γ is the ratio of the partial pressure of oxygen to the pressure of a mixture of gases, P is the pressure (in mm) of the mercury column.

The calculation results for this dependence for two values of γ presented in Figure 5.

Figure 5. Areas of existence of a cold flame of carbon disulfide oxidation solid lines - calculation by the infinite reaction zone method; dotted lines - experimental data (Semyonov, Voronkov, 1939);

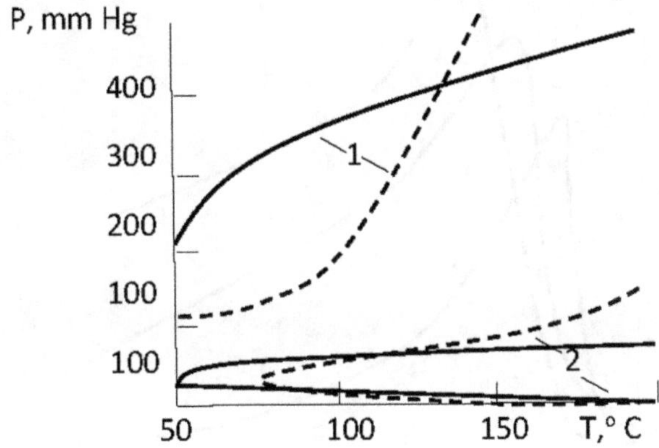

1 - γ = 0.2; 2 - γ = 1.

From Figure 5 it can be seen that the regions of existence of a cold flame of carbon disulfide oxidation, found experimentally and by the method of an infinite reaction zone based on a one-component model, are qualitatively consistent with each other. Figure 6 shows the results of calculating the concentration profile of the intermediate reactant in a cold flame wave for three values of parameter β.

Figure 6. Intermediate concentration profiles in a one-component cold flame model for second-order autocatalysis and zero intermediate reagent concentration at the end of the wave

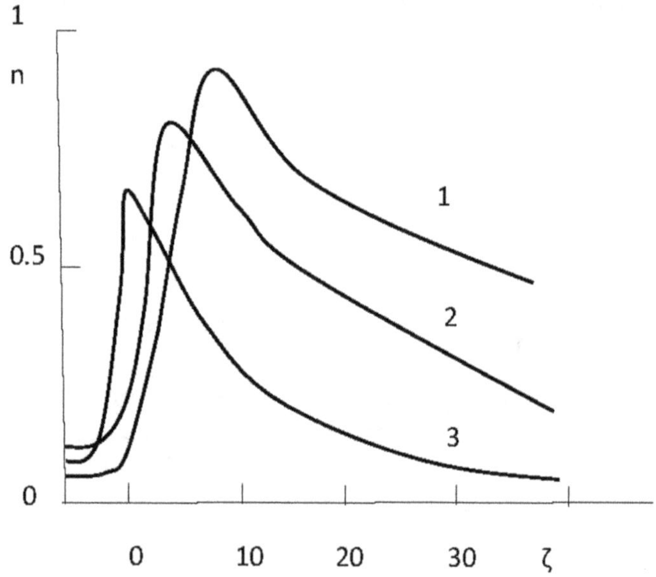

1 - β=0.0164; 2 - β=0.0312; 3 - β=0.0465

From Figure 6 it can be seen that a decrease in the parameter b from the critical value leads to a significant stretching of its part to the right of the top, while the value of the concentration of the intermediate reagent at the maximum increases.

Thus, using the infinite reaction zone method, it was possible to prove (Blinded for Review, 2004) that the one-component model of a cold flame with second-order autocatalysis (Semyonov and Voronkov, 1939) qualitatively correctly describes the observed conditions for disruption of the stationary regime of cold flame propagation.

5.4. TWO-COMPONENT COLD FLAME MODEL WITH SECOND-ORDER AUTOCATALYSIS

At the beginning of the fifth chapter, we said that in order to describe a cold flame with second-order autocatalysis, two components of the reaction medium must be taken into account. Such a two-component cold flame model was constructed by N.N. Semyonov and V.G. Voronkov (1939). However, they were unable to solve the system of equations of this model and turned to simplified models. In this section, we will show that the infinite reaction zone method can also cope with this difficult task. Let us turn to the original non-stationary diffusion equations for the concentrations of the intermediate reagent and fuel. With the adopted kinetic scheme, these equations have the form

$$\frac{\partial[\xi]}{\partial t} = D_\xi \frac{\partial^2[\xi]}{\partial x^2} + \frac{k_2 k_3}{k}[A][\xi]^{\,2} - k_4[\xi][B][A], \tag{5.43}$$

$$\frac{\partial[A]}{\partial t} = D_A \frac{\partial^2[A]}{\partial x^2} - \frac{k_2 k_3}{k}[A][\xi]^{\,2}. \tag{5.44}$$

In the analysis this time we will use dimensionless variables, whose scales consist only of the parameters included in the original non-stationary differential equations. Such non-dimensionalization of variables expands the possibilities of qualitative analysis and is widely used in the group study of differential equations (Ovsyannikov, 1978).

Let us introduce the wave coordinate $z=ut-x$ and dimensionless variables

$$\zeta = z(k_4/D_\xi)^{\,1/2}, w = u(k_4 D_\xi)^{\,-1/2}, n = [\xi]/\varsigma, r = [A]/\varsigma, \varsigma = (kk_4)^{\,1/2}(k_2 k_3)^{\,-1/2}.$$

Then the system (5.43), (5.44) is transformed to the form

$$w\frac{dn}{d\zeta} = \frac{d^2 n}{d\zeta^2} + nr(n-\beta), \tag{5.45}$$

$$w\frac{dr}{d\zeta} = \gamma\frac{d^2 r}{d\zeta^2} - n^2 r, \tag{5.46}$$

where $\gamma=D_A/D_\xi$, $\beta=[B]/\varsigma$.

This system can be written as four first-order equations

$$dp/d\zeta = wp + nr(\beta - n), \tag{5.47}$$

$$dV/d\zeta = \gamma^{-1}wV + \gamma^{-1}n^2 r, \tag{5.48}$$

$$dn/d\zeta = p, \tag{5.49}$$

$$dr/d\zeta = V. \tag{5.50}$$

The equilibrium positions of the system (5.47) - (5.50) form the r axis in four-dimensional space (p, V, n, r), i. e. the line p = 0, V = 0, n = 0. From the discriminant of the system (5.47) - (5.50) in the vicinity of an arbitrary point on the r axis, i. e. points r_s, we find the eigenvalues of singular points

$$\lambda_1 = \gamma^{-1} w, \lambda_{2,3} = 0.5w \pm \sqrt{0.25 w^2 + r_s \beta}, \tag{5.51}$$

those. equilibrium positions are of saddle-node type. The wave solution can leave the initial equilibrium position ($r_s = r_-$) only with a positive eigenvalue, i.e. either with λ_1 or

$$\lambda_2 = 0.5w + \sqrt{0.25 w^2 + r_- \beta}. \tag{5.52}$$

The wave solution can enter the final equilibrium position only with a negative eigenvalue, i.e., with

$$\lambda_3 = 0.5w - \sqrt{0.25 w^2 + r_+ \beta}. \tag{5.53}$$

Along the wave solution, the fuel concentration only decreases, since no fuel is created again in any of the reactions. Moreover, as we approach the initial and final equilibrium positions, the derivative d r/d ζ tends to zero. Therefore, in its middle part, the profile r(ζ) has an inflection point. In addition, since we are interested in a soliton-type solution, we will assume that somewhere in the middle part of the wave solution there is a maximum in the profile n(ζ).

In accordance with the infinite reaction zone method, we divide the cold flame wave into three zones. In the first zone ($-\infty<\zeta<0$), we consider the solution of the system (5.47) - (5.50) to be fair in the vicinity of the initial equilibrium position

$$r = r_- - (r_- - r_f)\exp(\lambda_{1,2}\zeta), \tag{5.54}$$

$$n = n_f \exp(\lambda_{1,2}\zeta), \tag{5.55}$$

where index f correspondent of variables at $\zeta = 0$.

In the second zone ($\Delta<\zeta<+\infty$), we consider the solution of the system in the vicinity of the final equilibrium position to be fair

$$r = r_+ + (r_\Delta - r_+)\exp(\lambda_3(\zeta - \Delta)), \tag{5.56}$$

$$n = n_\Delta \exp(\lambda_3(\zeta - \Delta)), \tag{5.57}$$

where index Δ correspondent of variables at $\zeta = \Delta$.

In the third zone ($0<\zeta<\Delta$), given that it contains an inflection point in the profile r (ζ) and has a finite length, we will accept a linear dependence r (ζ)

$$r = r_f - \frac{r_f - r_\Delta}{\Delta}\zeta. \tag{5.58}$$

Equation (5.58) is obtained from the condition of continuity of change r at the zone boundary. We note that in this case, due to the increase in the number of variables and the increase in the dimension of the phase space, the length of the buffer zone does not tend to zero. Its length is not given but is determined from the solution of an algebraic system of equations, remaining finite. This is the development of the infinite reaction zone method for phase spaces of dimension greater than three. Assuming, that zone 3 contains the maximum of the profile n(ζ), we expand the right-hand side of equation (5.49) in the vicinity of the maximum in a Taylor series, restricting ourselves to the first non-zero term of the series

$$\frac{dn}{d\zeta} = n_m r_m (\beta - n_m)(\zeta - \zeta_m). \tag{5.59}$$

Integrating (5.59), we get

$$n = n_m + 0.5 r_m n_m (\beta - n_m)(\zeta - \zeta_m)^2. \tag{5.60}$$

On the boundaries of zone 3 from (5.60) we have

$$n_f = n_m + 0.5 r_m n_m (\beta - n_m)\zeta_m^2, \tag{5.61}$$

$$n_\Delta = n_m + 0.5 r_m n_m (\beta - n_m)(\Delta - \zeta_m)^2. \tag{5.62}$$

From the condition of continuity of the first derivatives $\frac{dr}{d\zeta}, \frac{dn}{d\zeta}$ at the boundary of zones 1 and 3 from (5.54), (5.58), (5.55), (5.59) for $\zeta = 0$ we have

$$\lambda_{1,2}(r_- - r_f) = \frac{r_f - r_\Delta}{\Delta}, \tag{5.63}$$

$$\lambda_{1,2} n_f = -r_m n_m (\beta - n_m)\zeta_m. \tag{5.64}$$

From the condition of continuity of the first derivatives $\frac{dr}{d\zeta}, \frac{dn}{d\zeta}$ at the boundary of zones 2 and 3 from (5.56), (5.58), (5.57), (5.59) for $\zeta = \Delta$ we have

$$\lambda_3(r_\Delta - r_+) = -\frac{r_f - r_\Delta}{\Delta}, \tag{5.65}$$

$$\lambda_3 n_\Delta = r_m n_m (\beta - n_m)(\Delta - \zeta_m). \tag{5.66}$$

At the vertex of the soliton, from (5.58) we obtain

$$r_m = r_f - \frac{r_f - r_\Delta}{\Delta}\zeta_m. \tag{5.67}$$

As a result of linearization, the second derivatives with respect to ζ from n and r at the boundaries of zones 1 - 3, 2 - 3 have a discontinuity. From the symmetry condition, we assume that the sum of the second derivatives $\frac{d^2r}{d\zeta^2}$ to the left of the point $\zeta = 0$ and to the right of the point $\zeta = \Delta$ is equal to zero. Using (5.54), (5.56) we get

$$(r_\Delta - r_+)\lambda_3^2 - (r_- - r_f)\lambda_{1,2}^2 = 0. \tag{5.68}$$

We set the second derivatives $\frac{d^2r}{d\zeta^2}, \frac{d^2n}{d\zeta^2}$ in the initial equations (5.45), (5.46) to zero at the point $\zeta = \Delta$

$$wn_\Delta\lambda_3 + n_\Delta r_\Delta(\beta - r_\Delta) = 0, \tag{5.69}$$

$$w\Delta^{-1}(r_\Delta - r_f) + n_\Delta^2 r_\Delta = 0. \tag{5.70}$$

Since in the resulting algebraic system the parameter γ enters only in λ_1, and the parameter γ must naturally enter the system and determine its solutions, then λ_1 is chosen from $\lambda_{1,2}$. Solving a closed algebraic system of equations for unknown quantities (Blinded for Review, 2002), we obtain one non-linear algebraic equation for calculating the wave velocity

$$w^2 = \frac{\beta(r_f - r_\Delta)r_\Delta(r_-4 - 3r_f - r_\Delta)(r_f{}^2 - r_\Delta^2)\gamma^2}{64r_\Delta(r_- - r_f)^3 + \gamma(r_f^2 - r_\Delta^2)(r_-4 - 3r_f - r_\Delta)}, \tag{5.71}$$

$$r_f = r_- - r_\Delta + w^2(1+\gamma)\gamma^{-2}\beta^{-1}, \tag{5.72}$$

$$r_\Delta = \frac{w^2}{2\gamma}\left(\frac{w^2}{\gamma} - \beta^2\right)^{-1}\left(\frac{w^2}{\gamma^2}\frac{1+\gamma}{\beta} - 2\beta + \frac{w}{\gamma}\sqrt{\frac{w^2}{\gamma^2}\frac{(1+\gamma)^2}{\beta} - 4}\right). \tag{5.73}$$

The remaining unknowns are determined by the equations

$$\begin{aligned}
&n_m = 0.25(\beta - w^2 r_\Delta{}^{-1}\gamma^{-1})(4r_- - 3r_f - r_\Delta)(r_- - r_f)^{-1},\ \lambda_1 = -\lambda_3 = -w/\gamma, \\
&r_m = 0.5(r_f + r_\Delta),\ \zeta_m = \Delta/2 = (r_f - r_m)\gamma(r_- - r_f)^{-1}w^{-1},\ r_+ = w^2(1+\gamma)\beta^{-1}\gamma^{-2}, \\
&n_f = n_\Delta = n_m - 0.5n_m r_m(n_m - \beta)\zeta_m^2.
\end{aligned} \tag{5.74}$$

After the division of all unknown parameters, the concentration profiles by zones are calculated according to equations (5.54) - (5.57), (5.58), (5.60). An analysis of the obtained solution shows that it has the property of similarity in parameter β, i. e. according to the concentration of the oxidant. It is expressed in the fact that if we divide all the values of the concentrations n and r, as well as the velocity w by the parameter β, and multiply the coordinate ζ by the parameter β, then the parameter β, as it were, disappears from the system, i. e. finding unknown values of these new parameters and constructing profiles of relative concentrations by zones does not require setting the parameter β. In this case, the ratio of the initial concentration of fuel and oxidizer r_-/β and the ratio of diffusion coefficients γ become the control parameters of the system.

The value of β/r_-, as it is easy to see, coincides with the parameter β introduced in the one-component model of a cold flame (see Section 5.3). Analysis of equations (5.71), (5.72) shows that the solution of equation (5.71) exists only for $\gamma < 1$, i. e. with a greater or equal diffusion coefficient of the intermediate reagent compared to the diffusion coefficient of the fuel. Indeed, since for any values of the parameters the relation $r_- - r_f \geq 0$ must be satisfied, and, consequently, taking into account (5.72), also the relation $r_\Delta \beta - w^2(1+\gamma)\gamma^{-2} \geq 0$, it follows from (5.73) that this is possible only with $\gamma \leq 1$. The influence of the ratio of the diffusion coefficients of the reagents is also significant in the region $\gamma < 1$. A decrease in γ in this region leads to a decrease in the critical value of the parameter β/r_-, at which the stationary regime of cold flame propagation is disrupted. This significantly distinguishes the two-component cold flame model from the one-component model. Figure 7 shows the results of calculating the velocity w/β as a function of a parameter β/r_- for two values of the ratio of diffusion coefficients $\gamma = 1$ and $\gamma = 0.5$.

Figure 7. Dependence of the wave velocity w/β in the two-component cold flame model on the parameter β/r

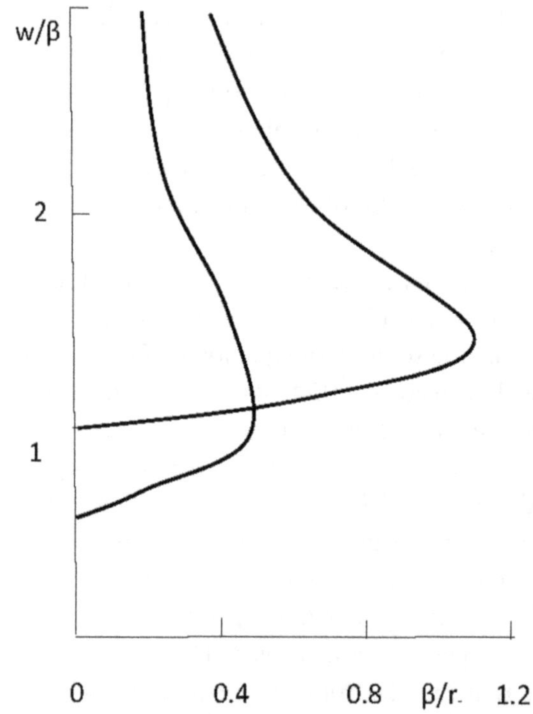

1 – γ = 0.5; 2 – γ = 1.

From Figure 7 it can be seen that the dependence of the wave speed on the parameter β/r_-, as for the one-component model, has two branches with high and low speed. Slow waves, as usual, seem to be unstable. Disruption of a stationary flame occurs at the value of the parameter $\beta/r_- = 0.114$ in case $\gamma = 1$, and $\beta/r_- = 0.054$ in case $\gamma = 0.5$.

Figure 8 shows, as an example, the concentration profiles of the fuel and the intermediate reactant in high - speed cold flame waves at a value of $\beta/r_- = 0.5$ and two diffusion coefficient ratios $\gamma = 1$ and $\gamma = 0.5$.

Comparison with the one-component model of a cold flame shows that the dependence of the velocity on the break parameter in the two-component model also differs in that the branches with high and low speed behave differently with a decrease in the break parameter. The branch of high speed with a decrease in the breakage parameter sharply goes into the region of large values w/β, and the branch of low values of speed tends to some finite values w/β.

Figure 8. Reagent concentration profiles in a two-component cold flame model with second-order autocatalysis

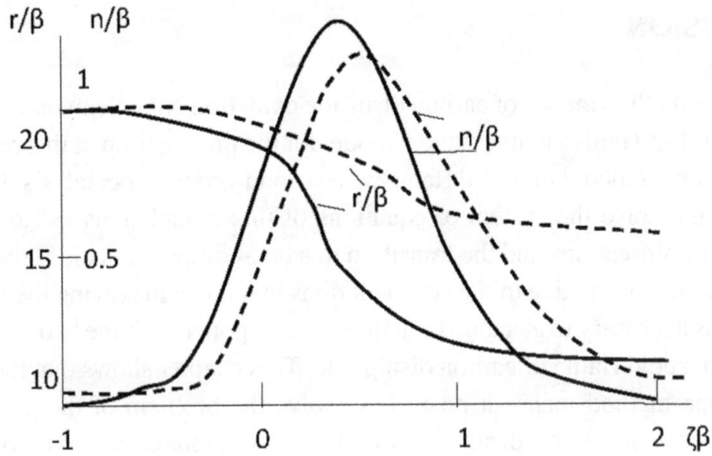

Solid line – $\gamma = 0$, dashed line – $\gamma = 1$.

Figure 8 shows that an increase in the diffusion coefficient of the intermediate reactant, other things being equal, leads to a sharpening of the soliton of the intermediate reactant and a deeper use of fuel in the wave. While the speed of propagation of the cold flame decreases.

B.V. Novozhilov and V.S. Posvyansky (Novozhilov, Posvyansky, 1973) numerically investigated the non-stationary equations of a cold flame in a two-component approximation. However, they believed that in the reaction of the interaction of an intermediate substance with the initial substance, the first interacts only with an oxidizing agent, the concentration of which was assumed to be constant. The exclusion of fuel with a variable concentration from the reaction of the interaction of an intermediate substance with the initial reagents leads to the fact that the model of a cold flame proposed by them is much closer to a one-component model than the model of N.N. Semyonova, V.G. Voronkov (equations (5.1), (5.2)). This explains the fact that the critical value of the interruption parameter and the corresponding velocity values corresponding to the disruption of the stationary regime of flame propagation, determined in the work of B.V. Novozhilova, V.S. Posvyansky, almost coincide with the corresponding values obtained in this monograph when analyzing a single-component model of a cold flame by the infinite reaction zone method. The corresponding critical value of the cutoff parameter for a two-component cold flame model, determined in this chapter by the infinite reaction zone method, turned out to be almost three times higher than the value of this parameter for a single-component flame model.

CONCLUSION

Considering the kinetics of carbon disulfide oxidation N.N. Semyonov and V.G. Voronkov in 1939 built a mathematical model for the propagation of this reaction as a two-component model of a cold flame with second-order autocatalysis. However, they could not solve the system of equations of this model. This led to the need for model simplifications and the transition to a one-component cold flame model. However, in the future, due to the contradictions that arose in solving the problems posed, it was necessary to abandon both the one-component and the two-component models of the cold flame of carbon disulphide. This chapter shows that the infinite reaction zone method makes it possible to solve the problem of the propagation of a cold flame of carbon disulfide in both one-component and two-component approximations. In all these cases, approximate analytical solutions were obtained, and calculations were carried out. The results of studies of one-component and two-component cold flame models are compared with each other and with known experimental data. The study confirmed the correctness of N.N. Semyonov and V.G. Voronkov on the existence of a soliton solution in a two-component model of carbon disulphide oxidation. In the case when the equation of a one-component model of a cold flame has an exact solution (solution by D.A. Frank-Kamenetsky), then the approximate solution obtained by the infinite reaction zone method practically coincides with the exact solution. This can be considered as a reference point for checking the correctness of the infinite reaction zone method, which proves the legitimacy of its use in more complex problems. It should be noted that the infinite reaction zone method in problems with a large number of variables is being further developed. If in problems with a small number of variables the buffer zone between two zones in the vicinity of the initial and final equilibrium positions tends to zero after the solution of the algebraic system of equations, then in systems with a large number of variables the buffer zone has a finite extent, and its size is determined from the condition for the solvability of the algebraic system of equations.

Obviously, the two-component cold flame model is a more general model than the one-component model. In the two-component model, more complex regularities in the propagation of a cold flame are found. In particular, it was found that an increase in the diffusion coefficient of an intermediate reactant in a flame, other things being equal, leads to a sharpening of the soliton of the intermediate reactant, a deeper use of fuel in the wave, and a decrease in the propagation velocity of a cold flame.

REFERENCES

Ayvazyan, R. G., & Sinelnikova, T. A. (1994). The mechanism of linear and quadratic branching of reaction mixtures and the role of HO2 radicals in chain combustion. *Kinetics and Catalysis, 35*(1), 22–29.

Azatyan, V. V. (2002). Fast reactions of substitution and displacement of atoms from polyatomic molecules and solid salts by thermally equilibrium atomic reagents. *Kinetics and Catalysis, 42*(2), 165–181.

Azatyan, V. V., ... (1976). Isothermal propagation of a rarefied nitrogen trifluoride flame. *Proceedings of the Academy of Sciences of the USSR. Chemical series*. 7, 1459 - 1461.

Azatyan, V. V., Novozhilov, B. V., Posvyansky, V. S., & Semyonov, N. N. (1976). Speed and limits of propagation of the cold flame of carbon disulphide oxidation. *Kinetics and Catalysis, 17*(6), 1386–1395.

Fedotov, V. G., & Chaikina, A. M. (1972). Investigation of the cold flame of the reaction of fluorine with hydrogen in a flow. *Reports of the Academy of Sciences of the USSR, 203*(2), 406–408.

Frank-Kamenetsky, D. A. (1967). *Diffusion and heat transfer in chemical chemistry*. Nauka.

Kolmogorov, A. N., Petrovsky, I. G., & Piskunov, N. S. (1937). The study of the diffusion equation, combined with the increase in the amount of matter and its application to one biological problem. *Bulletin of Moscow State University. Series A.Mathematics and Mechanics, 1*(6), 1–26.

Loskutov, A. Yu., & Mikhailov, A. S. (1990). *Introduction to synergetics*. Nauka.

Neumann, M.B. (1938). The role of peroxides in the formation of cold and hot hydrocarbon flames. *Good luck with chemistry*. 7(7), 341-384.

Novozhilov, B.V., & Posvyansky, V.S. (1973). On the speed of propagation of a cold flame. *Physics of combustion and explosion*. 9(2), 225-230.

Ovsyannikov, L. V. (1978). *Group analysis of differential equations*. Nauka.

Voronkov, V. G., & Semenov, N. N. (1939). Cold flame propagation in combustible mixtures containing 0.03% carbon disulfide. *Journal of Physical Chemistry, 13*(12), 1695–1727.

Zeldovich, Ya. B., Barenblatt, G. I., Librovich, V. B., & Makhviladze, G. M. (1980). *Mathematical theory of combustion and explosion*. Nauka.

Chapter 6
Waves of Hot Flame

ABSTRACT

The methods of the non-asymptotic theory of combustion are used to study the regularities of diffusion-thermal waves of a flame, taking into account the non-zero variable length of the zone of chemical reactions. It is proved that wave solutions under these conditions have a number of fundamental features that escape from analysis when using assumptions about a narrow reaction zone (including in the method of matched asymptotic expansions). These features include the existence of several wave solutions (with unique solutions in the narrow reaction zone method) for the same values of regime parameters, for example, the same wave propagation velocity. Bifurcation diagrams of wave regimes are constructed under these conditions. It has been found that the length of the zone of chemical reactions in diffusion-thermal nonadiabatic waves is the internal degree of freedom of these waves, which ensures the stability of the combustion regime with heat losses to the environment.

INTRODUCTION

If the chemical reaction of combustion is essentially exothermic, and at the same time the processes of diffusion of molecules are important, which is where burning of gases takes place, then to take into account the distribution of the region of the region of chemical reactions in space, it is generally impossible to use only one of the considered methods of an infinite or semi-infinite reaction zone. Then it is necessary to apply both methods at the same time. One of the special cases was considered in Chapter 3. This was an example of the adiabatic approximation, when the change in the concentration of the reactant in the flame is consistent with the change in temperature in the flame. However, the possibility of using the adiabatic approximation requires special proofs. In the general case, it is necessary to use both

DOI: 10.4018/978-1-6684-7745-8.ch006

methods (infinite and semi-infinite reaction zone) simultaneously, while additional conditions must be set for the solvability of the system of equations.

The problem of a diffusion-thermal combustion wave in a gas was considered in many papers under different conditions (Zel'dovich, 1948, 1961; Zel'dovich, Barenblatt. Librovich, Makhviladze, 1980; Novikov, Ryazantsev, 1966; Bachelis and Melamed, 1966; Bush and Fendell, 1970 ; Berman, Ryazantsev, 1972). Fundamental here were the assumptions of the narrow reaction zone method. These assumptions were modified into the method of matched asymptotic expansions (Bush and Fendell 1970; Berman and Ryazantsev 1972; Williams 1971). According to this method, in order to determine the wave velocity, there is no need to introduce an unknown front temperature, up to which the reaction rate is null. The solution for the wave velocity is obtained as a result of expanding the functions in a series in terms of a small parameter, which in this case is the ratio RT_+/E (R is universal gas constant, T_+ is temperature at the end of the wave, E is activation energy of the chemical reaction), and matching the solutions in the vicinity of the maximum temperature and away from it. To as already mentioned, the assumptions of the narrow reaction zone method, including the assumptions of matched asymptotic assumptions, are often not met and asymptotic methods conflict with known results. In this chapter, non-asymptotic research methods are applied to the analysis of diffusion-thermal combustion waves. At the beginning, a diffusion-thermal combustion wave is considered for an arbitrary source form. Then a power law is adopted for the dependence of the reaction rate on the mass fraction of the reagent. One of the most important issues in the theory of combustion is the question of the effect of heat losses from the burning medium on the laws of flame propagation and the disruption of the stationary regime of flame propagation. For technical devices using combustion, this issue is important for ensuring a stationary combustion regime. In the fight against fires, this issue is important for determining the conditions for extinguishing fires. In this chapter, a model of a combustion wave of a thermal nature is constructed under conditions of heat removal. Non-asymptotic methods for studying waves are applied. In all these cases, approximate analytical solutions are obtained, and their numerical studies are presented.

6.1. DIFFUSION-THERMAL WAVES OF COMBUSTION IN A GAS WITH AN ARBITRARY SOURCE

Let us consider a combustion wave model taking into account the distribution of the chemical reaction zone with an arbitrary dependence of the combustion reaction rate on temperature and the mass fraction of the initial reagent, assuming that the combustion reaction rate can be expressed as a function of temperature and of the mass fraction of one of the components.

The non-stationary one-dimensional equations of the laws of conservation of energy and mass under the accepted conditions have the form

$$c_p\rho\frac{\partial T}{\partial t} = \frac{\partial}{\partial x}\lambda\frac{\partial T}{\partial x} + \rho QW, \tag{6.1}$$

$$\rho\frac{\partial C}{\partial t} = \frac{\partial}{\partial x}\rho D\frac{\partial C}{\partial x} - \rho W. \tag{6.2}$$

In contrast to equations (3.1), (3.2) of the third chapter, here in the mass conservation equation (6.2) a term with a derivative with respect to the spatial variable of the mass fraction of the reagent C, associated with the diffusion of the reagent, has been added. In addition, since the working medium is a gas, it is necessary to take into account the dependence of thermophysical parameters (density ρ, diffusion coefficient D, thermal conductivity coefficient λ) on temperature. This significantly complicates the problem of obtaining an approximate analytical solution.

As in Chapter 3, we introduce the wave coordinate $z = ut - x$ (u is wave velocity), assuming the existence of a stationary solution for the combustion wave propagation. Then equations (6.1), (6.2) take the form

$$\frac{d}{dz}\lambda\frac{dT}{dx} - c_p\rho u\frac{dT}{dz} + \rho QW = 0, \tag{6.3}$$

$$\frac{d}{dz}\rho D\frac{dC}{dz} - \rho u\frac{dC}{dt} - \rho W = 0. \tag{6.4}$$

We set the coefficients of thermal conductivity and diffusion by the following power functions of temperature (Lovachev, Basevich, Kogan, Posvyansky, 1970)

$$\lambda = \lambda_0(T/T_0)^{0,67}, D = D_0(T/T_0)^{1,67}, \tag{6.5}$$

where subscript 0 corresponds to the values of the parameters at the beginning of the wave ($z = -\infty$).

The specific heat capacity of the gas, in view of its weak dependence on temperature, will be considered constant and equal to its average value along the wave. The specific mass flow rate ρu is also constant, as follows from the mass conservation condition. We consider the gas to be ideal, so its density is given by the relation

$$\rho = \rho_0 T_0/T. \tag{6.6}$$

We represent the reaction rate as a product of two functions

$$W = k\varphi'(C)\phi'(T), \tag{6.7}$$

where k is the proportionality factor and $\phi'(0) = 0$.

Equations (6.3), (6.4) have the first integral

$$\lambda dT/dz + Q\rho DdC/dz = u\rho c_p(T - T_+) + Qu\rho C. \tag{6.8}$$

The integration constant in (6.8) was determined from the boundary condition

$$T = T_+, C = 0, dT/dz = 0, dC/dz = 0, z = +\infty . \tag{6.9}$$

From (6.8) under the boundary condition

$$T = T_0, C = C_0, dT/dz = 0, dC/dz = 0, z = -\infty \tag{6.10}$$

we have $C_0 = (T_+ - T_0)c_p/Q$. (6.11)

Let us introduce dimensionless variables and parameters

$$n = C/C_0, (0 \le n \le 1);\ \eta = (T - T_+)/(T_+ - T_0), (-1 \le \eta \le 0);$$
$$\zeta = \int \frac{u\rho c_p}{\lambda_0}\left(\frac{T_0}{T}\right)^{0.67} dz; w = u\rho\left(\frac{c_p}{\rho_0 k\lambda_0}\right)^{1/2}. \tag{6.12}$$

Equations (6.4), (6.8), taking into account (6.7), (6.11), (6.12), can be written as a system

$$dn/d\zeta = p, \tag{6.13}$$

$$d\eta/d\zeta = \eta + n - pLe, \tag{6.14}$$

$$Ledp/d\zeta = p + w^{-2}C_0^{-1}\varphi(n)\phi(\eta), \tag{6.15}$$

where

$$\varphi(n) = \varphi'(C_0 n), \phi(\eta) = (T_0/T)^{0.33}\phi'[\eta(T_+ - T_0) + T_+], \varphi(0) = 0.$$

The system of equations (6.13) - (6.15) is autonomous and its solutions can be represented by trajectories in the phase space (*n*, *η*, *p*). The negative octant of the phase space $p \le 0$, $\eta \le 0$ makes sense physically. The location of the trajectories in this octant is determined by zero surfaces (Figure 1)

$F_1{:}p \;=\; 0$, on which $dn/d\zeta \;=\; 0$;
$F_2{:}n \;=\; pLe - \eta$, on which $d\eta/d\zeta=0$;
$F_3{:}p \;=\; -w^{-2}C_0^{-1}\varphi(n)\phi(\eta)$, on which $dp/d\zeta=0$.

Figure 1. The phase space of system (6.13) – (6.15)

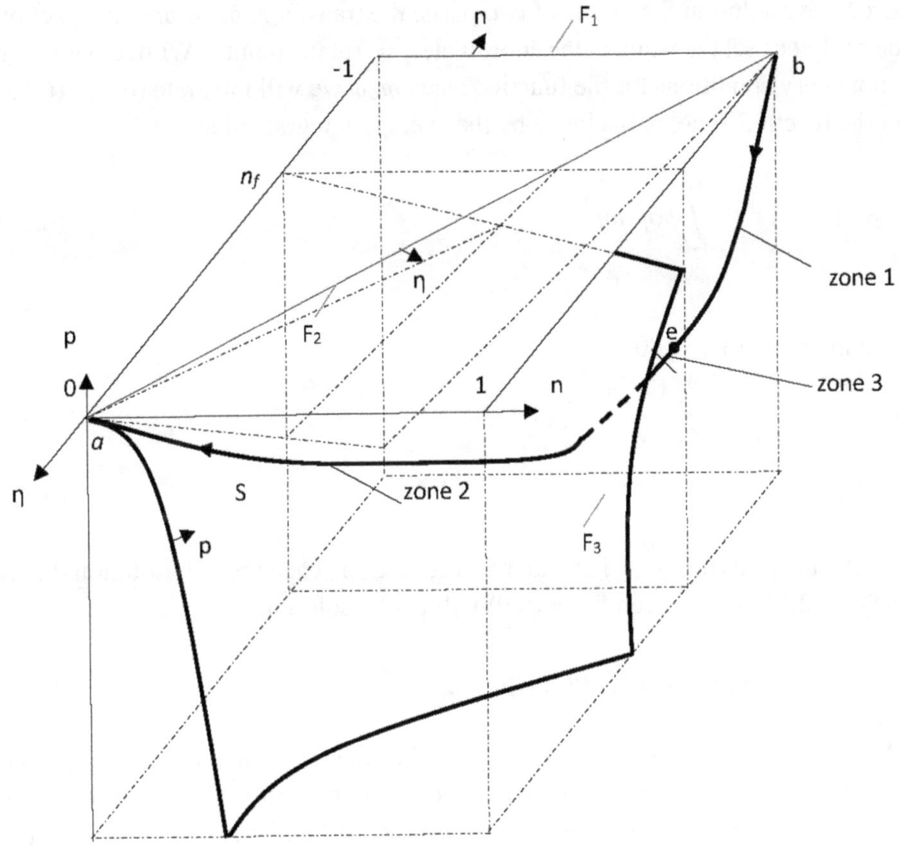

The arrows at the zero surfaces in Figure 1 indicate the areas where the values of the corresponding phase coordinates increase. The point *a* with coordinates p = 0, n = 0, η = 0 is the equilibrium position of the system. It corresponds to the state at the end of the wave. Point *b* with coordinates p = 0, n = -1, η = -1, if $\phi(-1)\neq0$, is not an equilibrium position, although the values of the phase coordinates at this point correspond to the state at the beginning of the wave. In order for the point *b* to become the equilibrium position, it is necessary to put $\phi(-1)=0$. Let us put $\phi(\eta)=0$ on the segment

$$-1 \le \eta \le -1 + h, h = (T_f - T_0)/(T_+ - T_0), T_f$$

is the temperature of the wave front, i.e. boundaries of the transfer zone and the reaction zone. Then, for $\eta=\eta_f=(T_f-T_+)/(T_+-T_0)$ the surface F_3 has a discontinuity; for $\eta < \eta_f$ the surface F_3 coincides with the surface F_1. In this case, a trajectory can leave position b and reach a certain point e on the plane $\eta = \eta_f$. After the point e, the trajectory, crossing the surface F_3, can enter the equilibrium position a. Thus, the desired wave solution S consists of two parts: the transfer zone *be* and the reaction zone *ea*. Let us set the value of the coordinate $\zeta = 0$ at the point e. Without violating the boundary conditions for the functions n, η, φ, ϕ we will integrate (6.13), (6.15) over the reaction zone, replacing φ by the average integral value

$$\langle\phi\rangle = (1-h)^{-1} \int_{-1+h}^{0} \phi(\eta)\,d\eta \tag{6.16}$$

on the interval $-1+h \le \eta \le 0$.

Let

$$n = F(\zeta; h, w) \tag{6.17}$$

be the integral (6.13), (6.15) under this condition. Then the only solution (6.14) satisfying the boundary condition (6.9) will be the solution

$$\eta = \exp(\zeta) \int [F(\zeta) - Le\,dF(\zeta)/d\zeta] \exp(-\zeta)\,d\zeta. \tag{6.18}$$

Let us introduce a buffer zone (3) $0 \le \zeta \le \Delta$ between the transfer zone (1) and the reaction zone (2), the extent Δ of which is small and, after the resolution of the system obtained after integration over the zones, will tend to zero. In view of the smallness of zone 3, the functions $\eta(\zeta)$, $n(\zeta)$ can be considered linear in this zone. We write the integral (6.18) in the Cauchy form

$$\eta = -\exp(\zeta) \int_{\zeta}^{\infty} \exp(-\tau) [F(\tau) - Le\,dF(\tau)/d\tau]\,d\tau, \tag{6.19}$$

using (6.14) we equate the derivatives $d\eta/d\zeta$ on both sides of zone 3

$$\eta_\Delta + n_\Delta - Le\,p_\Delta = \eta_f + n_f - Le\,p_f. \tag{6.20}$$

By virtue of the assumption that $n(\zeta)$ is linear in zone 3

$$p_\Delta = p_f \tag{6.21}$$

and (6.20) have a form

$$\eta_\Delta + n_\Delta = \eta_f + n_f \tag{6.22}$$

From (6.17), (6.20) it follows that

$$\begin{aligned} \eta_\Delta &= -\exp\Delta \int_\Delta^\infty \exp(-\tau)[F(\tau;h,w) - LedF(\tau;h,w)/d\tau]d\tau, \\ n_\Delta &= F(\Delta;h,w), \eta_f = h-1. \end{aligned} \tag{6.23}$$

Let us find n_f in (6.22) by integrating (6.13) - (6.15) with $\phi(\eta)$=0 over the transfer zone. From (6.13), (6.15) for $\phi(\eta)$=0 we have

$$dn/dp = Le \tag{6.24}$$

Integrating (6.24) under condition (6.10), we find

$$n = pLe + 1 \tag{6.25}$$

Substituting (6.25) into (6.14) and integrating, we get

$$\eta = (\eta_f + 1)\exp\zeta - 1. \tag{6.26}$$

Integrating (6.15) with $\phi(\eta)$=0, we have

$$p = p_f \exp(\zeta/Le). \tag{6.27}$$

Substituting (6.27) into (6.25), we obtain an expression for the concentration profile in the transfer zone

$$n = Lep\exp(\zeta/Le) + 1. \tag{6.28}$$

From (6.25), (6.28) it is seen that

$$n_f = p_f Le + 1. \tag{6.29}$$

The value of p_f in (6.29) is expressed in terms of the integral (6.17), taking into account (6.21)

$$p_f = dF(\zeta; h, w)/d\zeta\big|_{\zeta=\Delta}. \tag{6.30}$$

Thus, from (6.23), (6.29), (6.30) it follows that both sides of equality (6.22) enter as unknowns h, w, Δ. Due to the linearity of the change n and η in zone 3, from (6.13), (6.15), (6.20), taking into account (6.23), (6.29), we have

$$n_\Delta = n_f - w^{-2} C_0^{-1} \varphi\left(\frac{n_f + n_\Delta}{2}\right) \phi\left(\frac{\eta_\Delta + h - 1}{2}\right), \tag{6.31}$$

$$\eta_\Delta = h\Delta - 1 + h. \tag{6.32}$$

Eliminate η_Δ from (6.32) using (6.23)

$$\Delta h - 1 + h = -\exp\Delta \int_\Delta^\infty \exp(-\tau)\,[F(\tau; h, w) - Le\,dF(\tau; h, w)/d\tau]\,d\tau. \tag{6.33}$$

Using (6.22), (6.32), (6.23) we can write

$$n_f - n_\Delta = \eta_\Delta - \eta_f = h\Delta. \tag{6.34}$$

Substituting (6.34) into (6.31), canceling by Δ, which is not yet equal to zero, we obtain

$$h = w^{-2} C_0^{-1} \varphi\left(\frac{n_f + n_\Delta}{2}\right) \phi\left(\frac{\eta_\Delta + h - 1}{2}\right). \tag{6.35}$$

Let us now let Δ tend to zero in (6.33), (6.35)

$$1 - h = \int_0^\infty \exp(-\tau)\,[F(\tau; h, w) - Le\,dF(\tau; h, w)/d\tau]\,d\tau, \tag{6.36}$$

$$w^2 = h^{-1} C_0^{-1} \varphi[F(0; h, w)]\,\phi(h - 1). \tag{6.37}$$

Equations (6.36), (6.37) constitute a closed system with respect to two unknowns h and w for an arbitrary form of the kinetic function

Thus, in this model, the existence of an exothermic chemical reaction is predominant. Here, the principle of cutting off the kinetic function to a certain, determined as a result of solving the problem, temperature of the combustion front has been preserved. The mass fraction of the reagent, in contrast to problems in which diffusion is not taken into account, is variable over the entire wavelength, including up to the reaction zone.

6.2. DIFFUSION-THERMAL WAVE OF COMBUSTION IN A GAS WITH A POWER-LAW DEPENDENCE OF THE REACTION RATE ON THE MASS FRACTION OF THE REAGENT

Let the reaction rate be determined by the expression

$$W = kC^m \exp\left(-\frac{E}{RT}\right), \varphi'(C) = C^m, \phi'(T) = \exp\left(-\frac{E}{RT}\right).$$

Then equation (6.15) is transformed

$$Le dp/d\zeta = p + w^{-2} C_0^{m-1} n^m \phi_1(\eta), \tag{6.38}$$

$$\phi_1(\eta) = (\eta\sigma^{-1} + \sigma^{-1})^{\,0.33} \exp\left(-\beta\frac{1+\sigma}{\eta+1+\sigma}\right), \sigma = \frac{T_0}{T_+ - T_0}, \beta = \frac{E}{RT_+}. \tag{6.39}$$

The mean integral value ϕ_1, according to (6.16), (6.39), has the form

$$\langle\phi_1\rangle = (1-h)^{-1} \int_{h-1}^{0} (\eta\sigma^{-1} + 1 + \sigma^{-1})^{-0.33} \exp\left(-\beta\frac{1+\sigma}{\beta+1+\sigma}\right) d\eta. \tag{6.40}$$

Expanding in (6.40) the exponential according to Frank-Kamenetsky, taking into account the weak dependence of the denominator in the integrand in (6.40) on η, we obtain

$$\langle\phi_1\rangle = (1-h)^{-1}\left(\frac{h+1}{2\sigma} + 1\right)^{-0.33} \frac{1+\sigma}{\beta}\left[1 - \exp\left(\beta\frac{h-1}{1+\sigma}\right)\right]\exp(-\beta). \tag{6.41}$$

Let the integral of equations (6.38), (6.13) with the replacement of ϕ_1 by $\langle\phi_1\rangle$ have the form (6.17). In this case, the system (6.36), (6.37) can be solved for $\phi(\eta)=\phi_1(\eta)$ only numerically. Consider the case of a first order reaction (m = 1). Integrating (6.13), (6.15) with the replacement of ϕ_1 by $\langle\phi_1\rangle$, using the boundary condition at $\zeta = 0$, we find

$$n = n_f \exp(\gamma\zeta) = F(\zeta), \tag{6.42}$$

$$\gamma = 0,5Le^{-1} - \sqrt{0.25Le^{-2} + Le^{-1}w^{-2}\langle\phi_1\rangle}. \tag{6.43}$$

From (6.29), (6.30) for Δ=0, taking into account (6.42), (6.17), we have

$$n_f = (1-\gamma)^{-1} \tag{6.44}$$

$$F(\zeta) = (1-\gamma)^{-1}\exp(\gamma\zeta). \tag{6.45}$$

Substituting (6.45) into (6.36), we get

$$1 - h = (1-\gamma)^{-2}(1 - Le\gamma). \tag{6.46}$$

Let us substitute into (6.37) the values of the functions $\varphi, \phi = \phi_{1,}$ at ζ=0

$$w^2 = h^{-1}(1-\gamma)^{-1}(h\sigma^{-1}+1)^{-0.33}\exp\left(-\beta\frac{1+\sigma}{h+\sigma}\right). \tag{6.47}$$

Substituting γ from (6.47) into (6.46), we obtain the equation for the wave velocity

$$\begin{aligned}(1-Le)w^4 + h^{-1}(h\sigma^{-1}+1)^{-0.33}Le\exp\left(-\beta\frac{1+\sigma}{h+\sigma}\right)w^2 - \\ -h^{-2}(h\sigma^{-1}+1)^{-0.66}(1-h)\exp\left(-2\beta\frac{1+\sigma}{h+\sigma}\right) = 0.\end{aligned} \tag{6.48}$$

From (6.48) follows three expressions for the wave velocity as a function of the Lewis number. If is Le = 1 (the similarity of the temperature and concentration profiles in the wave), so

$$w^2 = h^{-1}(h\sigma^{-1}+1)^{-0.33}(1-h)\exp\left(-\beta\frac{1+\sigma}{h+\sigma}\right). \tag{6.49}$$

If is Le < 1, so

$$w^2 = \frac{-Le + \sqrt{Le^2 - 4(Le-1)(1-h)}}{h(h/\sigma+1)^{0.33}2(1-Le)}\exp\left(-\beta\frac{1+\sigma}{h+\sigma}\right). \tag{6.50}$$

If is Le > 1, so

$$w^2 = \frac{-Le + \sqrt{Le^2 + 4(Le-1)(1-h)}}{h(h/\sigma+1)^{0.33}2(1-Le)}\exp\left(-\beta\frac{1+\sigma}{h+\sigma}\right). \tag{6.51}$$

Figure 2 shows the results of calculating the values of wave speed w and front temperature h depending on the Lewis number Le at $\beta = 3$, $\sigma = 0.144$, determined by equations (6.49) - (6.51).

Figure 2. Calculated dependences of the velocity and temperature of the combustion wave front on the Lewis number ($\beta = 3$, $\sigma = 0.144$) according to the model with a distributed chemical reaction zone

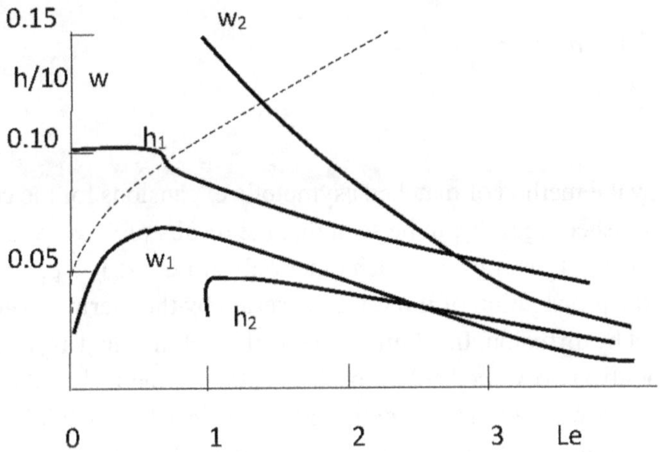

1 - slow burning wave, 2 - fast burning wave, dotted line - solution by the method of matched asymptotic expansions

From Figure 2 it can be seen that at Le < 1 there is one wave solution that continuously passes into the solution at Le > 1. At the same time at Le > 1, there are solutions with different velocities in the region. Moreover, the new solution, which is not a continuation of the solution from area Le < 1, has a much higher speed, especially in the vicinity of Le = 1. As follows from (6.51), the rate limit of the new solution as the Lewis number tends to unity is equal to infinity, and the limit h is equal to zero. This corresponds to wave degeneration, in which the temperature along the wave remains constant, equal to the initial temperature. In the area L > 1, where the influence of diffusion is significant, with an increase in the Lewis number, the velocities of both wave solutions decrease, approaching each other in the region of numbers Le > 5. In this case, the wave expands due to diffusion and the difference between the temperature of the wave front and the initial temperature of the gas decreases (strong diffusion equalizes the distribution of concentrations over space and weakens the wave).

Calculations show that an increase in the number β leads to a sharp decrease in the velocities of both wave solutions and to an increase in the temperature of the wave front.

It should be noted that the solution (6.49) obtained under the condition that the temperature and concentration profiles are similar (Le = 1), coincides with the limits of the wave solutions (6.50), (6.51) as Le tends to unity, respectively, from the left and from the right.

In Figure 2, as a comparison, the dependence of the speed of the diffusion-thermal wave on the Lewis number is shown, calculated by the formula

$$w^2 = 2Le\beta^{-2}(1+\sigma)^{1.67}\sigma^{0.33}e^{-\beta}, \quad (6.52)$$

obtained by the method of matched asymptotic expansions for the conditions of this problem (dashed line). It can be seen that the results of wave velocity calculations by both methods are close to each other only at Le << 1, i. e. when the main influence on the propagation of the wave is exerted by the thermal conductivity of the gas, and not by diffusion. In addition, the method of matched asymptotic expansions does not give two-valued solutions. The nature of the influence of the Lewis number on the wave speed in the region Le greater than 1, which follows from the method of the semi-infinite reaction zone and the method of matched asymptotic expansions, is directly opposite: in the first case, as the Lewis number increases, the wave speed decreases, in the second case, the wave speed increases. All this is a consequence of the incorrectness of the method of matched asymptotic expansions associated with the neglect of the length of the zone of chemical reactions.

6.3. NON-ADIABATIC COMBUSTION WAVES

One of the most important issues in the theory of combustion is the question of the effect of heat losses from the burning medium on the laws of flame propagation and the disruption of the stationary regime of flame propagation. For technical devices using combustion, this issue is important for ensuring a stationary combustion regime, since there is always heat loss from a high-temperature environment. In firefighting, this issue is important for determining the conditions for extinguishing a fire. Zel'dovich was the first to construct a theory of a combustion wave under conditions of heat removal (Zel'dovich, 1941), and approximate expressions were obtained for the velocity and limits of the existence of such a wave. The main assumptions characteristic of this author were the assumptions of a narrow reaction zone and smallness of heat loss to the environment. Later, the question of the effect of heat loss on the combustion wave was considered in a simplified way in the case of a power-law approximation of the chemical reaction rate function (Spalding, 1957).

F. Williams took into account the length of the zone of chemical reactions in a combustion wave with heat losses (Williams, 1971), but he assumed that the rate of a chemical reaction is constant along the entire length of the reaction zone, which cannot be. Non-adiabatic combustion waves were considered by the method of matched asymptotic expansions, which is based on the assumption of a narrow reaction zone (Joulin and Klavin, 1976; Bookmaster, 1976). At present, interest in the analysis of combustion waves under conditions of heat removal is associated, in particular, with the need to fight various kinds of natural fires, in which the propagating zones of chemical reactions occupy significant spatial regions and the assumptions about the negligibly small length of the chemical reaction zone clearly do not correspond to reality (Grishin, 2002, 2004). In the process of flame propagation in the natural environment, not only heat exchange takes place, but also the mass exchange of the burning medium with the environment. Under these conditions, it is necessary to take into account the possibility of suction of a combustible reagent from the external environment. In addition, in peat fires, filtration of burning gases through the soil is possible.

In this section, a model of a combustion wave of a thermal nature with a finite (not small) length of the chemical reaction zone under heat removal conditions is constructed. The finite extent of the zone of chemical reactions in this case is associated with the need to apply a double cutoff of the kinetic function when the wave coordinate tends to infinity in both directions. As shown, it is under this condition that the existence of a wave solution as a trajectory going from one equilibrium position to another (final) is possible. An approximate analytical solution is obtained, calculations are carried out. The correctness of the analytical solution is proved by comparing it with the numerical solution of the original system of differential equations. Calculations have shown, and this is significant, that the velocity of a stationary nonadiabatic combustion wave can be many times less than the velocity of an adiabatic combustion wave for equal kinetic and initial parameters of the system. This indicates that the combustion wave, due to its open ability to change its structure, and, above all, to change the length of the chemical reaction zone, has an increased ability to maintain a stationary combustion regime. This ability is absent in models that neglect the extent of the chemical reaction zone (or assume a constant reaction rate), in which the minimum velocity of a stationary nonadiabatic wave is 0.6 of the adiabatic wave velocity (Williams, 1971). It has been found that an increase in heat loss to the environment (when trying to extinguish a flame) leads to an increase in the length of the chemical reaction zone (flame self-regulation) and, as a result, to maintaining a stationary flame regime.

The effect of diffusion on the regularities of the nonadiabatic combustion wave is studied. The analysis showed that the inclusion of diffusion transfer processes in the system in the presence of heat exchange with the environment, the existence of a wave solution requires the introduction of a term describing the process of mass transfer of the mixture with the environment into the system of equations. It was found that the processes of diffusion transfer lead to an effect characteristic of exothermic chemical reactions in stirred volumes. It turned out that at the same combustion wave velocity, the existence of from one to three different combustion waves is possible. These waves differ significantly in the temperatures at the soliton apex. According to the value of this temperature, one can distinguish high-temperature, medium-temperature and low-temperature wave regimes. The discovered effect is a consequence of the finite length of the zone of chemical reactions in the combustion wave and diffusion mixing. The domains of existence of wave solutions under these conditions are determined.

6.4. NON-ADIABATIC COMBUSTION WAVES OF A THERMAL NATURE

The non-stationary equation of heat conduction in a medium with a chemical exothermic combustion reaction and taking into account heat exchange with the environment in a one-dimensional approximation has the form

$$c_p\rho\frac{\partial T}{\partial t} = \frac{\partial}{\partial x}\lambda\frac{\partial T}{\partial x} + \rho QW - \alpha(T - T_s). \quad (6.53)$$

We will consider the thermophysical characteristics of the medium (density ρ, heat capacity c_p, thermal conductivity coefficient λ, thermal effect of the reaction Q, heat transfer coefficient α, ambient temperature T_S) to be constant. The rate of a chemical reaction W is related to the change in the mass fraction C of the combustible reagent by the equation

$$\frac{\partial C}{\partial t} = -W \quad (6.54)$$

and is given by the equality

$$W = k_0 C\exp(-E/(RT)), \quad (6.55)$$

corresponding to a first order chemical reaction.

Assuming the existence of a stationary wave solution, we introduce the wave coordinate z = ut - x, where u is the wave velocity. Then the system (6.53) - (6.55) is reduced to the form

$$c_p\rho u\frac{dT}{dz} = \lambda\frac{d^2T}{dz^2} + \rho QkC\exp\left(-\frac{E}{RT}\right) - \alpha(T - T_s). \tag{6.56}$$

$$u\frac{dC}{dz} = -kC\exp\left(-\frac{E}{RT}\right). \tag{6.57}$$

The assumption about the existence of a stationary wave solution in this case requires clarification. The wave propagates through the medium due to the transfer of heat by thermal conduction and, in the mathematical sense, is the trajectory of the system (6.56), (6.57), leaving one equilibrium position and entering another equilibrium position of this system. Heat losses to the environment lead to the fact that in any equilibrium position of the system (6.56), (6.57) the temperature of the reaction medium must be equal to the ambient temperature. Only under this condition will equilibrium be established in the system. At the entire wavelength, with the exception of infinitely distant points, the temperature is higher than the ambient temperature. The wave can propagate in a stationary manner under these conditions, if the heat released in the chemical reaction is sufficient for the heat transferred by thermal conduction to be sufficient to

Introducing dimensionless variables and parameters

$$\theta = \frac{RT}{E}, \zeta = z\left(\frac{\rho QRk}{\lambda E}\right)^{1/2}, w = uc_p\left(\frac{E\rho}{\lambda QRk}\right)^{1/2}, K_1 = \frac{\alpha E}{\rho QRk}, K_2 = \frac{Ec_p}{QR}, \tag{6.58}$$

let us bring the system (6.56), (6.57) to the form

$$w\frac{d\theta}{d\zeta} = \frac{d^2\theta}{d\zeta^2} + C\exp\left(-\frac{1}{\theta}\right) - K_1(\theta - \theta_s), \tag{6.59}$$

$$w\frac{dC}{d\zeta} = -K_2C\exp\left(-\frac{1}{\theta}\right). \tag{6.60}$$

Equation (6.59) can be written as two first order equations

$$\frac{d\theta}{d\zeta} = \eta, \tag{6.61}$$

$$\frac{d\eta}{d\zeta} = w\eta - C\exp(-\frac{1}{\theta}) + K_1(\theta - \theta_s). \tag{6.62}$$

Autonomous system solutions (6.60) - (6.62) can be represented by trajectories in three-dimensional phase space (η, θ, C).

In this case, in order for equilibrium positions to exist in the system, it is necessary to put exp(-1/θ) zero in the vicinity of the temperature θ_s at $\theta > \theta_s$. Obviously, in these areas, the mass fraction of the combustible reagent is constant. At the beginning, the mass fraction is equal to the initial one $C = C_0$, at the end of the wave - the mass fraction is some final $C = C_+$. Solutions to system (6.61), (6.62) under these conditions are represented by trajectories in the phase plane (η, θ) (Figure 3).

Figure 3. Phase plane of system (6.61), (6.62) at zero rate of chemical reaction

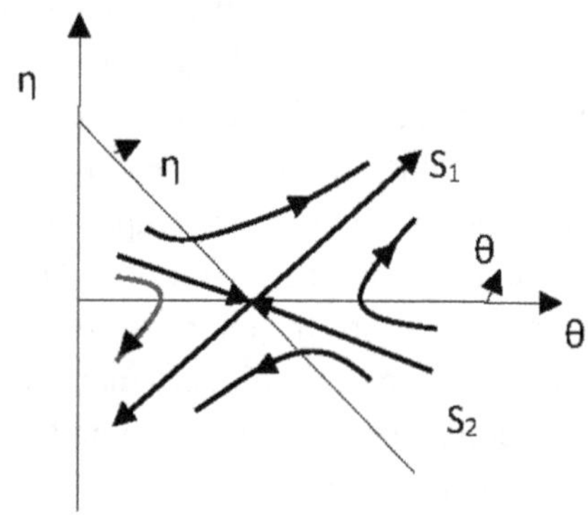

It has a saddle-type equilibrium position ($\eta = 0$, $\theta = \theta_s$) with a separatrix S_1 leaving the equilibrium position in the positive quadrant of the plane, and separatrix S_2 entering the equilibrium position from the negative quadrant of the plane. The eigenvalues are found from the characteristic equation of the system (6.61), (6.62) with exp(-1/θ) = 0

$$\xi^2 - w\xi - K_1 = 0,$$
$$\xi_{1,2} = 0.5w \pm \sqrt{0.25w^2 + K_1}. \quad (6.63)$$

The first characteristic number corresponds to the separatrix S_1, the second - to the separatrix S_2. In the three-dimensional phase space, the equilibrium positions form a straight line parallel to the C axis. Between the neighborhoods of the initial and final equilibrium positions in the three-dimensional phase space, there are parts of the trajectories that transfer the phase point $a(\eta = 0, \theta = \theta_s, C = C_0)$ and to the neighborhood of the final equilibrium position, the point $b(\eta = 0, \theta = \theta_s, C = C_+)$. In this region of the phase space, there is a zero surface F: $\eta = Cw^{-1}exp(-1/\theta)-K_1 w^{-1}(\theta - \theta_s)$, on which $d\eta/d\zeta = 0$, and a zero plane $\eta = 0$, on which $dC/d\zeta = 0$. The desired trajectory leaves the equilibrium position *a* along the separatrix S_1 and enters the equilibrium position *b* along the separatrix S_2 crossing the null surface F and the plane $\eta = 0$. The corresponding integral curves are shown in Figure 4

Figure 4. Integral curves of the system (6.60) - (6.62)

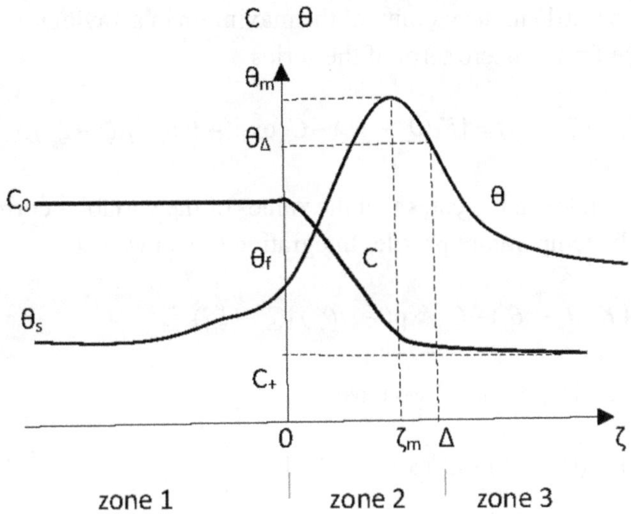

Three zones can be distinguished in the wave solution (Figure 6.4). In the first and third zones (vicinities of the initial and final equilibrium positions) there is no chemical reaction and equation (6.59) has solutions:

in the first zone

$$\theta = \theta_s + (\theta_f - \theta_s)\exp(\xi_1\zeta), C = C_0, \tag{6.64}$$

in the third zone

$$\theta = \theta_s + (\theta_\Delta - \theta_s)\exp(\xi_2(\zeta - \Delta)), C = C_+. \tag{6.65}$$

Here, θ_f is the temperature of the leading edge of the reaction, θ_Δ is the temperature of the trailing front of the reaction, Δ is the width of the chemical reaction zone, θ_m is the maximum temperature in the wave. All parameters included in equations (6.64), (6.65), except for the initial mass fraction C_0 and ambient temperature θ_s, are to be determined.

Due to the finiteness of zone 2 and the content of an inflection point in it in the mass fraction profile, we will accept a linear change in the mass fraction in zone 2

$$C = C_0 - (C_0 - C_+)\Delta^{-1}\zeta. \tag{6.66}$$

Since zone 2 contains the temperature profile maximum, we expand the right side of equation (6.61) in the vicinity of the maximum in a Taylor series, restricting ourselves to the first nonzero term of the series

$$d\theta/d\zeta = (d\eta/\zeta)_m(\zeta - \zeta_m) = [K_1(\theta_m - \theta_s) - C_m\exp(-1/\theta_m)](\zeta - \zeta_m). \tag{6.67}$$

In (6.67) the index m indicates that the values of the variables correspond to the maximum of the temperature profile. Integrating (6.67) we get

$$\theta = \theta_m + 0.5[K_1(\theta_m - \theta_s) - C_m\exp(-1/\theta_m)](\zeta - \zeta_m)^2. \tag{6.68}$$

At the boundaries of zone 2 we have

$$\theta_f = \theta_m + 0.5[K_1(\theta_m - \theta_s) - C_m\exp(-1/\theta_m)]\zeta_m^2, \tag{6.69}$$

$$\theta_\Delta = \theta_m + 0.5[K_1(\theta_m - \theta_s) - C_m\exp(-1/\theta_m)](\Delta - \zeta_m)^2. \tag{6.70}$$

From the condition of continuity of the first derivatives $d\theta/d\zeta$ at the boundaries of zone 2, we have

$$(\theta_f - \theta_s)\xi_1 = -[K_1(\theta_m - \theta_s) - C_m\exp(-1/\theta_m)]\zeta_m, \tag{6.71}$$

$$(\theta_\Delta - \theta_s)\xi_2 = [K_1(\theta_m - \theta_s) - C_m\exp(-1/\theta_m)](\Delta - \zeta_m). \tag{6.72}$$

At the top of the soliton

$$C_m = C_0 - (C_0 - C_+)\Delta^{-1}\zeta_m. \tag{6.73}$$

We set the second derivative $d^2\theta/d\zeta^2$ in equation (6.59) to zero at the points $\zeta = 0$ and $\zeta = \Delta$

$$w(\theta_f - \theta_s)\xi_1 = C_0\exp(-1/\theta_f) - K_1(\theta_f - \theta_s), \tag{6.74}$$

$$w(\theta_\Delta - \theta_s)\xi_2 = C_+\exp(-1/\theta_\Delta) - K_1(\theta_\Delta - \theta_s). \tag{6.75}$$

The slope of the straight - line C (ζ) in the reaction zone is determined from equation (6.60) by the average values of the variables for zone 2 on the right side of (6.60)

$$\frac{C_0 - C_+}{\Delta} = \frac{K_2}{w} 0.5(C_0 - C_+)\exp(-\frac{4}{2\theta_m + \theta_f + \theta_\Delta}). \tag{6.76}$$

As a result of linearization, the second derivatives $d^2 \theta / d \zeta^2$ at the boundaries of zone 2 have a discontinuity. we assume that the sum of the second derivatives $d^2 \theta / d \zeta^2$ on both sides of the boundary at $\zeta = 0$ is equal to zero

$$(\theta_f - \theta_s)\xi_1^2 + [K_1(\theta_m - \theta_s) - C_m \exp(-1/\theta_m)] = 0. \tag{6.77}$$

Solving the resulting system of algebraic equations, we obtain one equation for the wave front temperature

$$0.5 K_2 = w\xi_1 \frac{\varphi_1 - (1.5K_1 + \xi_1^2)(\theta_f - \theta_s)}{\varphi_1 + (\xi_2 w + K_1)\varphi_2} \exp\left(\frac{1}{\theta_f + 0.25\varphi_2}\right) = Y, \tag{6.78}$$

$$\varphi_1 = C_0 \exp s\left(-2/(3\theta_f - \theta_s)\right), \varphi_2 = (\theta_f - \theta_s)(\sqrt{1 + 3\xi_2^2/\xi_1^2} - 1)\xi_1^2/\xi_2^2,$$
$$w^2 = \frac{\theta_f - \theta_s}{C_0 \exp(-1/\theta_f)}\left(K_1 - \frac{C_0 \exp(-1/\theta_f)}{\theta_f - \theta}\right)^2. \tag{6.79}$$

After finding θ_f using (6.78), the wave velocity using (6.79), and the eigenvalues using (6.63), the values of the remaining unknown parameters are determined (Zhizhin, 2008).

Numerical studies have shown that the function $Y(\theta_f)$ given by equations (6.78), (6.79) is a curve with maximum. Figure 5 shows the results of calculating this function for C_0=1, θ_s=0.05 and various values of the parameter K_1, which is proportional to the heat transfer coefficient.

Figure 5. Plots of $Y(\theta_f)$ function (Equation 6.78)

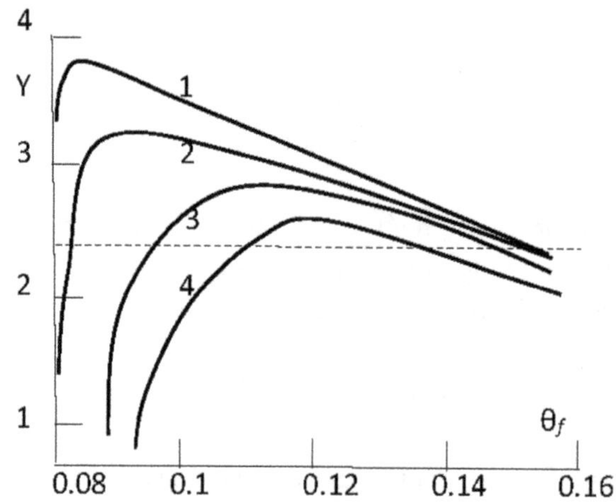

It can be seen that as the value of the parameter K_1 increases, the maximum value of the function $Y(\theta_f)$ decreases and the function itself shifts to higher temperatures θ_f. The points of intersection of the function $Y(\theta_f)$ with the horizontal $Y(\theta_f)=K_2/2$ are the roots of equation (6.78) and determine the values of the temperature of the leading edge of the wave. It follows from this that for a relatively small thermal effect of the reaction (large K_2) and a given value of the parameter K_1, there may be no stationary wave solutions. With an increase in the thermal effect of the reaction (a sufficient amount of heat), two stationary wave solutions can appear - with a high and low temperature of the reaction front. As an example, Figure 6 shows the results of calculating the dependences of the temperature of the leading front of the reaction and the wave velocity on the parameter K_1 at K_2 =4.8 and the accepted values of C_0, θ_s. These dependencies have two mating branches.

Figure 6. Dependences of the velocity and temperature of the leading edge of the combustion wave on the parameter K,

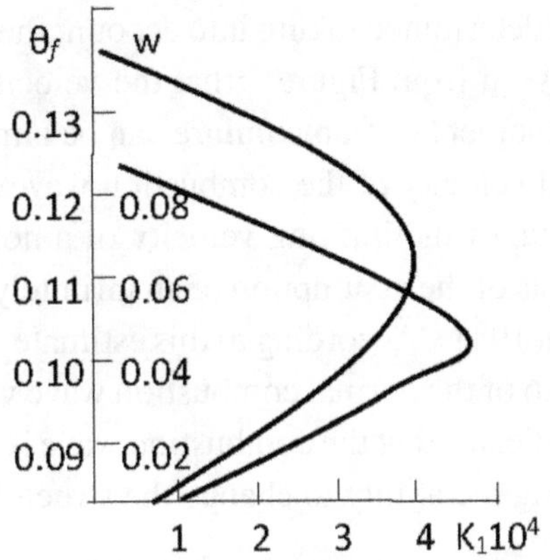

The upper branches correspond to the high-speed high-temperature wave solution, the lower branches correspond to the low-temperature low-velocity wave solution. However, the lower branches of these functions contradict the physical meaning, since on them with an increase in heat loss to the environment, the temperature of the reaction medium increases. Therefore, the lower branches of the functions can be discarded without proving their instability. Figure 7 shows the results of calculating the dependence of the limiting speed on the parameter K_1.

Figure 7. Dependences of the characteristics of the combustion wave at the point of disruption of the stationary regime on the parameter K,

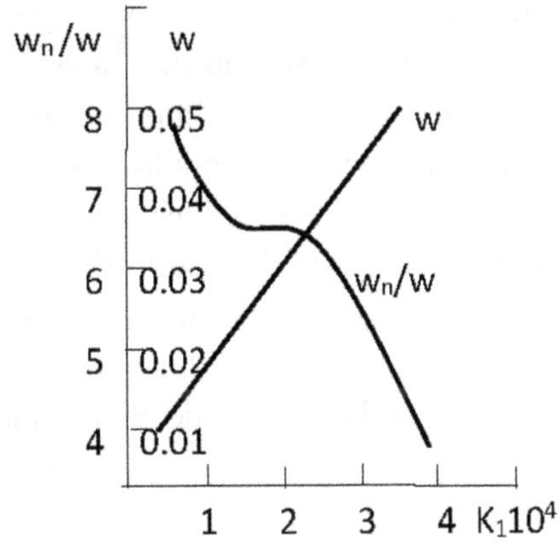

In addition, as a comparison, Figure 7 shows the results of calculating the ratio of the normal velocity of an adiabatic combustion wave, determined by the narrow reaction zone method (Zeldochich et al., 1980), to the limiting velocity of a nonadiabatic combustion wave, determined taking into account the length of the chemical reaction zone. It can be seen from Figure 7 that the velocity of the non-adiabatic combustion wave at the moment of flame failure can be almost an order of magnitude less than the normal velocity of the combustion wave. This is fundamentally different from the estimate of the limiting velocity of a nonadiabatic combustion wave, obtained on the basis of the assumption of an infinitely small extension of the reaction zone (Zel'dovich, 1941). According to this estimate, the limiting velocity of a nonadiabatic flame is 0.6 of the normal combustion wave velocity under adiabatic conditions. This result indicates that the combustion wave is much more resistant to external influences, having the ability to change the structure of the wave.

6.5. NON-ADIABATIC DIFFUSION-THERMAL COMBUSTION WAVE

If in a non-adiabatic combustion wave, not only the processes of heat transfer by thermal conduction are significant, but also the processes of diffusion transfer of the reactant, then the equation of the law of conservation of mass must be supplemented with the corresponding term describing the diffusion transfer

$$\frac{\partial C}{\partial t} = \frac{\partial}{\partial x} D \frac{\partial C}{\partial x} - W \quad , \qquad (6.81)$$

where the rate of the chemical reaction is still given by the equation (6.55).

It follows from (6.81), (6.55) that the equilibrium in the combustion wave under these conditions can be achieved only when the initial reagent is completely burned out. However, since we are considering a non-adiabatic wave, i.e. assuming the openness of the wave as a system for external influences, then the assumption of complete burnout of the initial reagent in the equilibrium position contradicts the possibility of finding a combustible reagent in the environment in some quantity. It follows that it is logical to supplement equation (6.81) with another term describing the mass transfer of the reaction medium with the environment

$$\frac{\partial C}{\partial t} = \frac{\partial}{\partial x} D \frac{\partial C}{\partial x} - W + \alpha_m (C_s - C). \qquad (6.82)$$

Here α_m is the coefficient of mass transfer of the reaction medium with the environment, C_s is the mass fraction of the combustible reagent in the environment.

Entering, as usual, the wave coordinate z = ut - x, u is the wave speed, we bring the equations (6.53), (6.55), (6.82) to the form

$$c_p \rho u \frac{dT}{dz} = \lambda \frac{d^2 T}{dz^2} + \rho Q k C \exp\left(-\frac{E}{RT}\right) - \alpha(T - T_s), \tag{6.83}$$

$$u \frac{dC}{dz} = D \frac{d^2 C}{dz^2} - kC \exp\left(-\frac{E}{RT}\right) - \alpha_m (C - C_s). \tag{6.84}$$

In addition to dimensionless quantities (6.58), we introduce parameters

$$K_3 = \frac{\alpha_m K_2}{k}, Le = \frac{c_p \rho D}{\lambda} \tag{6.85}$$

and write equations (6.83), (6.84) in dimensionless form

$$w \frac{d\theta}{d\zeta} = \frac{d^2\theta}{d\zeta^2} + C \exp\left(-\frac{1}{\theta}\right) - K_1(\theta - \theta_s), \tag{6.86}$$

$$w \frac{dC}{d\zeta} = Le \frac{d^2 C}{d\zeta^2} - K_2 C \exp\left(-\frac{1}{\theta}\right) - K_3(C - C_s). \tag{6.87}$$

System (6.86), (6.87) can be written as a system of four first-order differential equations

$$\frac{d\theta}{d\zeta} = \eta, \tag{6.88}$$

$$\frac{d\eta}{d\zeta} = w\eta - C \exp(-\frac{1}{\theta}) + K_1(\theta - \theta_s), \tag{6.89}$$

$$\frac{dC}{d\zeta} = p, \tag{6.90}$$

$$\frac{dp}{d\zeta} = wp/Le + K_2 C Le^{-1} \exp(-\frac{1}{\theta}) + K_3 Le^{-1}(C - C_s). \tag{6.91}$$

In order for the system (6.88) - (6.91) to have the equilibrium position necessary for the existence of a wave solution, it is necessary to set $\exp(-\frac{1}{\theta})$ zero in the vicinity of the temperature θ_s at $\theta \geq \theta_s$. In these areas we have a system

$$\frac{d\theta}{d\zeta} = \eta, \tag{6.92}$$

$$\frac{d\eta}{d\zeta} = w\eta + K_1(\theta - \theta_s), \tag{6.93}$$

$$\frac{dC}{d\zeta} = p, \tag{6.94}$$

$$\frac{dp}{d\zeta} = wp/Le + K_3 Le^{-1}(C - C_s). \tag{6.95}$$

It can be seen that the system (6.92) - (6.95) splits into two pairs of equations (6.92), (6.93) and (6.94), (6.95), i. e. the four-dimensional space (θ, η, C, p) in the vicinity of the equilibrium positions is the direct product of the phase planes (θ, η), (C, p). The equilibrium position of the system (6.92), (6.93) has coordinates $\eta = 0$, $\theta = \theta_s$. The eigenvalues in this equilibrium position are found from the discriminant of the system (6.92), (6.93)

$$\xi_{1,2} = 0.5w \pm \sqrt{0.25w^2 + K_1}. \tag{6.96}$$

The equilibrium position of the system (6.94), (6.95) has coordinates p = 0, C = C_s. The eigenvalues in this equilibrium position are found from the discriminant of the system (6.94), (6.95)

$$\xi_{3,4} = Le^{-1}(0.5w \pm \sqrt{0.25w^2 + K_3}). \tag{6.97}$$

From (6.96), (6.97) it follows that in both phase planes the equilibrium positions are of saddle type and the phase portraits of systems (6.92), (6.93) and (6.94), (6.95) are shown in Figure 8a and Figure 8b.

Figure 8. Behavior of the trajectories of the system (6.88) - (6.91) in the neighborhood of equilibrium positions

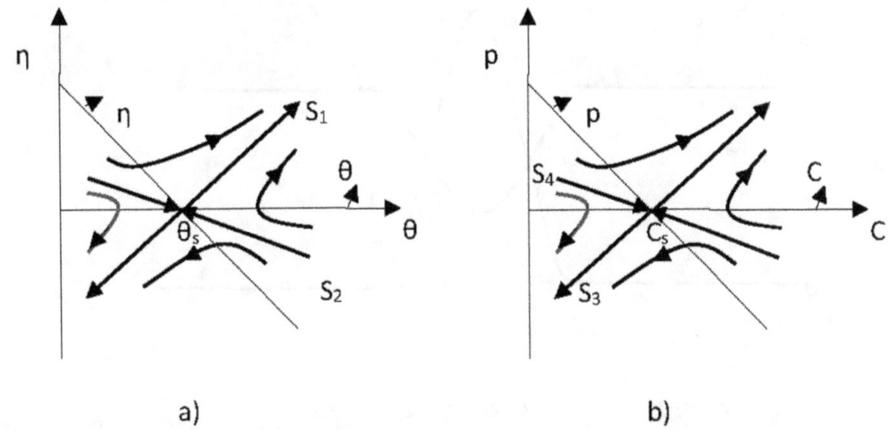

a) in the phase plane (η, θ), system (6.92), (6.93);
b) in the phase plane (p, C), system (6.94), (6.95).

The desired wave solution leaves the initial position of equilibrium (η = 0, θ = 0, p = 0, C = C_s) in the phase plane (η, θ) along the separatrix S_1, and in the phase plane (p, C) along the separatrix S_3. The desired wave solution enters the final position of equilibrium (η = 0, θ = $θ_s$, p = 0, C = C_s) along the separatrix S_2 in the phase plane (η, θ), and in the phase plane (p, C) along the separatrix S_4. Between the neighborhoods of the initial and final equilibrium positions, the trajectories pass in the four-dimensional phase space (η, θ, p, C), crossing zero surfaces

$$F_1{:}w\eta - C\exp(-1/\theta) + K_1(\theta - \theta_s) = 0,$$
$$F_2{:}wp - K_2 C\exp(-1/\theta) + K_3(C - C_s) = 0,$$

on which, respectively, the derivatives dη /dζ, dp /dζ change sign. After crossing these zero surfaces, the trajectories cross the zero surfaces η = 0, p = 0 (coordinate planes), changing the sign of the derivatives dθ /dζ, dC /dζ accordingly. This is accompanied by the transition of the trajectories from a part of the phase space η > 0, p < 0 to a part of the phase space η < 0, p > 0. The general view of the integral curves of the wave solution is shown in Figure 9.

Figure 9. The integral curves of system (6.88) – (6.91)

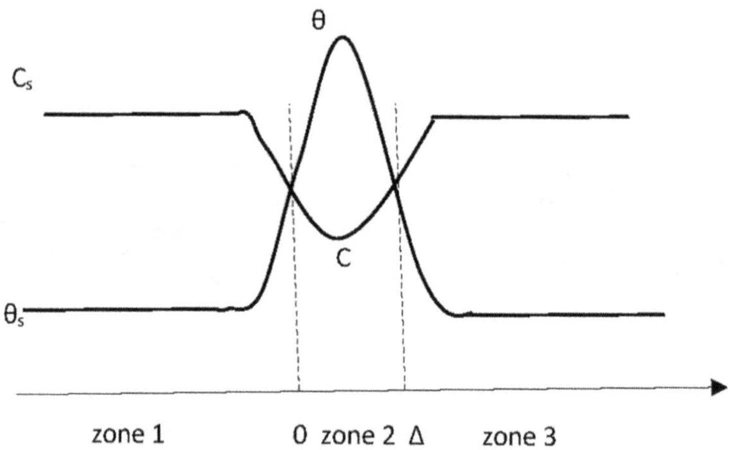

Three zones can be distinguished in the combustion wave structure (Figure 6.9). Zones 1 and 3 are neighborhoods of equilibrium positions. In these neighborhoods, chemical reactions do not occur. Zone 2 contains profile extremums of profiles θ(ζ), C(ζ). By linearizing the solutions in the vicinity of the equilibrium positions, expanding the functions in the Taylor series in the vicinity of their extremums, fulfilling the continuity conditions of the functions, one can obtain a closed system of algebraic equations for unknown values of variables at the boundaries of the zones. Solving this algebraic system of equations, an equation for the combustion wave velocity was obtained (Zhizhin, 2008)

$$\xi_4 \xi_2^{-1}(\xi_3^{-1} + 0.5\,\xi_1^{-1}) - 1.5\,\xi_1^{-1} = 0.5\,\xi_4^2 \xi_1^{-1}(\xi_1^{-1} - \xi_3^{-1})^{\,2}(\xi_4 \xi_2^{-1} - 1)^{\,-1}, \tag{6.98}$$

where ξ_i, $i = 1, 2, 3, 4$ are the functions of velocity at equations (6.96), (6.97).

After finding the velocity, the temperature at the maximum of the soliton is found by the equation

$$\frac{C_s}{(\theta_m - \theta_s)(K_1 + 2\xi_1^2/3)} = \exp(-\frac{1}{\theta_m}) + \frac{K_2 Le^{-1} \xi_1^{-1}(\xi_3^{-1} + 0.5\,\xi_1^{-1})}{1 + K_3 Le^{-1} \xi_1^{-1}(\xi_3^{-1} + 0.5\,\xi_1^{-1})}. \tag{6.99}$$

The mass fraction of the reagent at $\zeta = 0$ is determined by the equation

$$C_f = C_s - \varphi_2/(\xi_3 \xi_1), \tag{6.100}$$

$$\varphi_2 = \frac{(\theta_m - \theta_s)(K_1 + 2\xi_1^2/3)K_2 Le^{-1}}{1 + K_3 Le^{-1}\xi_1^{-1}(\xi_3^{-1} + 0.5\xi_1^{-1})}.$$

After that, the mass fraction of reagent is found at the extremum

$$C_m = C_f - 0.5(C_s - C_f)^2 \xi_3^2/\varphi_2, \qquad (6.101)$$

and also, at $\zeta = \Delta$

$$C_\Delta = (\xi_4\xi_2^{-1} - 1)^{-1}[(\theta_m - \theta_s)\varphi_2\varphi_1^{-1} - C_m + C_s\xi_4/\xi_2], \qquad (6.102)$$

$$\varphi_1 = K_1(\theta_m - \theta_s) - C_m \exp(-1/\theta_m).$$

The temperature at $\zeta = \Delta$ is determined by the equation

$$\theta_\Delta = \theta_m + \varphi_1(C_\Delta - C_m)/\varphi_2, \qquad (6.103)$$

and the width of the reaction zone from the equality

$$\Delta = (\theta_\Delta - \theta_s)\varphi_1{}^{-1}\xi_2 + (C_s - C_f)\xi_3/\varphi_2. \qquad (6.104)$$

The temperature at $\zeta = 0$ and the extremum coordinate are found from the equations

$$\theta_f = \theta_m + \varphi_1(C_f - C_m)/\varphi_2, \qquad (6.105)$$

$$\xi_m = \xi_3(C_s - C_f)/\varphi_2. \qquad (6.106)$$

The equation for finding the combustion wave velocity (6.98) can be represented as the ratio of two positive functions equal to unity

$$Y_1/Y_2 = 1 \qquad (6.107)$$

$$Y_1 = \frac{0.5 + \xi_1/\xi_2}{Le^2\xi_2^2} + \frac{0.5}{Le^2\xi_1\xi_3} + \frac{3}{2Le^2\xi_4^2}, \quad Y_2 = \frac{1 + \xi_1/\xi_3}{Le^2\xi_2\xi_3} + \frac{1}{Le^2\xi_1^2} + \frac{1}{Le^2\xi_3^2}.$$

Calculations show that the ratio Y_1/Y_2 is an increasing function of speed w. Therefore, if at w=0 the ratio $Y_1/Y_2 > 1$, then there are no wave solutions. Letting w tend to zero in (6.107), we obtain an equation for the boundaries of the regions of possible existence of wave solutions

$$Le_{1,2} = (2 \pm \sqrt{3})\sqrt{K_3/K_1}. \qquad (6.108)$$

These boundaries in the plane (Le, $\sqrt{K_3/K_1}$) are two rays emanating from the origin (Figure 10)

Figure 10. Bifurcation diagram of equation (6.107) for the combustion wave velocity

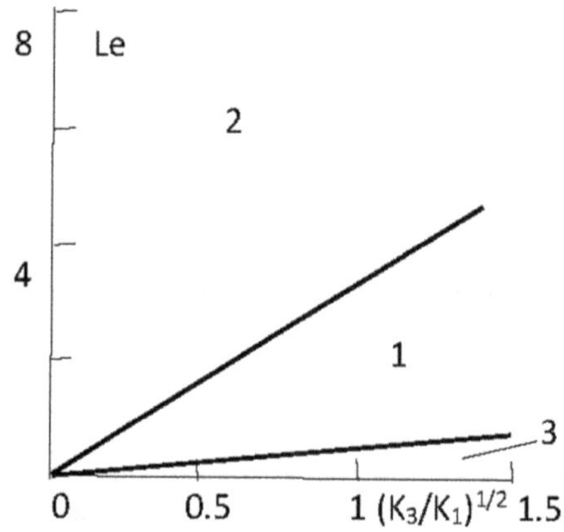

In region 1 between the rays, there are no solutions to equation (6.98). Thus, the region of existence of wave solutions consists of two parts: for large Lewis numbers (region 2) and for small Lewis numbers (region 3). To further solve the problem, it is necessary to determine the domain of existence of solutions to Eq. (6.99) with respect to the temperature at the maximum of the soliton. We represent equation (6.99) as

$$\Psi_1 = \Psi_2, \qquad (6.109)$$

$$\Psi_1 = \exp(-1/\theta_m), \Psi_2 = \frac{(\theta_m - \theta_s)(K_1 + 2\xi_1^2/3)}{C_s - (\theta_m - \theta_s)K_2 Le^{-1}\xi_1^{-1}(0.5\xi_1^{-1} + \xi_3^{-1})(K_1 + 2\xi_1^2/3)[1 + K_2 Le^{-1}\xi_1^{-1}(0.5\xi_1^{-1} + \xi_3^{-1})]^{-1}}.$$

It follows from the analysis of functions Ψ_1,Ψ_2, that they always have either one or three intersections. Thus, for one value of the combustion wave velocity, if there is a solution to equation (6.98), there are one or three values of the maximum temperature in the wave, i.e. from one to three possible temperature profiles and the mass fraction of the reagent in the wave. The critical condition for the birth or destruction of a pair of solutions to equation (6.109) is the tangency of the functions Ψ_1,Ψ_2. Differentiating the functions Ψ_1,Ψ_2, with respect to θm, solving equations (6.109) together with the equation $d\Psi_1/d\theta_m = d\Psi_2/d\theta_m$, we obtain two equations relating the values of the parameters at the points of contact of these functions

$$\Phi = \frac{2y\theta_s + 1 + \varphi}{2 - y\theta_s - y\varphi} \exp\frac{2(1+y)}{2y\theta_s + 1 + \varphi}, \tag{6.110}$$

$$\Phi = \frac{2y\theta_s + 1 - \varphi}{2 - y\theta_s + y\varphi} \exp\frac{2(1+y)}{2y\theta_s + 1 - \varphi}, \tag{6.111}$$

$$\Phi = C_s(K_1 + 2\xi_1^2/3)^{-1}, \varphi = \sqrt{1 - 4\theta_s^2 y - 4\theta_s}, y = \Omega/\Phi,$$

$$\Omega = \frac{K_2(\xi_3^{-1} + 0.5\xi_1^{-1})(K_1 + 2\xi_1^2/3)}{[Le\xi_1 + K_3(\xi_3^{-1} + 0.5\xi_1^{-1})](K_1 + 2\xi_1^2/3)}.$$

The temperature values θ_m in equalities (6.110), (6.111) are given by the equation

$$\theta_m = 0.5(1+y)^{-1}(2y\theta_s + 1 \pm \varphi), \tag{6.112}$$

where the sign "+" before φ corresponds to equation (6.110), and the sign "-" before φ corresponds to equality (6.111). Equations (6.110), (6.111) in the three-dimensional space (y, θ_s, Φ) of parameters define two surfaces, respectively F_1 and F_2, mating along a common line L (Figure 11).

Figure 11. Bifurcation diagram of equation (6.99) for the maximum temperature in the combustion wave

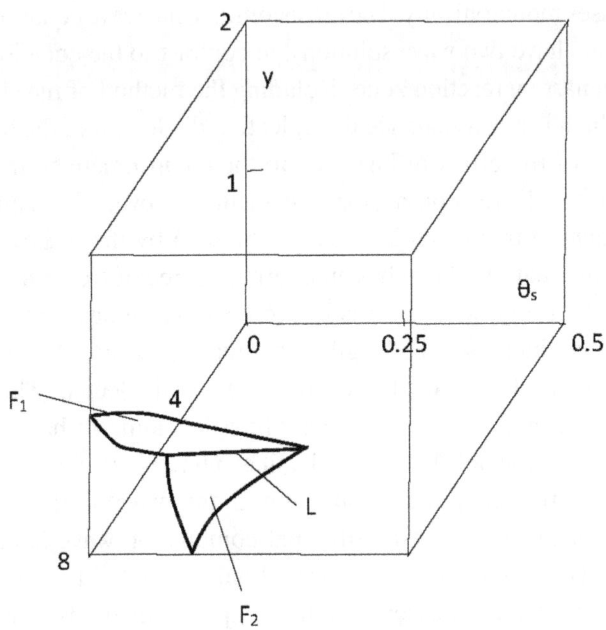

This line represents the locus of assembly-type catastrophe points. Under the surface, composed of sheets (6.110), (6.111), there is an area of parameter values , for which equation (6.109) has three roots. Outside this region, equation (6.109) has one root; at large θ_s, to the right of the F_2 surface, this is a high-temperature root (large θ_m), above the F_1 surface, this is a low-temperature root (small θ_m).

CONCLUSION

Combustion waves, in which both diffusion processes and the release of heat of exothermic chemical reactions are simultaneously important, are studied. Previously, such waves were considered, as a rule, in the approximation of a narrow reaction zone. Since this approximation often contradicts known experimental data, in this chapter these waves are considered by non-asymptotic methods (the assumption of a narrow reaction zone is not used). This made it possible to establish fundamentally new features of the laws governing the propagation of hot flame waves. It has been established that for Lewis numbers greater than one, the nature of the dependence of the combustion wave velocity on the Lewis number in the non-asymptotic theory is directly opposite to the nature of this dependence in the theory using the assumption of a narrow reaction zone. According to asymptotic methods, the velocity of the combustion wave increases indefinitely with an increase in the Lewis number, and according to non-asymptotic methods decreases monotonically. The diffusion-thermal wave equations in this range of Lewis numbers have two wave solutions, in contrast to the equations that use the assumption of a narrow reaction zone (including the method of matched asymptotic expansions), which is a consequence of neglecting the length of the reaction zone.

The question of the effect of heat loss to the environment from a wave of hot flame is important. Here, non-asymptotic methods reveal fundamental features of flame propagation that essentially cannot be seen by the narrow reaction zone theory. It turns out that the flame has an internal degree of freedom, which ensures the stability of the flame when heat is lost to the environment. To compensate for heat loss, the flame increases the length of the reaction zone. Of course, this effect cannot be seen if the length of the reaction zone is neglected. This is especially important when extinguishing a fire by cooling the flame. It has been found that for the same value of the hot flame velocity, it can have several wave solutions with different values of the maximum temperature in the wave. In the space of parameters, the equations of the diffusion-thermal combustion wave have a continuous set of assembly-type catastrophe points. It should be noted that in order to solve multiparameter hot flame problems, non-asymptotic methods require additional conditions that allow one to obtain approximate analytical solutions. For all the solutions obtained, their correctness was confirmed by the numerical solution of the original differential equations.

REFERENCES

Bachelis, R. D., & Melamed, V. G. (1966). On the non-uniqueness of the stationary solution of the system of equations of the theory of combustion. *Applied Mathematics and Mechanics*, *2*, 368–374.

Berman, V. S., & Ryazantsev, Yu. S. (1972). Application of the theory of matched asymptotic expansions to the calculation of stationary thermal propagation of an exothermic reaction front in a condensed medium. *Journal of Applied Mechanics and Technical Physics*, *5*, 106–112.

Buckmaster, J. (1976). The quenching of deflagration waves. *Combustion and Flame*, *26*(2), 151–162. DOI: 10.1016/0010-2180(76)90068-7

Bush, W. B., & Fendell, F. E. (1970). Asymptotic analysis of laminar flame propagation for general Lewis number. *Combustion Science and Technology*, *1*(6), 421–428. DOI: 10.1080/00102206908952222

Frank-Kamenetsky, D. A. (1967). *Diffusion and heat transfer in chemical chemistry*. Nauka.

Grishin, A.M. (2002). General mathematical models of forest and peat fires and their application. *Mechanical advances*. 12, 41 - 89.

Grishin, A.M. (2004). General mathematical model of steppe fires. *Ecological systems and devices*.12, 25 - 29.

Joulin, G., & Clavin, P. (1976). Asymptotic analysis of conditions of extinctions for luminary flames. *Acta Astronautica*, *3*(3/4), 223–240. DOI: 10.1016/0094-5765(76)90049-7

Lovachev, L. A., Basevich, V. Ya., Kogan, A. M., & Posvyansky, V. S. (1970). To the calculation of the stationary reaction zone on the example of the H2 + O2 flame. *Proceedings of the Academy of Sciences of the USSR. Chemical series*. 6, 1256 - 1263.

Novikov, S. S., & Ryazantsev, Yu. S. (1966). On the existence and uniqueness of the solution of the system of the thermal theory of combustion. *Applied Mathematics and Mechanics*, *2*, 368–374.

Spalding, D.B. (1957). A theory of inflammability limits and flame – quenching. *Combust. Proc. Roy. Soc. London.* V.A 240, 1220, 83 – 100.

Williams, F. A. (1971). *Theory of combustion*. Nauka.

Zeldovich, Ya. B. (1941). The theory of the propagation limit of a silent flame. *Journal of Experimental and Theoretical Physics*, *11*(1), 159–168.

Zeldovich, Ya. B. (1948). On the theory of flame propagation. *Journal of Physical Chemistry*, *1*, 27–48.

Zeldovich, Ya. B. (1961). Chain reactions in gas flames, approximate theory of flame speed. *Kinetics and Catalysis*, *11*(3), 305–374.

Zeldovich, Ya. B., Barenblatt, G. I., Librovich, V. B., & Makhviladze, G. M. (1980). *Mathematical theory of combustion and explosion*. Nauka.

Zhizhin, G. V. (2008). *Combustion waves with distributed zones of chemical reactions: (Non -asymptotic theory of combustion)*. Werner Regen Publishing House.

Chapter 7
Waves of Detonation

ABSTRACT

The non-asymptotical theory of waves of a detonation of chemical reactions is constructed. It, as against the classical theory of detonation, takes into account kinetic of the chemical reactions, changes of structure of the waves during reactions and extend of the chemical reactions. It has been established that the limitation of the gas velocity at the end of the detonation wave by the local value of the speed of sound, accepted in the classical theory of detonation, is at the same time a limitation on the depth of the chemical reaction so that the depth of the chemical reaction does not exceed 30% on average. It has been proven that the estimates of the parameters of adiabatic detonation waves according to the Chapman-Jouguet theory are very far from analytical solutions according to the non-asymptotic theory of detonation and numerical solutions of differential equations.

INTRODUCTION

Combustion waves moving at supersonic speeds, called detonation waves, cause catastrophic damage to the environment. Examples are well-known explosions in mines, accompanied by the propagation of detonation waves. Detonation waves are a powerful energy phenomenon. Natural desire to use it for human needs. It has long been used in various kinds of explosive work and the creation of weapons. Currently, in many countries, work is underway to create an engine for aircraft using continuous or pulsed detonation processes. According to modern concepts, a detonation wave is a complex of a traveling shock wave and a zone of chemical combustion reactions located immediately behind it. The medium through which the detonation wave propagates can be gaseous, liquid, solid, and can also be a homogeneous or heterogeneous mixture of components, at least one of whose components is chemically active and propagates the combustion reaction. A shock wave is a sharp disturbance, i.e. a sharp change in the main characteristics of the medium

DOI: 10.4018/978-1-6684-7745-8.ch007

(pressure, density, temperature), arising and propagating through the medium as a result of various kinds of atmospheric, underground, and underwater explosions. In the case of strong shock waves characterized by a large temperature difference, the propagation of a shock wave, due to a sharp increase in temperature behind the wave front, can cause a combustion reaction in the medium. In certain cases, the speed of propagation of the shock wave complex and the zone of bitterness will greatly increase and become supersonic. This creates a detonation wave. Very strong shock waves can cause nuclear reactions. This happens in the stars and, in particular, on the closest star to the Earth - the Sun. During periods of solar activity during flares in the solar atmosphere, powerful streams of contaminated particles are formed from the surface of the Sun, accompanied by strong magnetohydrodynamic shock waves (Pikelner, 1961). Moreover, due to the large loss of heat by radiation, nuclear reactions behind the shock wave are not initiated by the thermal motion of plasma particles but occur due to the powerful acceleration of particles in the magnetohydrodynamic shock wave (Syrovatsky, 1975 a, b). It should be noted that the idea that nuclear reactions can take place not only in the core of the Sun (thermonuclear reactions), but also in its atmosphere, was first expressed by I.S. Shklovsky (Shklovsky, 1955). Later, the mechanism of particle acceleration in the vicinity of the shock wave was elucidated, leading to nuclear reactions in the solar atmosphere. It is associated with the existence of a zero line of the magnetic field and the formation in the vicinity of this line of a neutral layer of electric current. When this layer is ruptured by a shock wave, there is a desire to heal the gap, and this happens with the help of particle acceleration. Moreover, the speed of movement of shock magnetohydrodynamic waves, accompanied by nuclear reactions, is supersonic. Taking into account that the temperature in the Sun's atmosphere reaches several million degrees, the speed of sound in the Sun's atmosphere reaches about 100 km/s, and the speed of propagation of shock waves is 500 km/s (Davis, Evans, 1974). The speed of movement of shock magnetohydrodynamic waves accompanied by nuclear reactions is supersonic. Thus, we can speak about the existence in nature of detonation waves of chemical and nuclear combustion.

Detonation waves of chemical combustion occur on the Earth. They can appear during bomb explosions; explosions of gas and dust-air mixtures in mines; explosions of gas cylinders; when large bodies fall, for example, meteorites; with powerful spark discharges; during volcanic eruptions and earthquakes, etc. A detonation wave can occur during fires when a combustion wave propagating at a subsonic speed (deflagration wave) is unstable and passes into a supersonic propagation mode.

Waves of chemical and nuclear detonation have various types of damaging effects on the ecological systems of the Earth. On the whole, detonation waves are of interest both from the point of view of safety and the reduction of their disturbing influence on the ecological systems of the Earth, and from the point of view of their fullest use. To be able to control detonation waves, it is necessary to know the quantitative patterns of their development. The most studied are the waves of chemical detonation in gases. However, in the works of the author (Zhizhin, 2008, 2009 a, b) it is shown that the existing theory of detonation in gases contains fundamental contradictions. Since the existing theory of detonation in gases is currently used under different conditions, it is necessary to present a new theory of detonation in gases. This chapter is dedicated to this.

7.1. NON-ASYMPTOTIC THEORY OF DETONATION IN GASES

The development of the theory of detonation reflects the stages in the development of the theory of combustion as a whole. The initial stage of the study of detonation is characterized by strong schematization and practical neglect of the length of the reaction zone in the wave (Zel'dovich, 1940, 1944; Zel'dovich and Kompaneets, 1955; Shchelkin and Troshin, 1963). In the classical theory of Chapman-Jouguet detonation (Chapman, 1899; Jouguet, 1905; Zel'dovich, 1944; Lewis and Elbe, 1968; Williams, 1971), it is assumed that a stationary adiabatic detonation wave is (as already mentioned) a complex of a shock wave and located behind her combustion zone. Moreover, it was believed that at the end of the combustion zone (chemical reaction zone) in coordinates associated with the detonation wave, the gas velocity is equal to the local speed of sound (Chapman-Jouguet condition). In addition, it was assumed that at the end of the chemical reaction zone, the depth of the chemical reaction is equal to one (everything that could burn burned down, Kompaneets, 1976). However, the adiabatic coefficient of the gas was assumed to be constant. It is easy to see that such assumptions contradict each other. First, if a chemical reaction has taken place in the gas, then the adiabatic coefficient has changed, since the composition of the gas has changed, especially if we assume that everything that could burn has burned down. This is especially important when determining the Mach number at the end of the chemical reaction zone, as the ratio of the velocity of gas to the speed of sound, the expression of which directly includes the adiabatic coefficient of the gas. Secondly, if we assume that at the end of the chemical reaction zone, the Mach number is equal to one, i.e. If a flow crisis sets in, which, as is well known (Vulis, 1950), occurs at a finite distance from the beginning of the flow, then there is no reason to believe that when the flow crisis was reached, everything that could burn out necessarily burned out.

Ya.B. Zeldovich (Zeldovich, 1940). However, in this work, the kinetics of chemical reactions was not taken into account, the heat capacity of the gas and its adiabatic coefficient were considered constant, and, moreover, when writing the conservation equations, an error was made, consisting in the fact that the differential of work against friction forces is included both in the energy conservation equation and into the equation of conservation of momentum. Although, as shown by L.A. Vulis, when establishing the law of reversal of influences a few years later (Vulis, 1946 a, b; 1947; 1950), the differential of work against friction forces enters the energy conservation equation twice: as the expenditure of system energy to perform work against friction forces and as thermal energy released when doing work against frictional forces. Moreover, these two differentials enter the energy conservation equation with opposite signs and compensate each other, i.e. in fact, the differential of work against friction forces disappears from the energy equation. But in the work of Ya.B. Zel'dovich, this differential is present in the energy conservation equation (this error was not corrected in later works).

Let us imagine the theory of detonation in gases without the listed conflicting requirements (Zhizhin, 2008).

According to existing ideas (Shchelkin, Troshin, 1963), in the laboratory (fixed) coordinate system, the detonation wave motion pattern, has the form shown in Figure 1a.

Figure 1. Graphs of changes in gas parameters in the laboratory coordinate system (moving detonation wave)

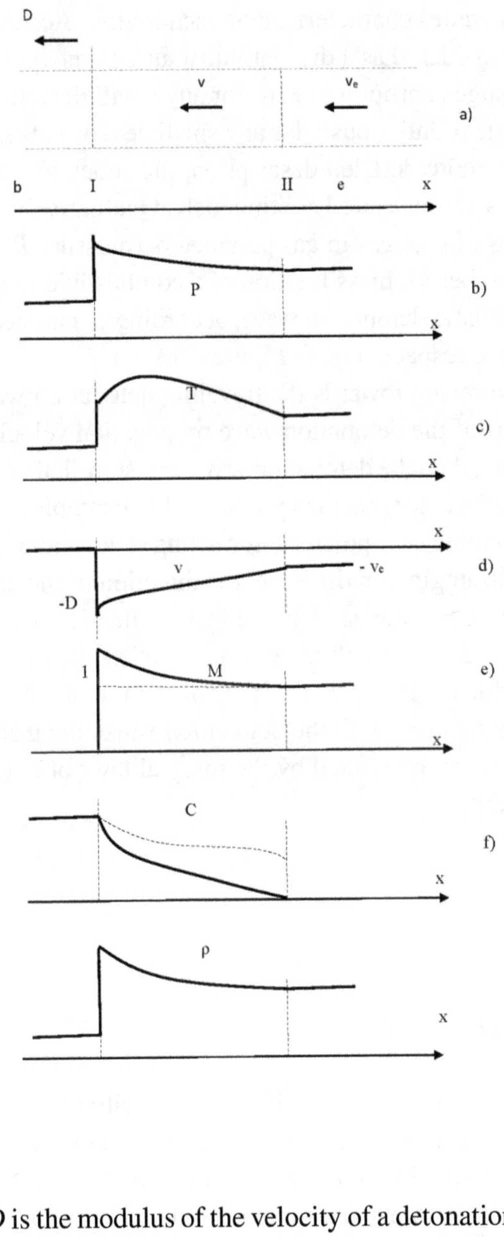

In this figure: D is the modulus of the velocity of a detonation wave propagating in the direction opposite to the direction of the spatial coordinate x; I there is a shock wave; II there is a trailing edge of the detonation wave; the space between sections I and II is the chemical reaction zone; b is the region of gas at rest in front of the detonation wave ($v_b = 0$) ; e there is a region of gas behind the detonation wave, where the gas moves at a constant speed v_e, entrained by the detonation wave. In the region 1-11, all gas parameters change along the spatial coordinate x (in compliance with the conservation laws), in contrast to the regions b and e, where all parameters have some constant values characteristic of each region. According to these ideas, the shock wave is considered as a discontinuity surface moving at a speed - D, and the gas velocity changes abruptly to zero for any small deviation from the surface to the left, and changes continuously for any small deviation from the surface to the right. Although in a more detailed description, the shock wave has a thin layer in which all parameters are measured continuously. Qualitatively, in the first approximation, the patterns of changes in gas parameters (pressure P, temperature T, gas velocity v, Mach number M, mass fraction of a combustible reagent C, gas density ρ) in a traveling adiabatic detonation wave, according to classical concepts, can be represented as shown, respectively, in Figures 1b) - f).

If a gas flow is directed towards the traveling detonation wave with a velocity equal to the modulus of the detonation wave propagation velocity in the gas in the initial stationary state, then the detonation wave will stop. This will make it possible to use the detonation wave for practical purposes. For example, a stopped detonation wave can be used to create a propulsive force acting on an aircraft that has a standing detonation wave in its engine. In this case, the direction of the driving force will be opposite to the direction of the gas furnace leaving the detonation wave.

Let us now proceed to a quantitative analysis of the processes in the adiabatic detonation wave within the framework of the non-asymptotic theory of combustion. In the coordinates associated with the detonation wave, the gas parameters before and after the shock wave are related by the integral laws of conservation of mass, momentum, and energy

$$\rho_1 u_1 = \rho_b u_b, \tag{7.1}$$

$$\rho_1 u_1^{\,2} + P_1 = \rho_b u_b^{\,2} + P_b, \tag{7.2}$$

$$c_{pA} T_1 + 0.5 u_1^2 = c_{pA} T_b + 0.5 u_b^2. \tag{7.3}$$

In equation (7.3)c_{pA} there is the specific heat capacity of the initial gas A at constant pressure. It is believed that during the passage of molecules through the shock wave, the chemical interaction of molecules does not have time to occur.

The gas parameters before and after the detonation wave are also related by the integral conservation laws

$$\rho_e u_e = \rho_b u_b, \tag{7.4}$$

$$\rho_e u_e^{\,2} + P_e = \rho_b u_b^{\,2} + P_b, \tag{7.5}$$

$$c_{pe} T_e + 0.5 u_e^2 = c_{pA} T_b + 0.5 u_b^2 + Q(1 - C_{Ae}). \tag{7.6}$$

In equation (7.6) Q there is a specific thermal effect of a chemical reaction; C_A is the mass fraction of the initial reagent A; C_{Ae} is the mass fraction behind the detonation wave, which must be found as a result of solving the problem and which, in the classical theory of detonation, is assumed to be equal to zero. Pressure, density and temperature in states *b, I, e,* are related by the equation of state of an ideal gas

$$P_i = \rho_i R_i T_i, \; i = b, I, e. \tag{7.7}$$

Assuming that when burning gas A turns into gas B, in equations (7.6), (7.7)

$$\begin{aligned} c_{pe} &= (c_{pA} - c_{pB}) C_{Ae} + c_{pB}, \\ R_e &= (R_A - R_B) C_{Ae} + R_B, \end{aligned} \tag{7.8}$$

In (7.8) $R_b = R_1 = R_A$ is the specific gas constant of the gas A, R_B is the specific gas constant of the gas B, $C_{Be} + C_{Ae} = 1$.

The system of equations (7.1) - (7.8) is not closed a priori for the unknown C_{Ae}. To close it, it is necessary to use the laws of kinetic processes in the zone of chemical reactions, and this is quite natural, since without chemical reactions, the detonation wave is reduced to a shock wave. For the zone of chemical reactions, the differential laws of conservation of mass, momentum, and energy have the form

$$\frac{d\rho}{\rho dz} + \frac{du}{udz} = 0, \tag{7.9}$$

$$\frac{dP}{dz} + \rho u \frac{du}{dz} = 0, \tag{7.10}$$

$$\frac{dh}{dz} + u \frac{du}{dz} = 0. \tag{7.11}$$

Here $h = (h_A - h_B) C_A + h_B$ is the specific enthalpy of the mixture of gases, h_A and h_B the specific enthalpies of the initial gas and the reaction product.

The system of equations (7.9) - (7.11), taking into account the equalities

$$\begin{aligned} P &= \rho R T, \; c_p = (c_{pA} - c_{pB}) C_A + c_{pB}, \\ R &= (R_A - R_B) C_A + R_B, Q = h_A - h_B \end{aligned} \tag{7.12}$$

can be solved with respect to the derivatives

$$(M^2 - 1)\frac{du}{udz} = K\frac{dC_A}{dz}, \tag{7.13}$$

$$(M^2 - 1)\frac{d\rho}{\rho dz} = -K\frac{dC_A}{dz}, \tag{7.14}$$

$$(M^2 - 1)\frac{dP}{Pdz} = -\gamma M^2 K\frac{dC_A}{dz}, \tag{7.15}$$

$$(M^2 - 1)\frac{dT}{Tdz} = -\frac{Q}{c_p T}(M^2 - 1 + K\frac{u^2}{Q})\frac{dC_A}{dz}, \tag{7.16}$$

where $K = \frac{Q}{c_p T} + \frac{R_A - R_B}{R}, \gamma = (\gamma_A - \gamma_B)C_A + \gamma_B$ is the coefficient of adiabatic gas, γ_A and γ_B are the coefficients of adiabatic gases A and B, $M^2 = \frac{u^2}{\gamma RT}$ is Mach number squared, γRT is square of the local adiabatic speed of sound in a gas.

Using (7.13) - (7.16), one can obtain an equation for the change in the square of the Mach number along the chemical reaction zone

$$(M^2 - 1)\frac{dM^2}{M^2 dz} = [(1 - \gamma M^2)K + (1 - M^2)\frac{\gamma_A - \gamma_B}{\gamma}]\frac{dC_A}{dz}. \tag{7.17}$$

Considering the chemical reaction irreversible, we write the equation for changing the mass fraction of the initial reagent in the form

$$\frac{dC_A}{dz} = -k_0 u^{-1} C_A^n \exp(-\frac{E}{R_0 T}), \tag{7.18}$$

where k_0 is the pre-exponent, n is the order of the reaction, E is the activation energy of the combustion reaction, R_0 is the universal gas constant.

By introducing a dimensionless temperature $\theta = \frac{R_0 T}{E}$, equations (7.16) - (7.18) can be reduced to an autonomous closed system of equations

$$\frac{dM^2}{d\zeta} = \left[(1 + \gamma M^2)(\frac{c_{pA}}{c_p \varepsilon\theta} + \frac{R_A - R_B}{R}) + (1 - M^2)\frac{\gamma_A - \gamma_B}{\gamma}\right]\frac{0.5 C_A^n \theta^{-1/2}}{1 - M^2}\exp(-\frac{1}{\theta}), \tag{7.19}$$

$$\frac{d\theta}{d\zeta} = \left[1 - M^2 - \frac{\gamma R}{c_{pA}} M^2 (\frac{c_{pA}}{c_p} + \varepsilon\theta\frac{R_A - R_B}{R})\right]\frac{c_{pA} C_A^n \theta^{-1/2}}{(1 - M^2)c_p M\varepsilon}\exp(-\frac{1}{\theta}), \tag{7.20}$$

$$\frac{dC_A}{d\zeta} = -\frac{C_A^n}{M\theta^{1/2}}\exp(-\frac{1}{\theta}), \varepsilon = \frac{Ec_{pA}}{R_0 Q}, \zeta = k_0(\frac{R_0}{\gamma RE})^{1/2}. \tag{7.21}$$

7.2. QUALITATIVE STUDY OF THE SYSTEM OF EQUATIONS (7.19) - (7.21)

Solutions of system (7.19) - (7.21) can be represented by trajectories in three-dimensional phase space (M,θ,C_A). Since the exothermic combustion reaction is a combination reaction, then

$$c_{pA} > c_{pB}, \gamma_A > \gamma_B, R_A > R_B. \tag{7.22}$$

From (7.19) - (7.21) under condition (7.22) and subsonic flow in the chemical reaction zone, it follows that in the part of the phase space (M,θ,C_A) that has a physical meaning $M \leq 1$, $C_A \leq 1$, the mass fraction of the initial reagent only decreases, and the Mach number only increases. In the open part of the phase space, there is a zero surface $F_\theta : M = \gamma^{-1/2}(1 + \frac{R_A - R_B}{c_{pA}} \varepsilon\theta)^{-1/2}$, at the points of which the derivative $d\theta / d\zeta$ changes sign, passing through zero. The equilibrium positions of the system, in which irreversible nonequilibrium processes occur, are located on the boundary of the physical part of the phase space (Zhizhin, 2004). In this problem the set of equilibrium positions of the system (7.19) - (7.21) coincides with the coordinate plane $C_A = 0$ of the phase space (M,θ,C_A). One of the boundaries of the physical part of the phase space is the singular plane $M = 1$, at the points of which the derivatives $dM / d\zeta$, $d\theta / d\zeta$ have a discontinuity $\pm\infty$. The set of initial points of the zone of chemical reactions form a plane $C_A = 1$, also limiting the physical part of the phase space. It is easy to see (Figure 2) that the solution manifold of system (7.19) - (7.21) can be divided into two subvarieties.

Figure 2. The phase space of the system (7.19) - (7.21)

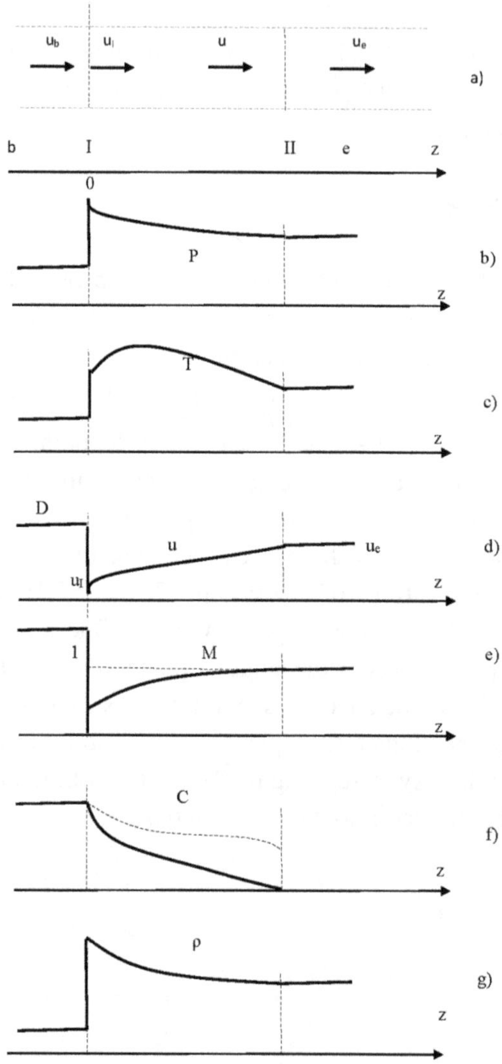

One of them includes trajectories starting from points of the plane $C_A = 1$ and ending at points of the singular plane $M = 1$ at $C_A > 0$ (trajectories *a*). Another submanifold includes trajectories starting at points of the plane $C_A = 1$ and ending at the equilibrium positions of the system - points of the plane $C_A = 0$ at $M < 1$.

These two submanifolds are separated by a separatrix surface F_s, consisting of trajectories S ending simultaneously with the complete burnout of the initial reagent and the achievement of the speed of movement of the value of the local speed of sound. Thus, the surface F_s passes through the straight-line M = 1, $C_A = 0$ and combines the solutions of the classical theory of detonation. The plane of initial states in the chemical reaction zone $C_A = 1$ is crossed by the separatrix surface along a certain line l_1. It is obvious that for solutions to the classical theory of detonation to exist, the line l_1 must lie in the Mach number interval from zero to one.

The zero surface F_θ intersects the plane of initial states along the line l_2 (Figure 2). Since it is known (Zel'dovich, 1940, 1944) that the temperature at the beginning of the zone of chemical reactions of the detonation wave increases, the line l_2, determined by equality $M = \gamma^{-1/2}(1 + \frac{R_A - R_B}{C_{pA}}\varepsilon\theta)^{-1/2}$, must lie above thl line l_1, since the temperature increases under the surface F_θ.

If the initial state on the plane of initial states is above the line l_2, (trajectories of type a_2), then the temperature decreases along the solution starting from this initial state. If the initial state on the plane of initial states is between the lines l_1 and l_2, then along the solution emanating from this initial state (trajectories of type a_1), the temperature first increases, and then, after crossing the zero surface F_θ, decreases. It was this nature of the temperature change in the reaction zone that was pointed out earlier (Zeldovich, 1940, 1944), which was reflected in Figure 1 and Figure 2. Solutions l_1 and l_2 end at the points of the singular surface M = 1 at a finite value of the mass fraction of the initial reagent. If the initial state on the plane of initial states is below the line l_1, then among them it is also possible to distinguish the initial states lying above some line l_3, from which solutions emerge (trajectories of type b_1), along which the temperature first increases and then decreases, and the initial states lying below the line l_3, from which solutions emerge (trajectories of type b_3), along which the temperature decreases. Both solutions b_1 and solutions b_2 end in the equilibrium positions of the system at M < 1.

7.3. APPROXIMATE ANALYTICAL AND NUMERICAL SOLUTIONS OF THE SYSTEM OF EQUATIONS (7.19) - (7.21)

Of the qualitatively different modes of the process found in the reaction zone of the detonation wave, solutions a_1, S deserve priority attention, since it is on these solutions that the nature of the change in temperature and velocity corresponds to the description of the processes in the zone of chemical reactions of the detonation wave contained in the works of Ya.B. Zeldovich and Figure 1, 2. Zel'dovich pointed out that combustion in the zone of a chemical reaction of a detonation wave has predominant features of the self-ignition mode. Therefore, to analyze the processes in the chemical reaction zone, one can use approximate estimates of the maximum temperature (T_*) and the corresponding mass fraction (C_*) of the reagent in the autoignition mode of the n - tph order chemical reaction (Zeldovich et al., 1980)

$$T_* = T_a(1 - nR_0T_a/E), \tag{7.23}$$

$$C_* = C_0 nR_0 T_0^2/(E(T_a - T_0)). \tag{7.24}$$

Here T_a is adiabatic temperature of boning, T_0 is initial temperature in autoignition mode regime.

Within this task $T_0=T_1$, $T_a=T_1+Q/c_{pA}$. The relationship between the maximum temperature in the reaction zone and other gas parameters in this state can be found from the zero-surface equation F_θ. Let index 2 denote the values of the parameters in the state with the maximum temperature in the zone of the chemical reaction of the detonation wave. Then equations (7.23), (7.24) and the equation for the maximum temperature in the reaction zone have the form

$$\frac{T_2}{T_1 + Q/c_{pA}} = 1 - \frac{nR_0}{E}(T_1 + Q/c_{pA}), \tag{7.25}$$

$$C_{A2} = \frac{nR_0c_{pA}}{EQ}(T_1 + Q/c_{pA})^2, \tag{7.26}$$

$$T_2 = \frac{1 - \gamma_2 M_2^2}{\gamma_2 M_2^2}\frac{\gamma_A}{\gamma_A - 1}\frac{Q/c_{pA}}{1 - R_B/R_A}. \tag{7.27}$$

For further analysis, it is convenient to represent the gas states at the characteristic points of the detonation wave as points in the plane (M^2, C_A) (Figure 3).

Figure 3. Scheme of the change in gas states in an adiabatic detonation wave

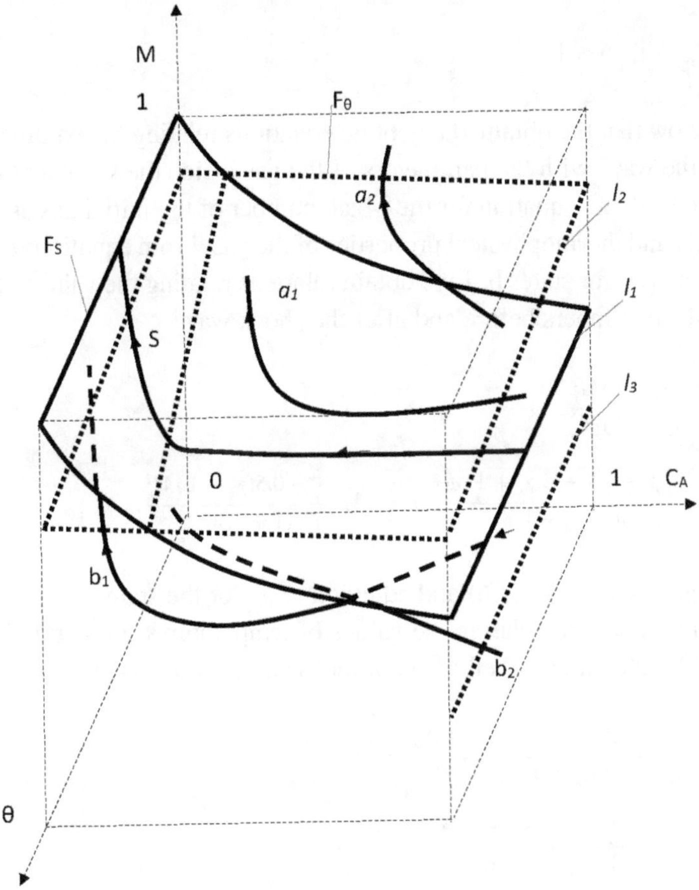

The gas from state *b* before the shock wave enters state 1 after the shock wave and moves through the chemical reaction zone in the self-ignition mode from state 1 to state 2, where the temperature reaches a maximum. Since, after reaching the maximum temperature, a relatively calm afterburning regime begins with a decrease in temperature, it can be assumed that this afterburning in the plane will take place along a straight - line segment from point 2 to the end point *e*. It should be noted that in the integral space the process between points 2 and *e* is by no means not linear. It is determined by the kinetics of the chemical reaction according to equation (7.18) The slope of the straight - line segment between points *e* and 2 in the plane is found from equation (7.17)

$$\left(\frac{dM^2}{dC_A}\right)_2 = -K_1, K_1 = M_2^2\left[\frac{1+\gamma_2 M_2^2}{1-M_2^2}\left(\frac{Q}{c_{p2}T_2}+\frac{R_A-R_B}{R_2}\right)+\frac{\gamma_A-\gamma_B}{\gamma_2}\right]. \qquad (7.28)$$

Since this straight - line segment passes through the point e, where M = 1, we get another algebraic equation

$$M_2^2 + K_1(C_{A2} - C_{Ae}) = 1. \tag{7.29}$$

Let us show that the obtained algebraic equations relating the parameters of the gas before the wave with the parameters of the gas inside the wave and after it are sufficient to find one equation for the Mach number of the initial gas as a function of the kinetic and thermophysical properties of the gas. From equations (7.1) - (7.3) and equation (7.7) for states b, 1 we obtain relations relating the values of temperatures and Mach numbers before and after the shock wave

$$\frac{T_b}{T_1} = \frac{1 + 0.5(\gamma_A - 1)M_1^2}{1 + 0.5(\gamma_A - 1)M_b^2}, \tag{7.30}$$

$$M_2^2 = \frac{1 - 2\gamma_A\phi - \sqrt{1 - 2(\gamma_A + 1)\phi}}{2\gamma_A^2\phi - \gamma_A + 1}, \phi = M_b^2\frac{1 - 0.5(\gamma_A - 1)M_b^2}{(1 + \gamma_A M_b^2)^2}. \tag{7.31}$$

From equations (7.4) - (7.6) and equation (7.7) for the states b, e, assuming M = 1, we obtain relations relating the values of temperatures, mass fractions of the reagent and Mach numbers before and after the detonation wave,

$$\frac{T_b}{T_e} = \frac{1 + 0.5(\gamma_e - 1)}{1 + 0.5(\gamma_A - 1)M_b^2 + \delta(1 - C_{Ae})}\frac{c_{pe}}{c_{pA}}, \delta = \frac{Q}{c_{pA}T_b}, \tag{7.32}$$

$$M_b^2\left(\frac{\gamma_A}{\gamma_e}\right)^2\frac{\gamma_e^2 - 1}{\gamma_A^2 - 1} = \frac{(1 + \gamma_A M_e^2)^2}{2 + (\gamma_A - 1)M_b^2 + 2\delta(1 - C_{Ae})^2}. \tag{7.33}$$

The reference for the calculation is equation (7.33), in which the unknown C_{Ae} is found from (7.29). The unknowns in (7.29) M_2^2, T_2, C_{A2} are found from (7.25) - (7.27), and the unknowns in (7.25) - (7.27) are in turn found from (7.30), (7.31).

In numerical studies, it was assumed that three atoms of gas A are combined into one molecule of gas B. Then γ_A=5/3, γ_B=4/3, n=3.

Figure 4 shows the results of calculating the dependence of the square of the Mach number of the detonation wave M_b on the activation parameter of a chemical reaction $\theta_b = \frac{R_0 T_b}{E}$ for various values of the heat release parameter $\delta = \frac{Q}{c_{pA}T_b}$.

Figure 4. Dependence of the squared Mach number of the detonation wave on the activation parameter θ_b of a chemical reaction for various values of the heat release parameter δ

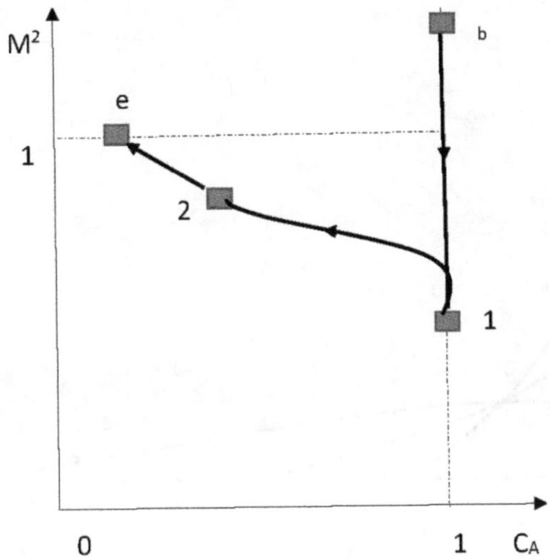

It can be seen that as δ increases, the steepness of the dependence $M_b^2(\theta_b)$ sharply increases, the dependence itself shifts to the region of small values of θ_b, while reaching very large ones M_b^2 . When moving from left to right along each dependence $M_b^2(\theta_b)$, the under burning of the initial reagent at the end of the detonation wave increases. Therefore, the dependence $M_b^2(\theta_b)$ for each δ has two limit points. The left limit point corresponds to the complete burnout of the initial reagent in the detonation wave, in which $C_{Ae} = 0$. The right limiting point corresponds to the absence of a chemical reaction, in it $C_{Ae} = 1$, $M_b^2 = 1$. This corresponds to the case of the absence of a shock wave initiating a chemical reaction. In the parameter plane (δ,θ_b), the limit points form two lines (Figure 5), limiting the area of existence of a stationary detonation wave (on line 1 – $C_{Ae} = 0$, on line 2 – $C_{Ae} = 1$).

Figure 5. Region of existence of a stationary detonation wave

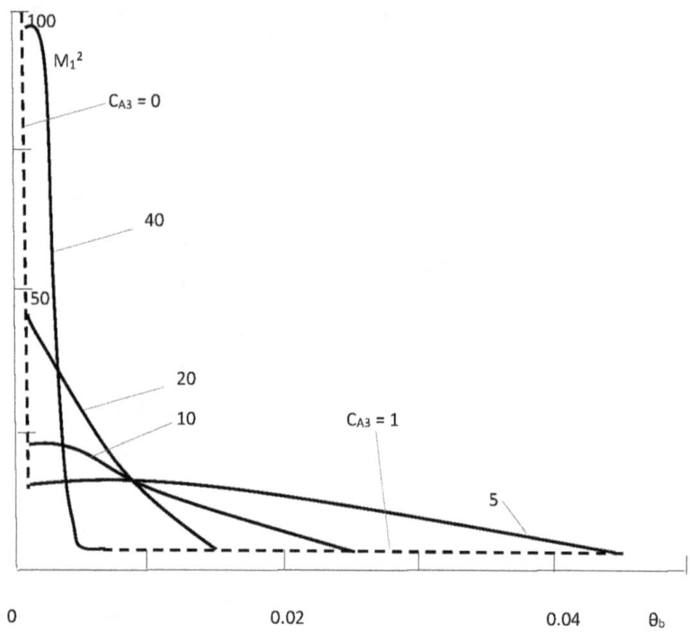

Thus, the classical S Chapman-Jouguet solutions corresponding to line 1 form only one of the boundaries of the region of existence of a stationary detonation wave with possible under burning of the initial reagent. Each point of the region of existence of a stationary detonation wave corresponds to the only values of regime parameters at the characteristic points of the detonation wave, including under burning of the reagent at the end of the wave. Figure 6 shows the results of calculating the dependence of the Mach number of the detonation wave on the parameters δ, θ_b in the form of a surface in three-dimensional phase space, clearly illustrating this dependence.

Figure 6. Dependence of the detonation wave Mach number on the chemical reaction activation parameter θ_b and on the heat release parameter δ

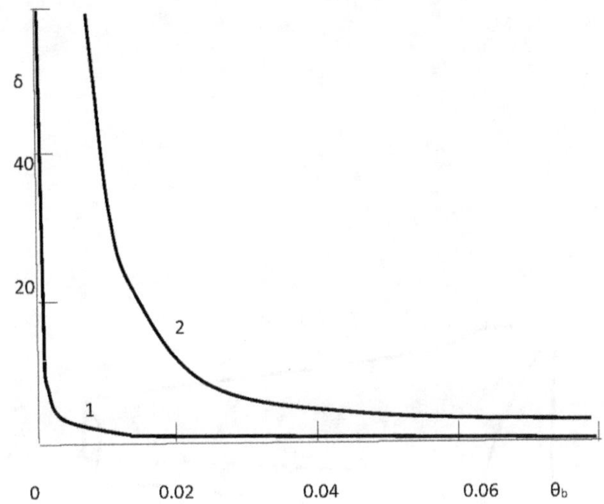

To check the correctness of the obtained analytical solution, numerical studies of the system of differential equations (7.19) - (7.21) were carried out. These equations describe the processes in the zone of chemical reactions of the detonation wave if the initial conditions correspond to the states after the shock wave preceding the zone of chemical reactions of the detonation wave. Numerical studies have shown that under these conditions only solutions a_1, S are realized.

Moreover, the numerical values of the length of the reaction zone as solutions a_1 approach solutions S greatly increase. This is due to the fact that S solutions exist for very small values of θ_b (Figure 7), leading to extremely small values of the exponent $\exp(-1/\theta_b)$ and, as a consequence, to a low rate of change of parameters along the trajectories. As an example, Figure 7 and Figure 8 show the results of calculating the system of differential equations (7.19) - (7.21) (solid lines) in the form of dependences of the Mach number of the detonation wave, the decimal logarithm of the length of the reaction zone of the detonation wave log L, the mass fraction of the initial reagent at the end of the reaction zone C_{Ae} on the activation parameter θ_b for values of the heat release parameter δ=20 (Figure 7).

Figure 7. Dependences of the parameters of the detonation wave on the activation parameter of the chemical reaction θ_b at δ=20, (1–C_{Ae}, 2–logL, 3–M_b)

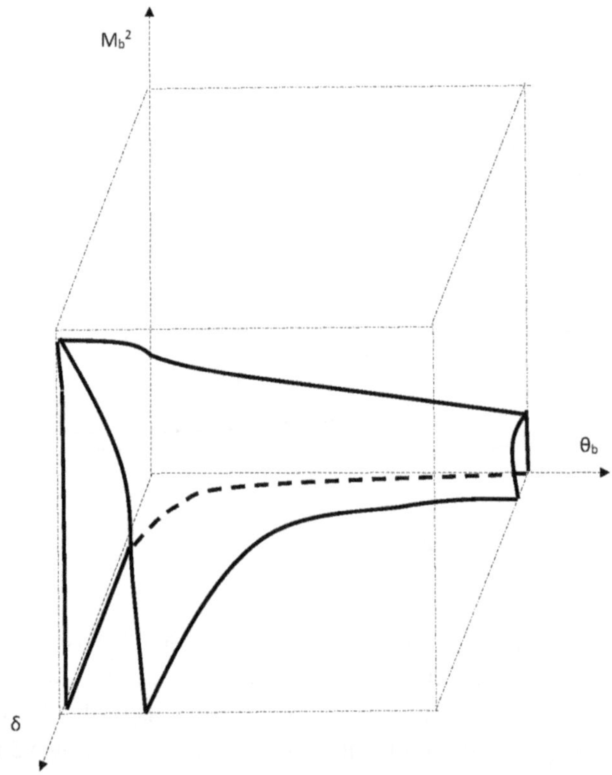

In the calculations for given parameters δ, θ_b, the Mach number M_b was varied until the temperature value at the maximum of the temperature profile over the reaction zone was compared with the temperature value T_2 determined from the estimate of the maximum temperature in the self-ignition mode (7.25). The values of the parameters along the wave obtained under this condition were taken as the values of the parameters of the desired wave solution with given values of δ, θ_b. In Figure 8, for comparison, the dependencies are shown as dotted lines $M_B(\theta_b)$ based on the analytical solution. From Figure 8 it follows that the analytical solutions and the numerical solution of the differential equations, in general, agree satisfactorily with each other. The greatest differences between the analytical solution and the numerical solution are observed when approaching the boundary of the region of existence of wave solutions corresponding to the value C_{Ae} = 0, i.e. complete burnout of the starting reagent. It is essential that with this approximation the length of the reaction zone increases sharply.

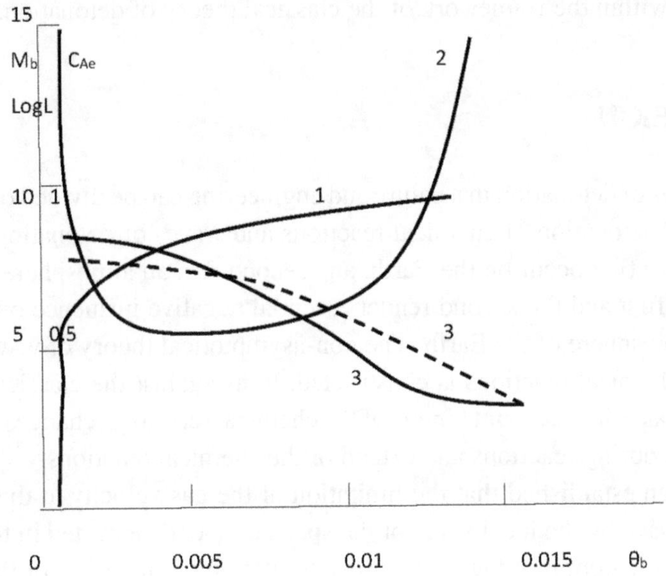

Only in the vicinity of the $C_{Ae} = 0$ boundary, the mass fraction of the initial reagent sharply decreases, remaining significant (more than 0.7) in the main part of the existence of wave solutions. The dependence of the length of the reaction zone on the parameter θ_b is interesting. It appears to have an extremal form. As θ_b decreases, the length of the reaction zone first decreases and then sharply increases. The obtained results of a numerical study of differential equations confirm the results of a qualitative and analytical analysis. Almost all wave adiabatic solutions end in a flow crisis with a significant under burning of the initial reagent (burnup no more than 30%). Wave solutions with significant burnup are characterized by very long chemical reaction zones and, therefore, such burnouts are practically unattainable. Thus, estimates of the parameters of detonation waves according to the Chapman-Jouguet theory are very far from numerical solutions.

They are also far from analytical solutions in the asymptotic theory of combustion, which describe processes in detonation waves. Even the qualitative regularities of the change in parameters along the detonation wave do not agree. For example, the graph of the change in the mass fraction of the reagent along the detonation wave, according to the studies carried out on the non-asymptotic theory of combustion, has the form of a dotted line in Figure 1 and Figure 2, and not a solid line with complete burnout, according to the assumption of the classical theory. In addition, the distance between sections I and II in Figure 1 and Figure 2, according to the results of studies on the non-asymptotic theory of combustion, sharply extremally depends on the activation parameter of the chemical reaction. This obviously cannot be detected within the framework of the classical theory of detonation.

CONCLUSION

The waves of detonation in a nature and engineering can be divided on two classes: waves of detonation of chemical reactions and waves of detonation of nuclear reactions. The first occur on the Earth, and second – in an atmosphere of the Sun, but also the first and the second render essential negative influence on ecological systems of biosphere of the Earth. The non-asymptotical theory of waves of a detonation of chemical reactions is constructed. It, as against the classical theory of detonation, takes into account kinetic of the chemical reactions, changes of structure of the waves during reactions and extend of the chemical reactions.

It has been established that the limitation of the gas velocity at the end of the detonation wave by the local value of the speed of sound, accepted in the classical theory of detonation, is at the same time a limitation on the depth of the chemical reaction so that the depth of the chemical reaction does not exceed 30% on average. This is a significant difference from the classical theory of detonation, which does not take into account the change in the composition of the gas along the detonation wave and at the same time assumes that everything that can burn is burned in the detonation wave.

It has been proven that wave solutions with significant burnup are characterized by very long chemical reaction zones and, therefore, such burnups are practically unattainable. The results of the analytical study are confirmed by the numerical solution of the systems of initial differential equations. It turned out that the estimates of the parameters of adiabatic detonation waves according to the Chapman-Jouguet theory are very far from analytical solutions according to the non-asymptotic theory of detonation and numerical solutions of differential equations. In addition to significant quantitative differences between calculations based on the Chapman-Jouguet theory and non-asymptotic theory, there are also fundamental qualitative differences.

In this regard, it is of interest (both from an energy and environmental point of view) to investigate the possibility of a continuous transition from subsonic gas velocities to supersonic velocities in a detonation wave after a shock wave, which transforms a supersonic gas flow into a subsonic flow. For the first time, the idea of the possibility of continuous gas acceleration to supersonic speeds behind the shock wave front, which is part of the structure of a nonadiabatic detonation wave, was put forward by Ya.B. Zel'dovich. At the same time, the conditions for the existence of such modes of motion have not been studied. Moreover, Ya.B. Zeldovich practically did not take into account the kinetics of chemical reactions, the adiabatic coefficient of the gas and its heat capacity were considered constant, and the equations of the conservation laws for the movement of gas under the influence of friction and heat transfer were written incorrectly. In second chapter, we shell study the conditions for a continuous transition from subsonic to supersonic velocities of gas in a detonation wave after a shock wave, taking into account the kinetics of chemical reactions, the variability of the adiabatic coefficient of the gas, and its heat capacity in the framework of the non-asymptotic theory of detonation. The results of these studies can be used to create energy-efficient high-power aircraft of a new generation using detonation waves. Implementation of the studied regimes in detonation waves would make it possible to significantly increase the depth of chemical reactions, i.e. to increase the degree of use of a combustible reagent to significantly increase the momentum imparted to the aircraft.

REFERENCES

Chapman, D. L. (1899).. . *Philosophical Magazine*, *47*, 90.

Davis, R., & Evans, J. C. (1974). Report on the Brookhaven Solar Neutrino Experiment. *Proceedings of the 6th Leningrad International Seminar "Particle Acceleration and Nuclear Reactions in Space"*.

Jouguet, E. (1905). *Jorn. De Mathematique*. 347.

Kompaneets, A. S. (1976). *Laws of physical statistics. Shock waves. Super dense substance*. Science.

Lewis, B., & Elbe, G. (1938). *Combustion*. Flames and Explosions of Gases.

Pikelner, S.B. (1961). *Fundamentals of space electrodynamics*. Moscow: State Publishing House of Phys. - mat. literature.

Shchelkin, K. I., & Troshin, Ya. G. (1963). *Gas dynamics of combustion*. Publishing House of the Academy of Sciences of the USSR.

Shklovsky, I. S. (1955). On the question of the content of lithium, beryllium and deuterium in the solar atmosphere. *Reports of the Academy of Sciences of the USSR*, *105*(5), 931–934.

Syrovatsky, S. I. (1975a). Characteristics of the current sheet and the thermal trigger of solar flares. *Letters to the Astronomical Journal*, *2*(1), 1–4.

Syrovatsky, S. I. (1975b). Acceleration of charged particles in processes like solar flares. *Proceedings of the Academy of Sciences of the USSR. Physical series*. 39(2), 359-374.

Vulis, L. A. (1946a). On the transition through the speed of sound in a gas flow. *Reports of the Academy of Sciences of the USSR*, (8), 669–672.

Vulis, L. A. (1946b). On the influence of friction on the transition through the speed of sound. *Reports of the Academy of Sciences of the USSR*, (9), 773–775.

Vulis, L. A. (1947). On the law of reversal of influences in a real gas flow. *Reports of the Academy of Sciences of the USSR*, (8), 799–801.

Vulis, L. A. (1950). *Thermodynamics of gas flows*. Gosenergoizdat.

Williams, F. A. (1971). *Theory of combustion*. Nauka.

Zeldovich, Ya. B. (1940). On the theory of detonation propagation in gaseous systems. *Journal of Experimental and Theoretical Physics*, *10*(5), 542–568.

Zeldovich, Ya.B. (1944). *Theory of combustion and detonation of gases*. Moscow - Leningrad: Publishing houses of the Academy of Sciences of the USSR.

Zeldovich, Ya. B., Barenblatt, G. I., Librovich, V. B., & Makhviladze, G. M. (1980). *Mathematical theory of combustion and explosion*. Nauka.

Zeldovich, Ya. B., & Kompaneets, A. S. (1955). *Theory of detonation*. Gostekhizdat.

Zhizhin, G.V. (2004). Synergetic of flows. *Life and safety*. 2 - 3a, 365 - 385.

Zhizhin, G. V. (2008). *Combustion waves with distributed zones of chemical reactions (non-asymptotic theory of combustion)*. Werner Regen Publishing House.

Zhizhin, G.V. (2009a). *Waves of detonation in nature and technology, their mathematical models and significance in violation of the ecological systems of the biosphere*. St. Petersburg: Preprint of the North -Western State Correspondence Technical University.

Zhizhin, G. V. (2009b). *Detonation wave in a gas with variable composition*. Polytechnic-service.

Chapter 8
Detonation Waves in a Dusty Gas

ABSTRACT

It is shown that the thermal effect of chemically inert particles on the gas, which acts simultaneously with the chemical effect on the gas (release of heat of a chemical reaction), can provide a continuous transition from subsonic flows to supersonic flows when gas moves in a dusty gas detonation wave. This eliminates the well-known sound limitation that exists in the classical theory of detonation. The structure of the dusty gas detonation wave is constructed, which consists of zones, in each of which qualitatively different sets of phase interaction processes occur. Approximate analytical solutions are obtained for each of the zones using the methods of non-asymptotic combustion theory and numerical studies of these solutions are carried out. It has been discovered that it is possible to increase many times the impulse created by the detonation wave of dusty gas to aircraft with such a detonation engine, while increasing the depth of reaction almost to unity.

INTRODUCTION

Since the limitation of the gas velocity at the end of the chemical reaction zone by the local value of the speed of sound significantly limits the depth of the reaction (Chapter 7), and this in turn leads to a limitation of the momentum created by the gas flow behind the standing detonation wave, loss of fuel and the presence of harmful products behind the detonation wave incomplete combustion, consideration should be given to overcoming this limitation. This can be achieved by creating conditions for a continuous transition through the speed of sound within the zone of chemical reactions of the detonation wave. Natural conditions for such a transition are created (Zhizhin, 2009 a, b) in a gas with chemically inert particles (dust). Interactions of gas with particles can be considered as external influences on the gas. Using the

DOI: 10.4018/978-1-6684-7745-8.ch008

methods of the non-asymptotic theory of combustion, we consider a detonation wave in a gas containing small particles of the condensed phase.

8.1. DIFFERENTIAL AND INTEGRAL CONSERVATION LAWS IN THE DETONATION WAVE OF A DUSTY GAS

It is assumed that dust particles are chemically inert and their content in the gas is small enough so that they do not collide with each other and occupy a negligible part of the flow cross section. Then the differential equations of the laws of conservation of mass, momentum, energy, and state in a one-dimensional stationary flow moving in the direction of a spatial variable z (Zhizhin, 1972, 2004; Vulis, Gusika, Zhizhin, 1972), taking into account the variability of the thermophysical properties of the gas and the release of heat in the process chemical reactions in gas have the form

$$\frac{d\rho_1}{\rho_1 dz} + \frac{du_1}{u_1 dz} = 0, \qquad (8.1)$$

$$\frac{dP}{\rho_1 dz} + u_1\frac{du}{dz} + \kappa u_1\frac{du_2}{dz} = 0, \qquad (8.2)$$

$$c_{p1}\frac{dT_1}{dz} + u_1\frac{du_1}{dz} + Q\frac{dC_A}{dz} + \kappa c\frac{dT_2}{dz} + \kappa u_2\frac{du_2}{dz} = 0, \qquad (8.3)$$

$$\frac{dP}{Pdz} = \frac{d\rho_1}{\rho_1 dz} + \frac{dT_1}{T_1 dz} + \frac{dR}{Rdz} = 0. \qquad (8.4)$$

Here ρ_1, u_1, T_1, P are the density, velocity, temperature and pressure of the gas, respectively; u_2, T_2 are speed and temperature of dust particles; c_{p1} is specific heat capacity of gas at constant pressure; c is specific heat capacity of particles; $\kappa = G_2/G_1$ is consumable content of particles; G_2 is particle consumption; G_1 is gas consumption; R is specific gas constant of a mixture of gases A and B; C_A is mass fraction of gas A; Q is specific thermal effect of a chemical reaction $A \rightarrow B$.

Equations (8.1) - (8.4) have complete integrals

$$\rho_1 u_1 = m, \qquad (8.5)$$

$$P = \rho RT, \qquad (8.6)$$

$$P + u_1 m + \kappa m u_2 = L_1, \qquad (8.7)$$

$$c_{p1} T_1 + u_1^2/2 + Q(1 - C_A) = L_2. \qquad (8.8)$$

Here m is specific gas consumption, ρ is the density of a mixture of gases A and B; L_1 and L_2 are integral constants.

Schemes of the detonation wave in a dusty gas in the laboratory coordinate system and in the wave coordinate system are shown in Figure 1 a,b. They coincide with the gas detonation patterns (Figure 1 and Figure 2 in Chapter 7). The only difference is that the velocity indices in Figure 1 take into account the presence of dust particles in the gas.

Figure 1. Scheme of a detonation wave in a dusty gas a) in the laboratory coordinate system, b) in the wave coordinate system.

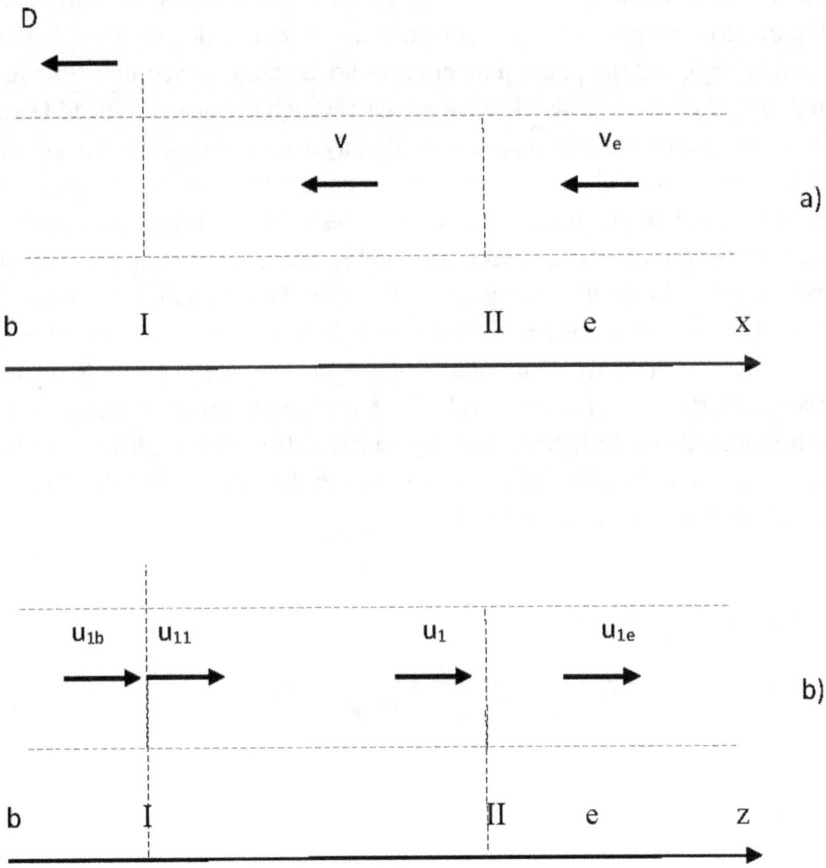

Section I corresponds to the shock wave. Section II corresponds to the state in the flow after the completion of the processes in the detonation wave. Conventionally, section II is depicted at a finite distance from section I since the completion of the processes of a chemical reaction and the interaction of particles and gas are asymptotic. Strictly speaking, section II must be attributed to an infinite distance from sections I or consider that the distance between sections I and II is the effective distance determined from graphs or analytical dependences of parameter changes, displaying the main part of these changes. Either the distance between sections I and II is determined not by the degree of completion of the processes in the detonation wave, but is limited by some other conditions, for example, the onset of a flow crisis, i.e. the gas velocity reaches the local value of the speed of sound when a continuous transition from subsonic to supersonic flows is impossible currents. Area b in Figure 1a corresponds to a gas at rest, i.e., a state in which the phase velocities are equal to zero, and the phase temperatures are equal to a (8combustible reagent is equal to 1, $C_A = 1$. The shock wave moves through the gas at a speed D against the direction spatial variable h, and immediately after the shock wave, simulating it with a shock wave with zero extent, the gas parameters change abruptly, and the values of the particle parameters are equal to the values of the particle parameters in state b. In the region between sections I and II, phase interaction processes and a chemical reaction in the gas take place. If the spatial coordinate is associated with the detonation wave (Figure 1b), then in area b the gas moves with velocity $u_{1b} = D$ in the direction of the wave coordinate z, and $u_{1b}=u_{2b}$, $T_{1b}=T_{2b}$, $C_A=1$. Reaching the cross section I, the gas parameters (velocity, temperature, pressure, density) change abruptly in accordance with the conservation laws and relations (7.30), (7.31), which, taking into account the designation of the gas and the state behind the shock wave by index 1, in this case have the form

$$\frac{T_{1b}}{T_{11}} = \frac{1 + 0.5(\gamma_A - 1)M_1^2}{1 + 0.5(\gamma_A - 1)M_b^2}, \tag{8.9}$$

$$M_1^2 = \frac{u_{11}^2}{\gamma_A R_A T_{1b}} = \frac{1 - 2\gamma_A\phi - \sqrt{1 - 2(\gamma_A + 1)\phi}}{2\gamma_A^2\phi - \gamma_A + 1}, \phi = M_b^2 \frac{1 - 0.5(\gamma_A - 1)M_b^2}{(1 + \gamma_A M_b^2)^{\ 2}}, \tag{8.10}$$

$$M_b^2 = \frac{u_{1b}^2}{\gamma_A R_A T_{1b}}. \tag{8.11}$$

In the region, the processes are described by the system of differential equations (8.1) - (8.4) and the equation for changing the mass fraction of the combustible reagent (7.18), which in this case should be written as

$$\frac{dC_A}{dz} = -k_0 u_1^{\ -1} C_A^n \exp(-\frac{E}{R_0 T_1}). \tag{8.12}$$

Between the parameters in state b and in state e, according to the integrals (8.5) - (8.8), there are relationships

$$\rho_{1e} u_{1e} = \rho_{1b} u_{1b}, \tag{8.13}$$

$$\rho_{1e} u_{1e}^{\ 2} + P_{1e} + \kappa \rho_{1e} u_{1e} u_{2e} = \rho_{1b} u_{1b}^{\ 2} + P_{1b} + \kappa \rho_{1b} u_{1b} u_{2b}, \tag{8.14}$$

$$c_{pe} T_{1e} + 0.5 u_{1e}^2 + Q(1 - C_{Ae}) = c_{pA} T_{1b} + 0.5 u_{1b}^2, \tag{8.15}$$

$$P_{1b} = \rho_{1b} R_A T_{1b}, \tag{8.16}$$

$$P_{1e} = \rho_{1e} R_B T_{1e}. \tag{8.17}$$

However, the values ρ_{1e}, u_{1e}, u_{2e}, P_{1e}, T_{1e}, T_{2e}, C_{Ae} are unknown. They can be determined from the calculations of processes in the region I - II. If we assume that in the region I - II all the processes of interaction of the components must be completed, then in the region *e*

$$u_{1e} = u_{2e}, T_{1e} = T_{2e}, C_{Ae} = 0. \tag{8.18}$$

At the same time, they remain unknown ρ_{1e}, u_{1e}, P_{1e}, T_{1e}. These unknowns are found from the system of equations obtained from (8.13) - (8.17) after substituting conditions (8.18) into them

$$M^2{}_{1e} = \frac{1}{(\gamma_B - 1)(\Phi / H_{1,2} - 0.5)},$$

$$\Phi = (\gamma_B - 1)^{-1} M_{1b}^{-2} + 0.5 - Q u_{1b}^{-2}, \Xi = 1 + \kappa - (\gamma_B - 1)(2\gamma_B)^{-1}, \tag{8.19}$$

$$H_{1,2} = \frac{1 + \kappa \gamma_A^{-1} M_{1b}^{-2}}{2\Xi} \pm \sqrt{\frac{(1 + \kappa + \gamma_A^{-1} M_{1b}^{-2})^{\ 2}}{4\Xi^2} - \frac{(\gamma_B - 1)(\Phi - 1)}{\gamma_B \Xi}},$$

$$u^2{}_{1e} = \frac{(\gamma_B - 1) M_{1b}^2 \Phi}{1 + 0.5(\gamma_B - 1) M_{1b}^2}, \tag{8.20}$$

$$T_{1e} = \frac{c_{PA} T_{1b} + 0.5(u_{1b}^2 - u_{1e}^2) - Q}{c_{pB}}. \tag{8.21}$$

From (8.19) it follows that solutions leading to the equilibrium position exist if

$$M^2{}_{1b} > (\gamma_A - 1)(0.5 + Q u^{-2}{}_{1b}). \tag{8.22}$$

8.2. MATHEMATICAL MODELING OF A DUSTY GAS DETONATION WAVE WITH THERMAL PHASE INTERACTION

In order to simplify the problem and the possibility of solving it with obtaining approximate analytical solutions with subsequent calculation, we will use these expressions to assume that the size of the dust particles is so small that the speed of moving dust grains quickly compares with the speed of the gas. The thermal interaction is not so fast, and it is this interaction that determines (limits). It is also reasonable to assume that the combustion reaction, accompanied by the appearance of a detonation wave, is characterized by a significant thermal effect of the chemical reaction. Then (Zhizhin, 2008; 2009 a, b) the system of equations describing the flow of dusty gas in the detonation wave after the shock wave has the form

$$(1 - M^2)\frac{du_1}{u_1 dz} = -\frac{Q}{c_p T_1}\frac{dC_A}{dz} + \frac{1}{T_1 c_p}\frac{dQ^e}{dz}, \tag{8.23}$$

$$(1 - M^2)\frac{dT_1}{T_1 dz} = (\gamma M^2 - 1)\frac{Q}{c_p T_1}\frac{dC_A}{dz} + (1 - \gamma M^2)\frac{1}{T_1 c_p}\frac{dQ^e}{dz}, \tag{8.24}$$

$$(1 - M^2)\frac{dM^2}{M^2 dz} = -(1 + \gamma M^2)\frac{Q}{c_p T_1}\frac{dC_A}{dz} + (1 + \gamma M^2)\frac{1}{T_1 c_p}\frac{dQ^e}{dz}, \tag{8.25}$$

$$\frac{1}{T_1 c_p}\frac{dQ^e}{dz} = W_T = -\kappa c\frac{\gamma - 1}{a^2}\frac{dT_2}{dz} = -\kappa c\frac{\gamma - 1}{a^2}\frac{T_1 - T_2}{l_T}, l_T = \frac{\rho_2 u_2 c d^2}{6\lambda Nu}. \tag{8.26}$$

Here, there W_T is the thermal interaction of phases; l_T is the thermal relaxation length (Zhizhin, 1972); d is the diameter of dust particles; λ is the thermal conductivity of the gas; Nu is the Nuselt number characterizing the interfacial heat transfer; $M = u_l/a$; $a^2 = \gamma RT_l$; the indexes of the variables refer 1 - to gas, 2 - to dust particles.

By introducing dimensionless variables $\upsilon = \frac{R_0 T_1}{E}, \theta = \frac{T_1}{T_2}, \chi = \int_0^z \kappa \frac{c}{c_p}\frac{dz}{l_T}$, equations (8.12), (8.24), (8.25), (8.26) are reduced to an autonomous closed system of equations

$$\frac{dM^2}{d\chi} = M^2\eta\frac{1 + \gamma M^2}{1 - M^2}, \tag{8.27}$$

$$\frac{d\theta}{d\chi} = \theta\frac{(1 - \gamma M^2)\eta - (1 - M^2)(\theta - 1)c_p(c\kappa)^{-1}}{1 - M^2}, \tag{8.28}$$

$$\frac{d\upsilon}{d\chi} = \upsilon\eta\frac{1 - \gamma M^2}{1 - M^2}, \tag{8.29}$$

$$\frac{dC_A}{d\chi} = -\frac{c_p\sigma}{c\kappa}C_A\exp(-\tfrac{1}{\upsilon}). \tag{8.30}$$

Here $W_c = \alpha\upsilon^{-1}C_A\exp(-1/\upsilon)$ is the chemical effect on the gas;

$\eta = W_c + W_T = \alpha\upsilon^{-1}C_A\exp(-1/\upsilon) + (1-\theta)/\theta$(8.31)

that is the sum of the effects on the gas,

$$\alpha = \frac{QR_0\sigma}{Ec\kappa}, \gamma = (\gamma_A - \gamma_B)C_A + \gamma_B, c_p = (c_{pA} - c_{pB})C_A + c_{pB}, \sigma = \frac{k_0 l_T}{u_1} = \frac{k_0 c d^2}{6\lambda Nu}.$$

Differentiating η, according to (8.31), using equations (8.28) - (8.30), we obtain an equation for changing the sum of actions

$$(1 - M^2)\frac{d\eta}{d\chi} = \left(\alpha C_A \frac{1-\upsilon}{\upsilon^2} e^{-1/\upsilon} - \frac{1}{\theta}\right)(1 - \gamma M^2)\eta + \\ +(1 - M^2)\frac{c_p}{c\kappa}\left(1 - \frac{1}{\theta} - \frac{\alpha}{\upsilon}\sigma C_A e^{-2/\upsilon}\right). \tag{8.32}$$

8.3. QUALITATIVE STUDY OF THE EQUATIONS OF DUSTY GAS FLOW IN A DETONATION WAVE AND IDEAS ABOUT THE STRUCTURE OF A DETONATION WAVE

Equations (8.27), (8.32) in the vicinity of the transition point through the speed of sound (*t* points (M = 0, η = 0)) can be written as

$$\frac{dM^2}{d\chi} = \eta\frac{1+\gamma_t}{1 - M^2}, \tag{8.33}$$

$$\frac{d\eta}{d\chi} = \frac{\left(\frac{1}{\theta_t} - \alpha C_{At}\frac{1-\upsilon_t}{\upsilon_t^2}e^{-1/\upsilon_t}\right)(\gamma_t - 1)\eta + (1 - M^2)\frac{c_{pt}}{c\kappa}\left(1 - \frac{1}{\theta_t} - \frac{\alpha\sigma}{\upsilon_t}C_{At}e^{-2/\upsilon_t}\right)}{1 - M^2}$$

. (8.34)

The coefficients at η and (1 – M²) in the numerator of the right side (8.34) can be transformed using the equality (8.31), given that η = 0 at the point t. Then equation (8.34) takes the form

$$\frac{d\eta}{d\chi} = \frac{\psi\eta + \varphi(1 - M^2)}{1 - M^2}, \tag{8.35}$$

$$\psi = (\gamma_t - 1)\left(1 - \frac{\theta_t - 1}{\theta_t \upsilon_t}\right), \varphi = \frac{c_{pt}}{c\kappa}\frac{\theta_t - 1}{\theta_t}\left(1 - \sigma e^{-1/\upsilon_t}\right). \tag{8.36}$$

It is possible to consider variants of the location of the trajectory plane (M^2, η) of the system (8.33), (8.35) depending on the signs of expressions (8.36), which are functions of the values of the variables included in equations (8.27), (8.32). This is of interest for determining the conditions necessary for a continuous transition from subsonic flows to supersonic flows in a detonation wave.

A change in ψ and φ leads to a change in the location of the zero isocline of equation (8.35) $\eta = \varphi\psi^{-1} (M^2 - 1)$ in the plane (M^2, η) and to a change in the sign of the derivative $d\eta / d\chi$ in the areas formed by dividing the plane by characteristic lines: this zero isocline, the straight line $M^2 = 1$ and the zero isocline of equation (8.33) $\eta = 0$. There are four options for combinations of signs of quantities ψ and φ: a) $\psi > 0$, $\varphi > 0$; b) $\psi > 0$, $\varphi < 0$; c) $\psi < 0$, $\varphi < 0$; d) $\psi < 0$, $\varphi < 0$. Each of these options corresponds to the relative positions of the characteristic lines and signs of the derivatives of the phase coordinates in the areas of the phase plane, indicated in Figure 2 a, b, c, d, e.

Figure 2. Phase diagrams of flows in a detonation wave of a dusty gas

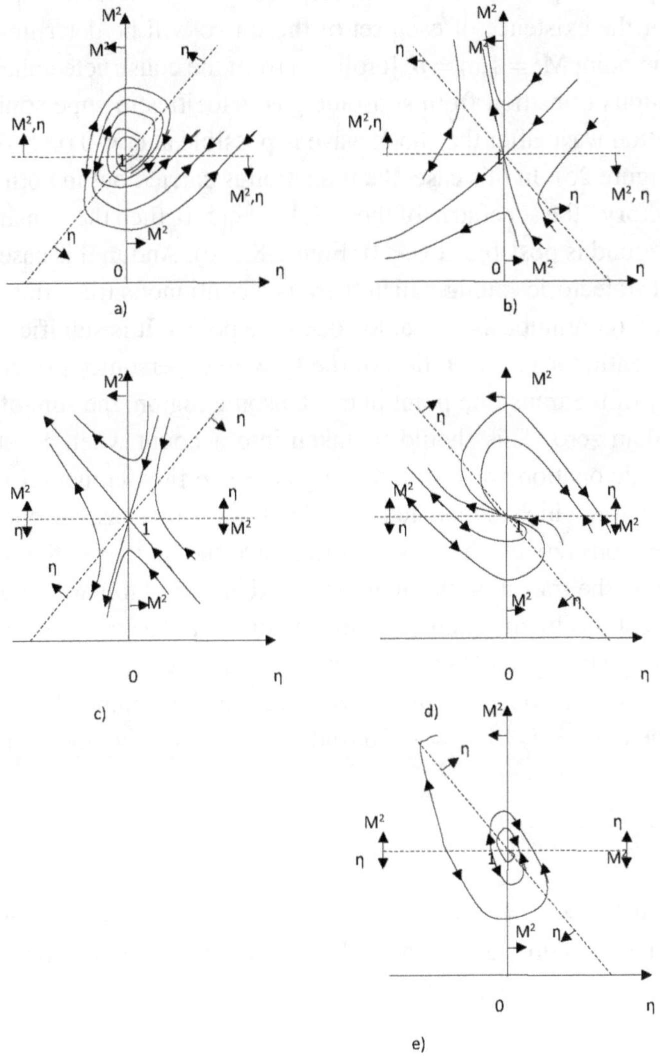

a) ψ > 0, φ > 0; b) ψ > 0, φ < 0; c) ψ < 0, φ < 0; d) ψ < 0, φ < 0.

The positive sign of the derivative of the phase coordinate is indicated by an arrow next to the corresponding characteristic line (zero or singular). The straight - line $M^2 = 1$ (singular) is the line of limit states, at its points the derivatives change sign, passing through infinity. The same figures show qualitatively different options for the location of trajectories in the phase plane.

Arrows on the trajectories indicate the direction of increase in the value of the independent variable. Sets of trajectories in Figure 2d and Figure 2e are different, although they correspond to the same combination of characters ψ and φ. The conditions for the existence of each set of these two will be determined from the analysis of the point $M^2 = 1$, $\eta = 0$. It follows from the constructed phase diagrams that a continuous transition from subsonic gas velocities to supersonic velocities in the detonation wave after the shock wave is possible at $\varphi < 0$ ($\psi > 0$, Figure 2b, and $\psi < 0$, Figure 2c). In this case, the transition is carried out in both cases along a single trajectory - the separatrix of the saddle. If $\varphi > 0$, then the transition through the speed of sound is possible at $\psi < 0$ (Figure 8.2. c). And in this case, there is an infinite set of trajectories along which there is a continuous transition through the speed of sound (continuously passable node type point). It is significant that in all cases where continuous acceleration of the flow to supersonic speeds is possible, in the vicinity of the transition point in the subsonic region, the sum of the actions η is greater than zero. This should be taken into account when constructing the structure of a detonation wave in a dusty gas. Since the detonation wave begins with a shock wave, which is modeled by a shock wave in a gas, the construction of subsequent sections (zones) in the detonation wave should lead to the section of the wave describing the transition through the speed of sound to have a positive value of the sum of actions in the vicinity of the transition. In the phase plane (M^2, η), the transition through the speed of sound from subsonic speeds to supersonic speeds of motion is carried out along a certain (own) direction determined by the derivative $dM^2 / d\eta$ at the point $M^2 = 1$, $\eta = 0$. To find it, we divide equation (8.33) by (8.35)

$$\frac{dM^2}{d\eta} = \eta\frac{1+\gamma_t}{(1-M^2)\varphi+\psi\eta}. \qquad (8.37)$$

In the point $M^2 = 1$, $\eta = 0$ the right side of (8.37) has an uncertainty of 0/0. Revealing this uncertainty according to L'Hopital's rule, we take the derivative of the numerator and denominator of the right side (8.37) with respect to η

$$\frac{dM^2}{d\eta} = \frac{1+\gamma_t}{\psi-\varphi dM^2/d\eta}. \qquad (8.38)$$

Solving the quadratic equation (8.38), we find

$$\left(\frac{dM^2}{d\eta}\right)_{1,2} = \frac{\psi \pm \sqrt{\psi^2-4\varphi(1+\gamma_t)}}{2\varphi}. \qquad (8.39)$$

From (8.39) and the phase diagrams in Figure 2, it follows that for the existence of a regime with a continuous transition from subsonic velocities to supersonic motions in a stopped detonation wave, a number of conditions must be met. If $\varphi < 0$ (Figure 2b, c), then the first direction $(\frac{dM^2}{d\eta})$ must be chosen as the proper direction, since the tangent of the slope of the eigendirection must be negative, and in this case

$$\left(\frac{dM^2}{d\eta}\right)_1 = \frac{-\psi - \sqrt{\psi^2 + 4|\varphi|(1+\gamma_t)}}{2|\varphi|} < 0.$$

If $\varphi > 0$ (Figure 2d), then in order for the derivative $dM^2 / d\eta$ to be real and negative, it is necessary that $\psi < 0$ and $\psi^2 > 4\varphi(1+\gamma_t)$. In this case, both eigendirections (8.39) will be negative. However, apparently, the second proper direction with a sharper slope should be chosen as the eigendirection

$$\left(\frac{dM^2}{d\eta}\right)_2 = \frac{\psi - \sqrt{\psi^2 - 4\varphi(1+\gamma_t)}}{2\varphi},$$

since the first eigendirection with a smaller angle of inclination is located close to the straight line $M^2 = 1$, and the state after the shock wave is characterized by small Mach numbers. However, the choice of one's own direction in this case will be determined by the model consistency condition.

Since the gas temperature sharply increases in the gas shock wave, while the temperature of the dust particles remains the same, immediately after the shock wave, the absolute value of the thermal effect of particles on the gas increases sharply, i.e. particles take heat from the gas (negative effect). In this case, the chemical reaction cannot start instantly, i.e. the chemical impact immediately after the shock wave is practically zero. Thus, the sum of the impacts on the gas flow immediately after the shock wave is negative and quite large in absolute value. In this regard, the established idea that a chemical reaction begins immediately after the shock wave, in this case for a dusty gas flow, requires clarification. If we assume that a chemical reaction begins immediately after the shock, then the gas temperature should increase even more, i.e. the absolute value of the negative thermal effect will increase even more, further removing the state of the medium from the possibility of passing through the speed of sound, where the effect on the flow should be equal to zero. However, it is well known that a significant increase in the rate of a chemical reaction requires an induction time (the induction time of a thermal explosion). Heat exchange processes practically begin immediately after the temperature difference occurs. Therefore, it should be assumed that immediately after the shock, the determining process is the process of heat transfer between the phases. This leads to a decrease in the absolute value of the negative sum of the effects on the gas (movement from the negative on η half-plane of the phase plane (M^2, η) to its positive half-plane). After the induction time has passed, a chemical reaction comes into play, and it becomes decisive. This brings the sum of effects on the gas even closer to the positive along m of the half-plane and, perhaps, transfers it to the positive half-plane. After that, the process of transition through the speed of sound begins. Here, the gas-dynamic effects of expansion become predominant, the gas temperature begins to fall due to expansion. The last section in the detonation wave is the relaxation section, on which the system tends to equilibrium: to the equalization of the phase temperatures and the complete combustion of the combustible reagent. The section is located in the supersonic region of the gas flow.

Thus, it seems that the detonation wave of a dusty gas consists of six qualitatively different zones: 1) a shock wave in a gas; 2) heat exchange zone between gas and dust particles; 3) zone of thermal explosion; the gas-dynamic zone of continuous acceleration of the gas flow from subsonic to supersonic speeds consists of two zones: 4) the subsonic part of the gas-dynamic zone; 5) supersonic part of the gas-dynamic zone; 6) relaxation zone - the tendency of the system to an equilibrium position, i.e. to the zero value of the sum of the effects on the gas due to the tendency of each of the effects to zero.

To confirm the assumption about such a structure of the dusty gas detonation wave, we construct mathematical models for each of the sections, obtain approximate analytical solutions, and confirm the consistency and consistency of the obtained solutions, including numerical calculations.

8.4. HEAT EXCHANGE ZONE IN THE DETONATION WAVE OF DUSTY GAS

In accordance with the above ideas about the structure of the detonation wave of a dusty gas, we will assume that immediately after the shock wave in the detonation wave there is a zone where heat exchange processes between gas and dust particles have a predominant effect.

Therefore, from the sum of actions η (8.31), we remove the chemical action, and we assume that the gas is ideal and has a constant adiabatic coefficient γ_A. Then Equations (8.27) - (8.29) have the form

$$\frac{dM^2}{d\chi} = M^2 \frac{1+\gamma_A M^2}{1-M^2} \frac{1-\theta}{\theta}, \tag{8.40}$$

$$\frac{d\theta}{d\chi} = (1-\theta)\left(\frac{1-\gamma_A M^2}{1-M^2} + \beta\theta\right), \tag{8.41}$$

$$\frac{d\upsilon}{d\chi} = \upsilon\frac{1-\theta}{\theta}\frac{1-\gamma_A M^2}{1-M^2}, \beta = \frac{c_{pA}}{c\kappa}. \tag{8.42}$$

Divide (8.41) by (8.40)

$$\frac{d\theta}{dM^2} = \theta\frac{1-\gamma_A M^2 + (1-M^2)\theta\beta}{M^2(1+\gamma_A M^2)}. \tag{8.43}$$

We introduce a variable $\zeta = 1/\theta$, then equation (8.43) is reduced to a first-order equation, which is easily integrated and, returning to the variable θ, we have

$$\theta = \frac{M^2}{J(1+\gamma_A M^2)^2 + \frac{\beta}{\gamma_A}(\frac{\gamma_A - 1}{2\gamma_A} - M^2)}. \tag{8.44}$$

We find the integration constant J in (8.44) from the condition that the solution passes through the initial point in the heat exchange zone, which coincides with the end point in the shock wave. If we denote the state at the end of the jump by the number 1, then we will have

$$J = \frac{\frac{M_1^2}{\theta_1} - \frac{\beta}{\gamma_A}(\frac{\gamma_A - 1}{2\gamma_A} - M_1^2)}{(1+\gamma_A M_1^2)^2}. \tag{8.45}$$

Dividing equation (8.40) by (8.42) and integrating, we obtain

$$\upsilon = \frac{FM^2}{(1+\gamma_A M^2)^2}. \tag{8.46}$$

We find the integration constant F in (8.46) from the condition for the solution to pass through point 1. Then

$$\upsilon/\upsilon_1 = \frac{M^2}{M_1^2}\frac{(1+\gamma_A M_1^2)^2}{(1+\gamma_A M^2)^2}. \tag{8.47}$$

Substituting (8.44) into (8.40), assuming that $\chi = 0$ at $M^2 = M_1^2$, we get the relationship of the coordinate χ with the variable M^2 in the heat exchange zone

$$\chi = \int_{M_1^2}^{M^2} \frac{(M^2-1)\,dM}{(1+\gamma_A M^2)\left(M^2(1+\frac{\beta}{\gamma_A}) - J(\gamma_A M^2+1)^2 - \frac{\beta}{2}\frac{\gamma_A - 1}{\gamma_A^2}\right)}. \tag{8.48}$$

In particular, at the end of the heat exchange zone, where $\chi = \chi_1$, $M^2 = M_2^2$ we have

$$\chi_1 = \int_{M_1^2}^{M_2^2} \frac{(M^2-1)\,dM}{(1+\gamma_A M^2)\left(M^2(1+\frac{\beta}{\gamma_A}) - J(\gamma_A M^2+1)^2 - \frac{\beta}{2}\frac{\gamma_A - 1}{\gamma_A^2}\right)}. \qquad (8.49)$$

The form of the integrals (8.48), (8.49) depends on the sign

$$\Delta = 4J\gamma_A^2 - (1 - 2\gamma_A J + \beta/\gamma_A)^2. \qquad (8.50)$$

If $\Delta < 0$, so

$$\chi = \ln\left\{ \left(\frac{1+\gamma_A M^2}{1+\gamma_A M_1^2}\right)^{B_1} \left(\frac{f_1(M^2)}{f_1(M_1^2)}\right)^{-B_1/2} \left(\frac{f_2(M^2) - \sqrt{-\Delta}\; f_2(M_1^2) + \sqrt{-\Delta}}{f_2(M^2) + \sqrt{-\Delta}\; f_2(M_1^2) - \sqrt{-\Delta}}\right)^{B_2} \right\}$$

, (8.51)

$$B_1 = \left(\frac{\gamma_A}{\gamma_A+1} + \frac{\beta}{2}\right)^{-1}, B_2 = \frac{B_1}{\sqrt{-\Delta}}\left(\frac{\gamma_A}{\gamma_A+1} - 2J\gamma_A + 0.5\right), f_2(M^2) = 2\gamma_A J - 1 +$$
$$2\gamma_A^2 J M^2, f_1(M^2) = M^4 J\gamma_A^2 - M^2(1 - 2\gamma_A J + \beta/\gamma_A) + J + \beta\frac{\gamma_A - 1}{2\gamma_A^2}. \qquad (8.52)$$

If $\Delta > 0$, so

$$\chi = \ln\left\{ \left(\frac{1+\gamma_A M^2}{1+\gamma_A M_1^2}\right)^{B_1} \left(\frac{f_1(M^2)}{f_1(M_1^2)}\right)^{-B_1/2} \right\}$$
$$+2\frac{B_2}{\sqrt{\Delta}}\left\{ arctg\left(\frac{f_2(M^2)}{\sqrt{\Delta}}\right) - arctg\left(\frac{f_2(M_1^2)}{\sqrt{\Delta}}\right) \right\}. \qquad (8.53)$$

The length of the heat exchange zone can be estimated from the induction time required for the occurrence of a thermal explosion. The time of adiabatic induction is determined by the equality (Zel'dovich, Barenblatt, Makhviladze, Librovich, 1980)

$$t_{ia} = \frac{R_0 T_o^2}{k_0 E \Delta T} \exp\frac{E}{R_0 T_0}$$

,

T_0 is initial temperature, $\Delta T \approx 25 \div 30^o$. In the terms of this chapter $t_{ia} = \frac{E\upsilon_1^2}{k_0 R_0 \Delta T}\exp\frac{1}{\upsilon_1}$. The dimensional extent of the induction zone in the adiabatic flow is equal to $z_{ia} = t_{ia} u_1 = \frac{l_T E \upsilon_1^2}{R_0 \Delta T \sigma}\exp\frac{1}{\upsilon_1}$. Since in the flow of dusty gases during the induction period, the gas gives off heat to the particles, the induction time increases. The nonadiabatic induction time can significantly exceed the adiabatic induction time. Let us assume that the induction time in a dusty gas exceeds the adiabatic induction time by a factor of two on average. Then, passing to the dimensionless length of the heat exchange zone, we obtain an approximate estimate of its length

$$\chi_1 = 2z_{ia}c\kappa/(c_{pA}l_T) \approx 60\frac{\upsilon_1}{\beta\sigma}exp\frac{1}{\upsilon_1}. \tag{8.54}$$

Equating (8.54), (8.50) taking into account the integral (8.51), we obtain an expression for calculating the square of the Mach number at the end of the heat exchange zone M^2

$$60\frac{\upsilon_1}{\beta\sigma}exp\frac{1}{\upsilon_1} = \ln\left\{\left(\frac{1+\gamma_A M^2}{1+\gamma_A M_1^2}\right)^{B_1}\left(\frac{f_1(M^2)}{f_1(M_1^2)}\right)^{-B_1/2}\left(\frac{f_2(M^2)-\sqrt{-\Delta}\,f_2(M_1^2)+\sqrt{-\Delta}}{f_2(M^2)+\sqrt{-\Delta}\,f_2(M_1^2)-\sqrt{-\Delta}}\right)^{B_2}\right\} \tag{8.55}$$

at $\Delta < 0$, and

$$60\frac{\upsilon_1}{\beta\sigma}exp\frac{1}{\upsilon_1} = \ln\left\{\left(\frac{1+\gamma_A M^2}{1+\gamma_A M_1^2}\right)^{B_1}\left(\frac{f_1(M^2)}{f_1(M_1^2)}\right)^{-B_1/2}\right\}$$

$$+2\frac{B_2}{\sqrt{\Delta}}\left\{arctg\left(\frac{f_2(M^2)}{\sqrt{\Delta}}\right) - arctg\left(\frac{f_2(M_1^2)}{\sqrt{\Delta}}\right)\right\} \tag{8.56}$$

at $\Delta > 0$.

The value of the square of the Mach number at the end of the heat exchange zone M_2^2, using the integrals (8.44), (8.45), we find the values of the gas temperature and the ratio of phase temperatures at the end of the heat exchange zone

$$\upsilon_2/\upsilon_1 = \frac{M_2^{\,2}(1+\gamma_A M_1^{\,2})^{\,2}}{M_1^2(1+\gamma_A M_2^{\,2})^{\,2}}, \tag{8.57}$$

$$\theta_2 = \frac{M_2^{\,2}}{J(1+\gamma_A M_2^{\,2})^{\,2}+\frac{\beta}{\gamma_A}(\frac{\gamma_A-1}{2\gamma_A}-M_2^{\,2})}. \tag{8.58}$$

8.5. NUMERICAL STUDY OF SOLUTIONS TO THE EQUATIONS OF A DUSTY GAS DETONATION WAVE WITH THERMAL PHASE INTERACTION

We will assume that combustible gas A is a monatomic gas, and the reaction product gas B is a diatomic gas, i.e. a chemical reaction $A \to B$ is a compound reaction, and the molecular weight of gas B is twice the molecular weight of gas A $\mu_B = 2\mu_A$. In this regard

$$c_{pA} = \frac{5}{2}R_A, R_A = \frac{R_0}{\mu_A}, c_{pB} = \frac{7}{2}R_B, R_B = \frac{R_0}{\mu_B},$$

$$\gamma_A = \frac{5}{3}, \gamma_B = \frac{7}{5}, c_{vA} = \frac{3}{2}R_A, c_{vB} = \frac{5}{2}R_B, \frac{c_p}{c_{pA}} = 0.3C_A + 0.7.$$

The main purpose of the numerical study is to prove the possibility of regimes with a continuous transition in the detonation wave of a dusty gas from subsonic flows to supersonic flows. Therefore, when choosing the values of the dimensionless parameters, which are included in the equations of the detonation wave of a dusty gas, we will proceed from the need to set such values of these parameters that would ensure continuous acceleration of the gas flow in the detonation wave from subsonic velocities to supersonic velocities. The determination of the specific values of the operating parameters that lead to such accelerations can be left to further stages. We will assume that if the dimensionless values corresponding to the continuous acceleration of the flow to supersonic speeds are found, then the values of the dimensional parameters corresponding to such an acceleration can also be found, due to the inclusion of a large number of dimensional parameters of different nature into the dimensionless parameters. As calculation options, we will take two values of the parameter characterizing the ratio of the heat capacity of the gas flow to the heat capacity of the particle flow: β = 1 (option 1); β = 0.5 (option 2). The dimensionless gas temperature before the shock wave in a stopped detonation wave must be significantly less than 1, since the activation energy of the combustion reaction in the detonation wave is of great importance. Let's take υ_b =0.01 The initial calculation is the Mach number, determined by the velocity of the detonation wave. Taking into account the constraint (8.22), it is clear that this Mach number should be significantly greater than 1. Let us choose the value of the Mach number M_b = 6. Then, in accordance with equations (8.9) - (8.11), we have the following values of the parameters after the shock wave

$$M_1^2 = 0.217876, \theta_1 = 12.11979, \upsilon_1 = \upsilon_b\theta_1 = 0.1211979.$$

The obtained solutions of the systems of equations in each of the zones of the detonation wave of a dusty gas make it possible to calculate the values of variables at the boundaries of the zones and plot graphs of their continuous change along the wave (Zhizhin, 2008, 2009 a, b; 2010 a, b; 2012).

Figure 3 shows the results of calculating the change in the parameters of the detonation wave at β = 1 (option 1).

Figure 3. Graphs of parameter changes along the detonation wave of dusty gas (calculation option 1, $M_h = 6$, $\beta = 1$)

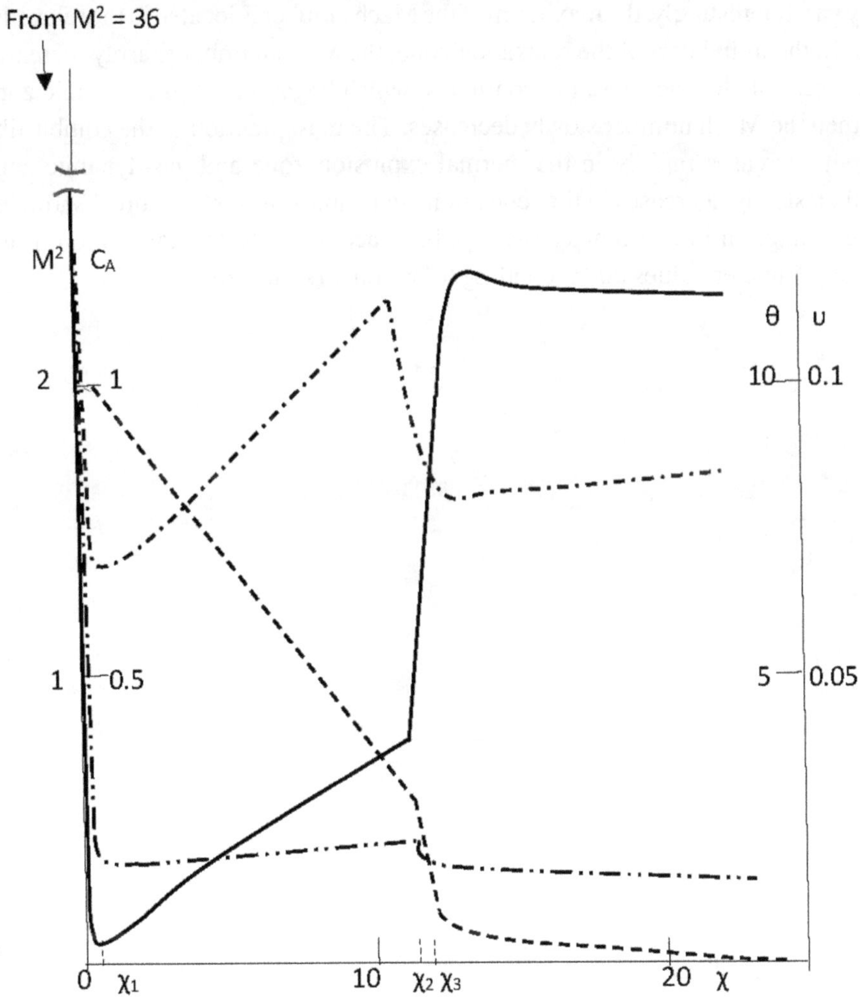

θ, υ, M², C_A

The graphs in Figure 3 demonstrate the strong non-linearity of the processes in the dusty gas detonation wave. Abrupt changes in parameters occur not only in the shock wave (for example, a sharp increase in gas temperature υ and a corresponding abrupt increase in the phase temperature ratio θ), but also in other zones of the detonation wave. Immediately after the shock wave in the heat exchange zone, there is a sharp decrease in the gas temperature and the ratio of phase temperatures.

In the thermal explosion zone, the gas temperature naturally increases, while in the gas-dynamic zone it drops sharply. In the relaxation zone, the gas temperature first continues to fall and then slowly increases. Thus, the existence of two maxima and two minima is found in the temperature profile along the detonation wave of a dusty gas. Interestingly, the maximum of the Mach number is located in the relaxation zone. In the initial part of the relaxation zone, the Mach number sharply increases; the process of flow acceleration continues, which began in the gas-dynamic zone, and then the Mach number slowly decreases. The mass fraction of the combustible reagent decreases rapidly in the thermal explosion zone and gas-dynamic zone, and then slowly decreases in the relaxation zone almost to zero. Figure 4 shows the process diagram in coordinates (M^2, η), built according to the calculation results for the parameter values corresponding to option 1 (solid lines).

Figure 4. Change of state in the detonation wave of dusty gas in the plane (M^2, η) (calculation variant 1, $M_{\text{н}}$ = 6, β = 1)

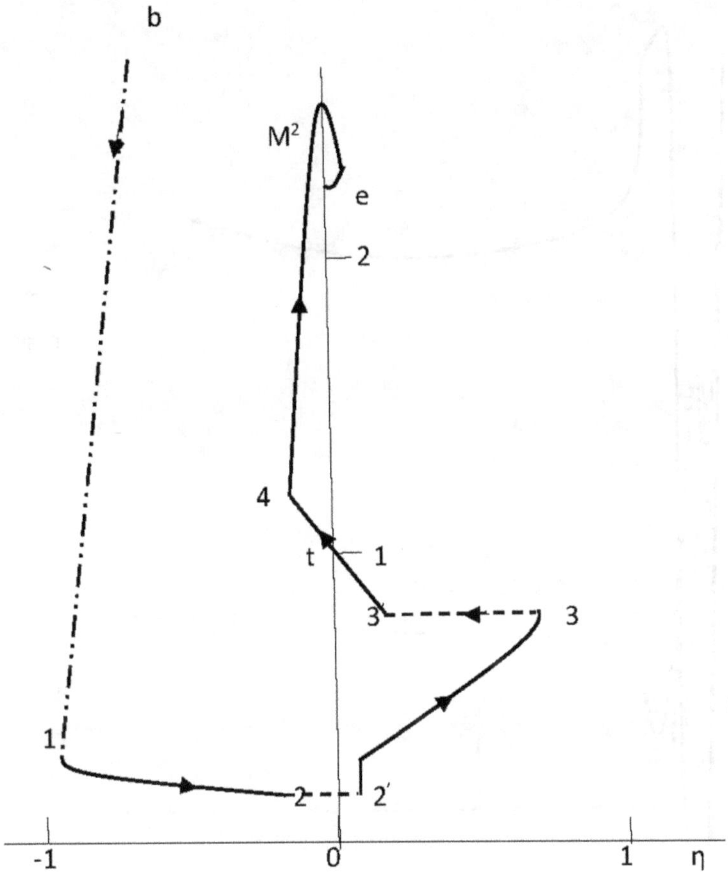

It should be noted that approaching the supersonic equilibrium in this case is accompanied by a change in the sign of the sum of actions η. This is due to a significant slowdown in the rate of chemical reactions when approaching the equilibrium position compared to the rate of heat exchange processes. Thus, in this version of the calculation, when moving along the detonation wave, a threefold transition through the zero value of the sum of actions η is observed. Figure 5 shows the results of calculating the change in parameters along the detonation wave of a dusty gas when the parameter value is halved (option 2).

Figure 5. Graphs of parameter changes along the detonation wave of dusty gas (calculation option 2, M_h = 6, β = 0.5)

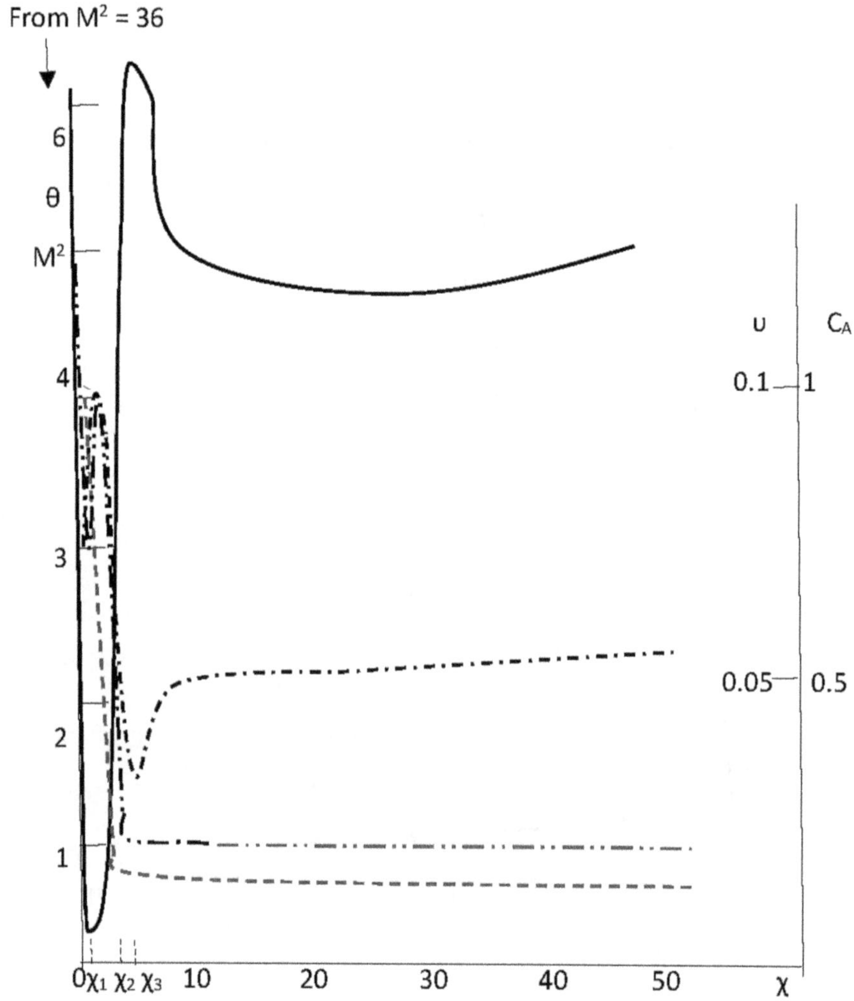

θ, υ, M^2, C_A

Here, first of all, a sharp increase in the characteristic detonation wavelength should be noted. Judging by the change in the mass fraction of the combustible reagent, the wavelength increased by several orders of magnitude compared to option 1 (β = 1). This once again testifies to the significant effect of slowing down the rate of chemical reactions in the "tail" of the wave, leading to a significant increase in the length of the wave precisely due to a decrease in this rate (i.e., the obvious inapplicability of the assumptions about the narrowness of the zone of chemical reactions).

The sharp nonlinearity of the processes in the detonation wave is preserved in this variant as well. Figure 6 shows the diagram of the process in the detonation wave in coordinates (M^2, η), built according to the calculation results for the parameters corresponding to option 2.

Figure 6. Change of state in the detonation wave of dusty gas in the plane (M^2, η) (calculation variant 2, M_h = 6, β = 0.5)

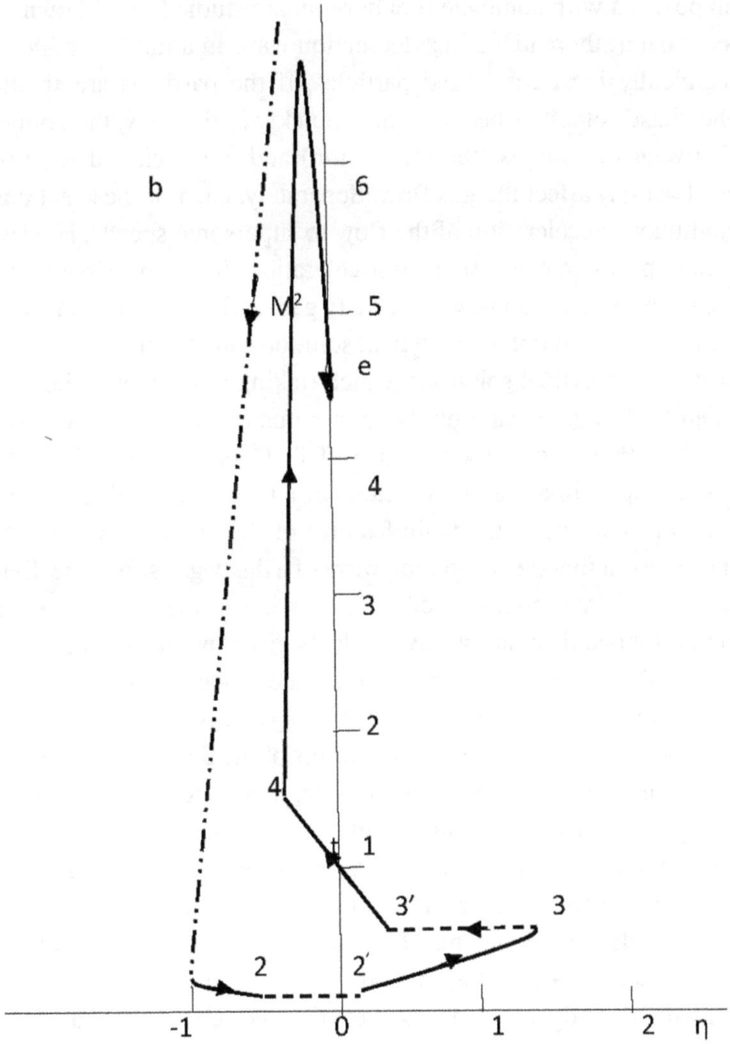

In this variant, the striving for a supersonic equilibrium position occurs without changing the sign of the sum of actions η.

CONCLUSION

The conditions for a continuous transition from subsonic to supersonic gas velocities in a detonation wave after a shock wave with the system tending to an equilibrium position with complete fuel burnout are studied. It is shown that these conditions are naturally realized in a detonation wave in a mixture of combustible gas and chemically inert condensed particles. If the particles are small enough such that the phase velocities become equal quickly in the flow, then only the heat exchange between the phases (thermal action) and the chemical reaction in the gas (chemical action) affect the gas flow. Separately, each of these actions cannot lead to a continuous acceleration of the flow to supersonic speeds. However, their simultaneous impact can cause such an acceleration. It was possible to determine the structure of the detonation wave of a dusty gas, taking into account all the main features of the process, to obtain analytical solutions and to study them numerically. When deriving analytical solutions, which (taking into account the complexity of the problem), of course, can only be approximate, we used the well-known in computational mathematics, proclaimed by P.L. Chebyshev (Chebyshev, 1947), the principle of approximating the functions of a real functional space by a set of simpler functions that reflect the main features of the process. When solving this problem, it turned out that the detonation wave of a dusty gas should be divided into six qualitatively different zones processes occurring in them. The first along the gas flow in the stopped detonation wave of dusty gas, obviously, is the shock wave, modeled by the shock wave in the gas. Immediately behind the shock wave there is a zone of heat exchange between the phases. Physically, this is quite understandable, since it is known that in order for the results of the development of a chemical reaction to become at least somehow noticeable, time is required - induction time. Heat exchange processes begin immediately, as soon as a temperature difference arises. Namely, after the shock wave, a sharp excess of the gas temperature over the temperature of the particles occurs. Behind the heat exchange zone is the zone of the predominant influence of chemical reactions - the zone of thermal explosion, since chemical reactions are exothermic. The zone of thermal explosion is followed by a zone in which gas-dynamic effects prevail, although heat exchange processes and chemical reactions are also taken into account. This is a zone of continuous transition from subsonic to supersonic gas motion. This zone should be divided into two more zones - the subsonic part and the supersonic part of the gas-dynamic zone.

Behind the gas-dynamic zone there is a relaxation zone - the zone of the system tending to the equilibrium position. Heat exchange processes and chemical reactions are also taken into account here. Only such ideas about the structure of the dusty gas wave turned out to be mathematically and physically consistent and led to a closed solvable system of equations and to a reasonable description of the detonation wave. Mathematical analysis and numerical studies revealed a complex structure of the dusty gas detonation wave, which is distinguished by sharply nonlinear dependences of the change in variables along the wave and the high sensitivity of the nature of these dependences to the varied dimensionless parameters of the problem. Numerical studies have revealed additional features of the processes in the detonation wave, which do not directly follow from the qualitative analysis. It has been established, for example, that a combustible reagent equal to the heat capacities of gas and particle flows almost completely burns out at a finite effective wavelength, tending to an equilibrium position in the supersonic flow region. It is interesting to note that in this version of the process, a threefold change in the sign of the sum of the effects on the gas is observed. It is significant that sharply nonlinear processes occur in different zones of the detonation wave of dusty gas. These features confirm the complete inapplicability of the assumptions of the narrow reaction zone method in the analysis of dusty gas detonation waves.

It is important that both analytical and numerical solutions prove the existence of flow regimes in the detonation wave of a dusty gas with continuous flow acceleration to supersonic gas velocities after the shock wave and the tendency of the system to an equilibrium position with complete burnout of the combustible reagent. This allows us to state that the detonation wave can be made controllable by stopping it, creating, in accordance with the calculation, the conditions for maintaining it in a stationary mode. In this case, it is possible to achieve high values of the momentum imparted by the outflowing flow to the aircraft. The existence of positions in the supersonic region of the flow of dusty gas removes the restriction on the gas flow velocity at the entrance to the stopped detonation wave. This means that with a stopped controlled detonation wave of a dusty gas, the flow velocity entering the detonation wave is tens of times greater than the flow velocity following from the classical theory of detonation. This will make it possible to achieve significantly greater values of the momentum imparted to the aircraft, moreover, with complete burnout of the reagent.

Even more striking results can be obtained if we assume the possibility of realizing a stopped controlled wave of nuclear detonation. Since the speed of sound in plasma (for example, in the atmosphere of the Sun), in which there can be a nuclear detonation wave (Pikelner, 1961), reaches 100 km / s, the speed of a hypothetical aircraft using a standing wave of nuclear detonation can be of the same order. The transition through the speed of sound in this device will occur due to the interaction of nuclear reactions and heat transfer between the plasma components.

REFERENCES

Chebyshev, P. L. (1947). *Mathematical analysis* (Vol. 2). The theory of mechanisms known as parallelograms. Complete works. Publishing House of the Academy of Sciences of the USSR.

Pikelner, S.B. (1961). *Fundamentals of space electrodynamics*. Moscow: State Publishing House of Phys. - mat. literature.

Vulis, L. A., Gusika, P. L., & Zhizhin, G. V. (1972). Two-phase flows in a channel of constant cross section (qualitative study). *Journal of Applied Mechanics and Technical Physics*, *5*, 143–156.

Zeldovich, Ya. B., Barenblatt, G. I., Librovich, V. B., & Makhviladze, G. M. (1980). *Mathematical theory of combustion and explosion*. Nauka.

Zhizhin, G. V. (1972). *Qualitative study of one-dimensional steady flows*. Diss... cand. Phys.-Math. Sciences. Leningrad.

Zhizhin, G.V. (2004). Synergetic of flows. *Life and safety*. 2 - 3a, 365 - 385.

Zhizhin, G. V. (2008). *Combustion waves with distributed zones of chemical reactions (non-asymptotic theory of combustion)*. Werner Regen Publishing House.

Zhizhin, G.V. (2009a). *Waves of detonation in nature and technology, their mathematical models and significance in violation of the ecological systems of the biosphere*. St. Petersburg: Preprint of the North -Western State Correspondence Technical University.

Zhizhin, G. V. (2009b). *Detonation wave in a gas with variable composition*. Polytechnic-service.

Zhizhin, G. V. (2010a). Nonequilibrium processes in the standing wave of dusty combustible gas. *Scientific Abstract Collection of III International Scientific and Technical Conference "AEROENGINES ofXXI Century", Electronic data, Moscow: Federal State Unitary Enterprise "Central Institute ofAviation Motors named after P.I. Baranov"*.

Zhizhin, G. V. (2010b). Mathematical model of a standing detonation wave in a jet of dusty gas. Jet, detached and unsteady flows. Anniversary seminar with international participation (collection of abstracts). St. Petersburg, 161 - 163.

Zhizhin, G. V. (2012). Supersonic motions of gas with condensed particles in the reaction zone of a standing detonation wave. International Scientific Conference on Mechanics: SIXTH POLYAKHOV`S READING. Book of abstract. Saint Petersburg: Saint Petersburg State University. 136.

Chapter 9
Waves of Solid Flame Combustion

ABSTRACT

A model for a combustion wave in condensed mixtures with autoinhibition by a high-melting reaction product is constructed with allowance for a reduction in the particle surface of the high-melting reactant during chemical interaction with the low-melting reactant. An approximate analytical solution is derived using the method of a semi-infinite reaction zone. The method is validated by qualitative and numerical studies of differential equations. The literature model of the process considered ignores the variation in the particle surface of the high-melting reactant, which leads to neglect of the key characteristics of all chemical reactions – a continuous decrease in the reaction rate during combustion of the reactants.

INTRODUCTION

In 1967, a small group of employees of the Department of the Institute of Chemical Physics of the USSR Academy of Sciences was lucky enough to make a scientific discovery (Merzhanov, A.G., Borovinskaya, I.P., Shkiro, V.M., 1967). Investigating the combustion mechanism of solid rocket fuels, they discovered a class of previously unknown combustion processes in which all substances (reagents, intermediate and final combustion products) are in This class of processes is called solid-flame combustion (Merzhanov, 2000, 2005).This scientific discovery is registered in the Register of State Discoveries of the USSR under the name "the phenomenon of wave localization of auto-braking solid-phase reactions" (Merzhanov, Shkiro, Borovinska-

DOI: 10.4018/978-1-6684-7745-8.ch009

ya, 1972, 1973 a, b, 1974, 1982). Later, due to the wide variety of such phenomena, processes in the broad sense of the word began to refer to solid-flame combustion.

This chapter deals with the combustion waves of solids, when processes in solids are limiting. An important representative of the class of such problems is solid-flame combustion in condensed mixtures with self-braking by a refractory reaction product. In this case, a chemical combustion reaction occurs on the surface of a refractory reagent surrounded in the initial state by a second low-melting reagent. The resulting product, as well as one of the reagents, it is refractory and it surrounds the particles of the refractory reagent, making it difficult for molecules of the low-melting reagent to access the refractory reagent, which are forced to diffuse through the growing layer of the refractory reaction product.

In this chapter, solid-flame combustion waves are considered taking into account the reduction in the surface of particles of a refractory reagent in the process of chemical interaction with a low-melting reagent (Zhizhin, 2004). The well-known model of the process under consideration (Aldushin, Martemyanova, Merzhanov et al., 1972) did not take into account the change in the surface of particles of a refractory reagent, which led to ignoring the fundamental characteristic of all chemical reactions - a continuous decrease in the reaction rate when the reagent burns out. To obtain an approximate analytical solution, the semi-infinite reaction zone method was used (Zhizhin, 1992, 1997). The correctness of the method is confirmed by a qualitative and numerical study of differential equations. It is shown that a stationary wave of solid-flame combustion exists at limited values of the parameter characterizing the self-stagnation of combustion by a solid-melting reaction product. This fundamentally distinguishes this model from the model proposed earlier (Aldushin, Martemyanova, Merzhanov et al., 1972), which does not take into account the change in the chemical interaction surface during the reaction. Moreover, in the work (Aldushin, Martemyanova, Marzhanov et al., 1972), it is for large values of the auto-braking parameter that an asymptotic expression for the velocity of a solid-flame combustion wave was obtained.

9.1. WAVES OF IDEAL SOLID-FLAME COMBUSTION WITH A VARIABLE SURFACE OF CHEMICAL INTERACTION

A model of a combustion wave in condensed mixtures was built, in which one of the reactants and reaction products are refractory substances (Aldushin, Martemyanova, Merzhanov and others, 1972). It was believed that a refractory reagent is particles distributed in a low-melting reagent, a chemical reaction occurs on the surface of a refractory reagent, while a refractory product grows on a particle of a refractory reagent, slowing down the reaction, since a low-melting reagent must diffuse through the product layer to enter into the reaction. This model is called ideal solid-flame combustion (Merzhanov, 2000). In the work (Aldushin, Martemyanova, Merzhanov and others, 1972) it was rightly assumed that the rate of heat release during the reaction is proportional to the area of the interface of the reagents, however, it was believed that the particles of the refractory reagent are flat, and one of the geometric dimensions of the particles is much smaller than the other two and at the same time constant. The change in the interface area of the reagents, but which the chemical reaction occurs, naturally decreasing as the refractory reagent burns out, was not taken into account. This was the contradiction of the model adopted in the work (Aldushin, Martemyanova, Merzhanov et al., 1972). It was assumed that the refractory reagent is consumed in the reaction, but neither the change in particle size nor their number was taken into account. The main attention in this work was paid to the analysis of the influence of a change in the reaction depth on the process of reaction deceleration due to an increase in the amount of a refractory product around the particles in the so-called oxidation laws. At the same time, a fundamental point in all chemical reactions fell out of consideration - a continuous decrease in the reaction rate (down to zero) due to the burning out of the reagents. In order to somehow smooth out this contradiction, the authors assumed that at a certain temperature in the vicinity of the adiabatic combustion temperature, the reaction rate suddenly vanishes. In this chapter, developing the model adopted in the work (Aldushin, Martemyanova, Merzhanov et al., 1967), the change in the surface area of the chemical interaction, which leads to a continuous decrease in the reaction rate, is taken into account. The influence of a refractory product on the reaction rate is taken into account using the laws of oxidation. To obtain an approximate analytical solution, the semi-infinite reaction zone method is used.

The system of one-dimensional equations describing the stationary propagation of an adiabatic wave of solid-flame combustion in the quasi-homogeneous approximation can be written as (Merzhanov, 2000)

$$\bar{c}\bar{\rho}u\frac{dT}{dz} = \bar{\lambda}\frac{d^2T}{dz^2} + \bar{\rho}QS_ak_0\varphi(\eta)\Phi(T), \tag{9.1}$$

$$u\frac{d\eta}{dz} = k_0S_a\varphi(\eta)\Phi(T). \tag{9.2}$$

Here $\bar{c},\bar{\rho},\bar{\lambda}$ are the averaged values, respectively, of the heat capacity, density and thermal conductivity of the medium; η is depth of chemical transformation; T is the temperature; u is the burning wave velocity; z is the wave coordinate; k_0 is the pre-exponent of the rate of a chemical reaction; ϕ(η) is the dependence of the rate of a chemical reaction on the depth of the chemical transformation, ϕ(1) 0; Φ(T) is the dependence of the rate of a chemical reaction on temperature; Q is the specific thermal effect of a chemical reaction (for example, the amount of thermal energy released during the combustion of a unit mass of a refractory reagent; S_a is the specific surface of a refractory reagent.

The rate of a chemical reaction in (9.1), (9.2) is written as $k_0 S_a \varphi(\eta) \Phi(T)$, i.e. products of a function of the depth of chemical transformation $S_a \varphi(\eta)$ and a function of temperature $k_0 \Phi(T)$. This type of reaction rate reflects, based on experimental data, ideas about thermal homogenization during solid-flame combustion (Merzhanov, 2000), according to which, during the reaction, the effective width of the reaction zone is spontaneously established, which significantly exceeds the particle size, so that the particles can be considered as point sources heat. The effect of particle size and shape is taken into account in S_a. The form of the function φ may change depending on the reactants (Brown, Dillimore, Galway, 1983).

If we consider the particles of a refractory reagent to be spherical with a radius r, then the volume of the particle is $v_0 \sim r^3$, the surface of the particle $s_0 \sim r^2$, i.e. $s_0 \sim v_0^{2/3}$. Let us determine the depth of transformation by the ratio of the mass of the reaction product to the mass of the mixture per unit volume $\eta = m_p/(m_h + m_p + m_f)$. Here m_h is the mass of the refractory reagent per unit volume of the mixture, m_p is the mass of the reaction product per unit volume of the mixture, m_f is the mass of the low-melting reagent per unit volume of the mixture. Considering that the refractory and low-melting reagents proportionally disappear in the reaction and that $m_h = n\rho v_0$ (ρ, n are density of particles and their number per unit volume of the mixture), we have

$$v_0 \sim (1-\eta)(m_h + m_p + m_f)/(n\rho).$$

Since $S_a = ns_0$, from the previous one we get

$$s_a \sim n\left((1-\eta)(m_h + m_p + m_f)/(\rho n)\right)^{2/3}.$$

Taking into account that the ratio $(m_h + m_p + m_f)/\rho$ and also n can be considered constants, taking into account the slight difference between 2/3 and 1, the last relation can be replaced in the first approximation by the equality $w = u(\overline{\rho}\overline{c})^{1/2}/(\overline{\lambda}k)^{1/2}$; $\varsigma = zu\overline{\rho}\overline{c}/\overline{\lambda}$. Where C is coefficient proportionally.

Let us introduce the proportionality coefficient C into the rate constant of a chemical reaction $k = k_0 C$ and introduce the following dimensionless variables and parameters: $\vartheta = (T - T_+)/(T_+ - T_0)$, T_+ is the temperature at the end of the wave (adiabatic combustion temperature), T_0 is the temperature at the beginning of the wave. Then equations (9.1), (9.2), together with the law of conservation of energy $Q/\overline{c} = T_+ - T_0$, take the form

$$w^2 \frac{d^2 \vartheta}{d\varsigma^2} - w^2 \frac{d\vartheta}{d\varsigma} + (1-\eta)\varphi(\eta)\Phi(\vartheta) = 0, \quad (9.3)$$

$$w^2 \frac{d\eta}{d\varsigma} - (1-\eta)\varphi(\eta)\Phi(\vartheta) = 0. \quad (9.4)$$

Equations (9.3), (9.4) taking into account the boundary condition

$$d\vartheta/d\varsigma = d\eta/d\varsigma = 0, \vartheta = -1, \eta = 0, at\ \varsigma = -\infty \quad (9.5)$$

have an integral

$$d\vartheta/d\varsigma = \vartheta + 1 - \eta. \quad (9.6)$$

9.2. QUALITATIVE AND NUMERICAL STUDY OF AN IDEAL SOLID-FLAME COMBUSTION WAVE WITH A VARIABLE CHEMICAL INTERACTION SURFACE

Solutions to system (9.4), (9.6) can be represented by trajectories in the phase plane (ϑ, η), where $0 \leq \eta \leq 1, -1 \leq \vartheta \leq 0$ (Figure 1).

Figure 1. The phase plane of equations for a wave of solid-flame combustion with a variable surface of chemical interaction

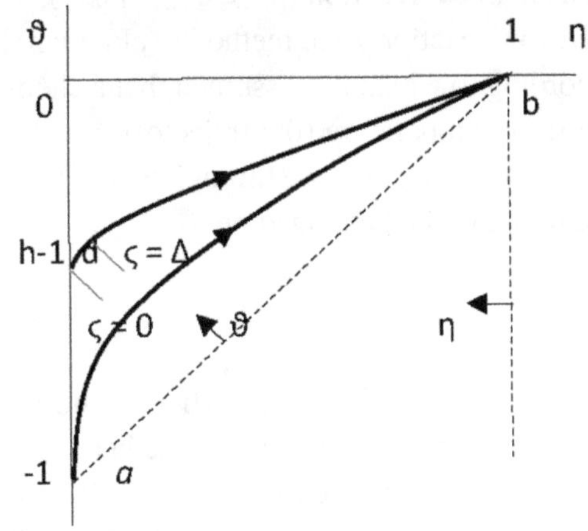

In Figure 1, the dotted line is the zero isocline of equation (9.6), at the points of which $d\vartheta / d\varsigma = 0$. Point *b* with coordinates (0,1) is special (the final equilibrium position), since the line $\eta = 1$ is the zero isocline $d\eta / d\varsigma = 0$. At this point, the boundary condition at the end of the wave is satisfied

$$d\vartheta / d\varsigma = d\eta / d\varsigma = 0, \vartheta = 0, \eta = 1, at\ \varsigma = +\infty$$

. (9.7)

The point *a* with coordinates $\vartheta = -1, \eta = 0$ is not an equilibrium position, since, strictly speaking, the derivative $d\eta / d\varsigma$ at it is non-zero. However, the system (9.4), (9.6) does not have wave solutions, which must be trajectories going from one equilibrium position to another. In order for the point *a* to be singular (the initial equilibrium position), it is necessary to set the function $\Phi(\vartheta)$ to zero at this point (in this case, the boundary condition (9.5) is satisfied). Then the desired wave solution is represented by a trajectory S going from point *a* to point *b*. In accordance with the method of the semi-infinite reaction zone, modeling the wave solution S, we will assume that the function $\Phi(\vartheta)$ is zero not only at the point *a,* but also on a certain segment *ad* of the ϑ axis, the length of which (the value of the source cut off) will be determined in the future. The segment *ad* hell corresponds to the heating zone $(-\infty \leq \varsigma \leq 0)$ in the exothermic reaction wave, and the temperature at point d $(\varsigma = 0)$ is the temperature of the reaction front T_f.

After the point d, the process in the wave develops along the trajectory db - a semi-infinite zone of a chemical reaction $(0 \leq \varsigma \leq +\infty)$. With such modeling, the limitation of the narrow reaction zone method (Zeldovich, Frank-Kamenetsky, 1938) is removed, according to which it is assumed that a chemical reaction occurs at an adiabatic combustion temperature (the trajectory S is replaced by a step *ao - ob*). Denoting the length of the segment *ad* through $h = (T_f - T_0)/(T_+ - T_0)$, from (9.6) at $\eta = 0$ we obtain that in the heating zone

$$\vartheta = -1 + h \exp\varsigma. \qquad (9.8)$$

The fundamental point for obtaining an analytical solution in the semi-infinite reaction zone method is the introduction between the heating zone and the chemical reaction zone of a certain buffer zone $(0 \leq \varsigma \leq \Delta)$ (Figure 9.1), the length of which Δ is small and, after resolving the Without violating the boundary conditions (9.5), (9.7), we will integrate (9.4) in the chemical reaction zone $(\Delta \leq \varsigma \leq +\infty)$, replacing the functions ϕ, Φ by their mean integral values over this zone. Then equation (9.4) takes the form

$$w^2 \frac{d\eta}{d\varsigma} = (1-\eta)\langle\varphi\rangle\langle\Phi\rangle \tag{9.9}$$

and has the integral

$$\eta = 1-(1-\eta_\Delta)e^{-\alpha(\varsigma-\Delta)}, \tag{9.10}$$

Where $\alpha = w^{-2}\langle\varphi\rangle\langle\Phi\rangle$, (9.11)

η_Δ is the η at $\varsigma = \Delta$.

Substituting (9.10) into (9.6) and integrating, we find

$$\vartheta = -(1-\eta_\Delta)(1+\alpha)^{-1}e^{-\alpha(\varsigma-\Delta)}. \tag{9.12}$$

In the buffer zone, due to its smallness, the dependences $\vartheta(\varsigma)$, $\eta(\varsigma)$ can be assumed to be linear. Taking this into account, using the continuity condition ϑ and η, we obtain

$$\vartheta = (\vartheta_\Delta + 1 - h)\varsigma/\Delta + h - 1, \tag{9.13}$$

$$\eta = \eta_\Delta \varsigma/\Delta. \tag{9.14}$$

The angle of inclination of the straight lines (9.13), (9.14) can also be determined on the basis of equations (9.4), (9.6) taking into account the average values for this zone in the right parts of these equations, i.e. we have the equations

$$(\vartheta_\Delta + 1 - h)/\Delta = 0.5(h - 1 + \vartheta_\Delta - \eta_\Delta)+1, \tag{9.15}$$

$$\eta_\Delta/\Delta = w^{-2}(1-0.5\eta_\Delta)\varphi(0.5\eta_\Delta)\Phi(0.5(h-1+\vartheta_\Delta)), \tag{9.16}$$

where $\vartheta_\Delta = \vartheta(\varsigma = \Delta)$is found from (9.12)

$$\vartheta_\Delta = -(1-\eta_\Delta)(1+\alpha)^{-1}. \tag{9.17}$$

Let us equate the derivatives $d\vartheta/d\varsigma$ at the boundary of the heating zone and the buffer zone. From (9.8), (9.13) we have

$$\vartheta_\Delta + 1 - h = h\Delta. \tag{9.18}$$

Comparing (9.18), (9.15), we get

$$h = \vartheta_\Delta + 1 - \eta_\Delta. \tag{9.19}$$

Eliminate ϑ_Δ from (9.18) using (9.17)

$$-(1 - \eta_\Delta)(1 + \alpha)^{-1} + 1 - h = h\Delta. \tag{9.20}$$

We exclude η_Δ from the left side of (9.16) with the help of (9.19), (9.18), canceling by Δ, which is not yet equal to zero

$$h = w^{-2}(1 - 0.5\eta_\Delta)\varphi(0.5\eta_\Delta)\Phi\big(0.5(h - 1 + \vartheta_\Delta)\big). \tag{9.21}$$

Let us now let Δ in (9.20), (9.21) tend to zero

$$(1 + \alpha_o)^{-1} = 1 - h, \tag{9.22}$$

$$w^2 = h^{-1}\varphi(0)\Phi(h - 1), \tag{9.23}$$

Where α_0 is the value α when calculating the mean integral values over the interval from 0 to infinity.

Equations (9.22), (9.23) set the required values of the quantity h (i.e. the temperature of the wave front) and the wave speed w. With the found values h, w the profile of the change in the depth of the chemical transformation is found on the basis of (9.10) (after the transition $\Delta \to 0$) according to the equation

$$\eta = 1 - \exp(-\alpha_o\varsigma), \tag{9.24}$$

and the temperature profile based on (9.12) according to the equation

$$\vartheta = -(1 + \alpha_o)^{-1}\exp(-\alpha_o\varsigma). \tag{9.25}$$

Let us pass from the general study of equations (9.1), (9.2) with rather arbitrary functions $\phi(\eta)$, $\Phi(\vartheta)$ to the analysis of these equations with specific functions $\phi(\eta)$, $\Phi(\vartheta)$ corresponding to ideal solid-flame combustion in the case of strong kinetic drag by a layer of refractory reaction product. Let us to be

$$\Phi(\vartheta) = \exp\left(-\frac{E/R_0}{\vartheta(T_+ - T_0) + T_+}\right). \tag{9.26}$$

The function $\phi(\eta)$ is set by an exponential law, in which the inhibition of the reaction is most pronounced (Merzhanov, 2000)

$$\varphi(\eta) = \exp(-m\eta), \tag{9.27}$$

m is the automatic braking parameter.

In order to use the general expressions (9.22), (9.24), (9.25) it is necessary to calculate the mean integral values of the functions $\phi(\eta)$, $\Phi(\vartheta)$ on the interval $0 \leq \varsigma \leq +\infty$ .. Transforming the exponent according to D.A. Frank-Kamenetsky, we get

$$\langle\Phi(\vartheta)\rangle_0 = (1-h)^{-1}\int_{h-1}^{\infty}\exp\left(-\frac{E/R_0}{\vartheta(T_+-T_0)+T_+}\right)d\vartheta = \frac{\gamma}{1-h}(1-e^{\frac{h-1}{\gamma}})e^{-\beta}, \tag{9.28}$$

$$\gamma = \frac{R_0T_+^2}{(T_+-T_0)E}, \beta = \frac{E}{R_0T_+}.$$

Using (9.27), we find

$$\langle\varphi(\eta)\rangle_0 = \int_0^1 \exp(-m\eta)\,d\eta = m^{-1}(1-\exp(-m)). \tag{9.29}$$

Thus, taking into account (9.11), (9.28), (9.29) we have

$$\alpha_0 = \gamma(1-h)^{-1}w^{-2}m^{-1}(1-\exp\frac{h-1}{\gamma})\exp(-\beta)(1-e^{-m}), \tag{9.30}$$

$$\varphi(0) = 1. \tag{9.31}$$

From (9.26) it follows that

$$\Phi(h-1) = \exp(-\frac{\gamma\beta^2}{h-1+\gamma\beta}). \tag{9.32}$$

Substituting (9.30) - (9.32) into (9.22) - (9.25), we obtain formulas for calculating the front temperature, wave velocity and its structure. Substituting (9.30) into (9.22) and (9.31), (9.32) into (9.23), respectively, we have

$$hw^2m = \gamma(1-\exp\frac{h-1}{\gamma})(1-e^{-m})e^{-\beta}, \tag{9.33}$$

$$w^2 = h^{-1}\exp(\frac{\gamma\beta^2}{1-h-\gamma\beta}). \tag{9.34}$$

We exclude w from (9.33) using (9.34) and get one equation for calculating h

$$(1 - y)Y_1 = (1 - y)Y_2, \tag{9.35}$$

$$y = 1 - h, Y_1 = m\gamma^{-1}\exp\frac{y\beta}{y - \gamma\beta}(1 - \exp(-m))^{-1}, Y_2 = 1 - e^{-y/\gamma}. \tag{9.36}$$

Equation (9.35) has one root y = 1, i. e. h = 0. However, this root should be discarded, since it leads to an infinite value of the combustion wave speed. The root of equation (9.35), which has a physical meaning, is determined by the intersection of the functions Y_1, Y_2 in the range from 0 to 1 (0 < y < 1). From the construction of graphs of these functions (Figure 2), it follows that such an intersection exists and is unique

$$1 - \exp(-1/\gamma) > m\gamma^{-1}\exp\frac{\beta}{1 - \beta\gamma}(1 - e^{-m})^{-1}. \tag{9.37}$$

Figure 2. Graphs of the functions $Y_1(y)$, $Y_2(y)$

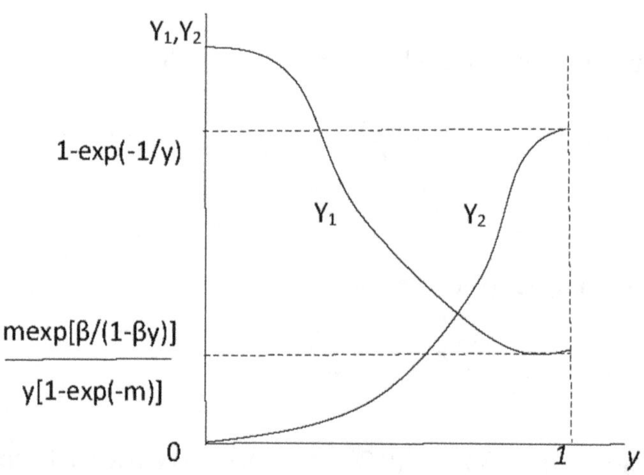

Equations (9.34), (9.35) in the absence of auto-braking go (after deleting the factors containing m) into equations for calculating the temperature of the wave front and its velocity, obtained earlier (Zhizhin, 1992) by the semi-infinite reaction zone method for first-order exothermic reactions. In the same way, from (9.24), (9.25), taking into account (9.30), equations are obtained for calculating the temperature profiles and the depth of chemical transformation in waves of first-order chemical reactions.

For further numerical investigation, it is convenient to introduce control parameters $\sigma = 1 + T_0\bar{c}/Q, \nu = RQ/(\bar{c}E)$. Then inequality (9.37) can be written as

$$\nu\sigma^2(1 - e^{-\nu^{-1}\sigma^{-2}})e^{-(\nu\sigma)^{-1}(\sigma-1)^{-1}} > m(1 - e^{-m})^{-1}. \quad (9.38)$$

The region of existence of the wave solution can be conveniently studied using the $T_1(\sigma, \nu) - T_2(m)$ diagram, where T_1 and T_2 are respectively the left and right parts of the inequality (9.38). Figure 3 shows the results of calculating the $T_1(\sigma)$ function for three values of the parameter ν (0.5, 1, 2) and the graph of the $T_2(m)$ function.

Figure 3. Definition of areas of existence of wave solutions ν = 0.5 - 1; ν =1 - 2; ν = 2- 3

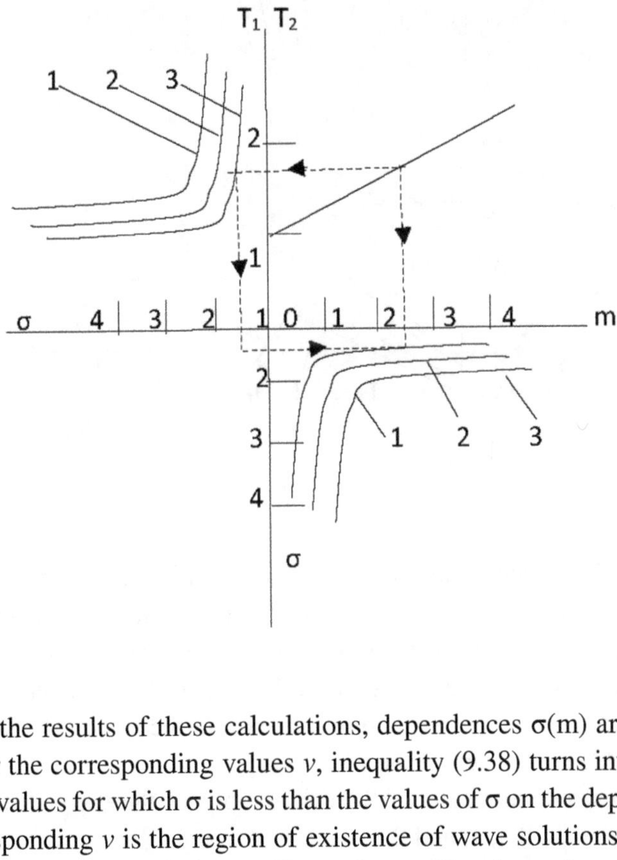

Based on the results of these calculations, dependences σ(m) are constructed, on which, for the corresponding values ν, inequality (9.38) turns into an equality. The range of values for which σ is less than the values of σ on the dependence σ(m) for the corresponding ν is the region of existence of wave solutions for a given ν. It follows from Figure 3 that the smaller ν, the smaller the region of existence of solutions. An increase in m (increasing auto-braking) reduces the range of values of the control parameter σ, for which the wave solution exists.

Figure 4 gives as an example the calculation results for the combustion wave velocity *w* and the magnitude of the source cutoff *h* for $\nu = 1$ and m = 0.5, 1, and 2 versus σ. It is evident that for any m, an increase in the initial temperature (i.e., σ) leads to a decrease in h, i.e., to the wave -front temperature approaching the initial temperature. The wave velocity thus increases. With approach to the boundary of the existence region of the wave solutions, the value h tends to the zero and the wave velocity increases without bound.

Figure 4. Wave velocity and the magnitude of the source cutoff versus the controlling parameter σ curve 1refers to ν = 1and m = 0.5; curve 2 refers to ν = 1 and m = 1; curves 3 refers to ν = 1 and m = 2, curve 4 refers to ν = 2 and m = 1

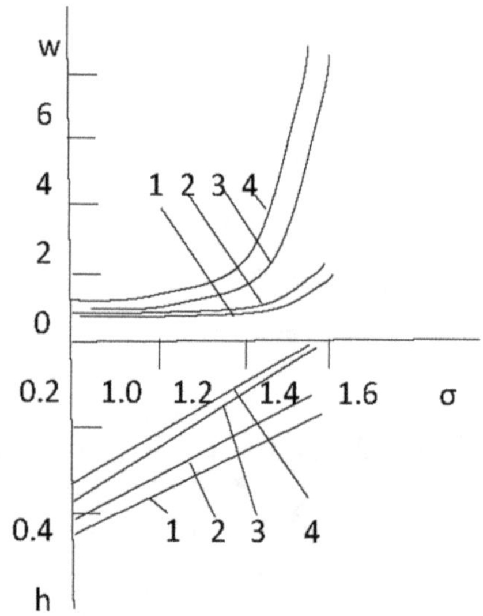

An increase in the autoinhibition parameter m increases the wave velocity and decreases the magnitude of the source cutoff. Growth in the parameter *ν*, i.e., the reaction activation energy leads to an increase in the wave velocity and a decrease in the magnitude of the source cutoff (curve 4 in Figure 4). A decrease in the source cutoff magnitude and the corresponding increase in the wave velocity are accompanied by a sudden extension of the wave.

Figure 5 presents the results of the calculations of the wave structure for σ = 1.6; *ν* = 1 and m = 0, 0.5, 1, and 2.

Figure 5. Structure of solid-flame combustion waves for m = 0 (curve 1), 0.5 (curve 2), 1 (curve 3), and 2 (curve 4), solid curves refer to η and dashed curves refer to ϑ

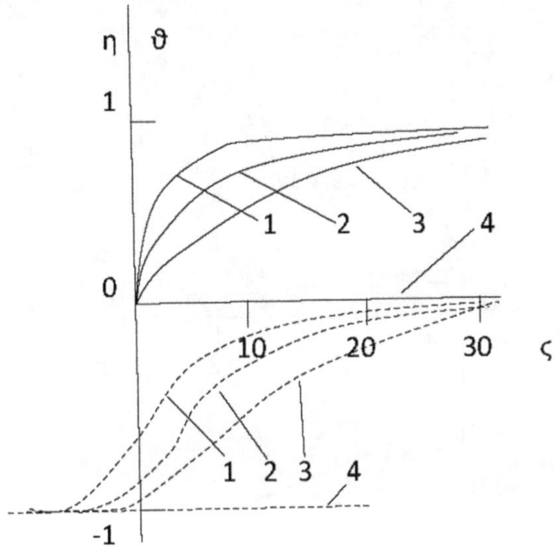

From Figure 5 it follows that despite the increase in the wave velocity with enhancement of autoinhibition, the wave is stretched so severely, that it can be used only for samples of very great length. In confined samples, the degree of chemical conversion is insignificant (curve 4 in Figure 5).

For comparison, we employ the narrow reaction-zone method to approximately calculate the velocity of a solid-flame combustion wave with exponential inhibition and a variable reaction surface. We divide equations (9.4) and (9.16) by each other with allowance for (9.26) and (9.27)

$$\frac{d\eta}{d\vartheta} = \frac{1-\eta}{\vartheta + 1 - \eta} w^{-2} exp\left(-m\eta - \frac{1}{\nu(\vartheta + \sigma)}\right). \tag{9.39}$$

In the denominator of the right side (9.39), we set $\vartheta = 0$ (i.e., $T = T_+$); then (9.39) is easily integrated. As a result, we obtain

$$w^2 = \nu\sigma^2 m(\exp m - 1)^{-1}\exp(-1/(\nu\sigma)). \tag{9.40}$$

Figure 6 gives calculated velocity of solid – flame combustion waves versus the autoinhibition parameter using the semi-infinite reaction zone method (curve 1) and the narrow reaction – zone method (curve 2) for σ = 1.6 and ν = 1.

Figure 6. Velocity and temperature of the wave front in solid-flame combustion versus the autoinhibition parameter

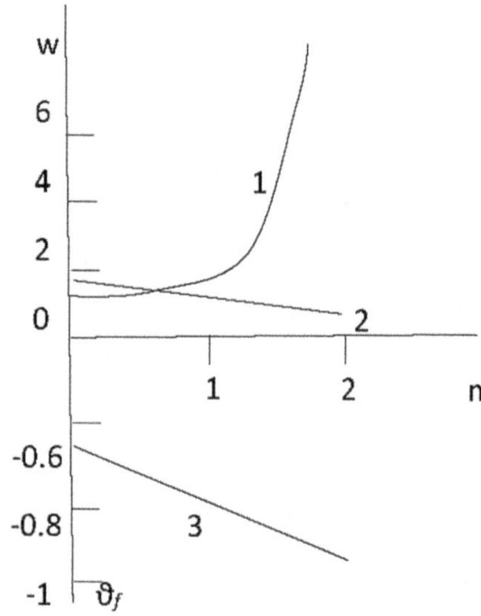

Curve 1 refers to the wave velocity determined by the semi-infinite reaction-zone method; curve 2 refers to the wave velocity determined by the narrow-reaction zone method; curve 3 refers to the wave-front temperature determined by the semi-infinite reaction-zone method and numerical integration of the differential equation (9.41)

It is obvious that the dependences of the wave velocity on the autoinhibition parameter computed using these methods are qualitatively different in nature and as m increases, the difference in the velocity values becomes very large.

To validate the semi-infinite reaction-zone method, we integrated equation (9.41) numerically. The calculation was begun the singular point in the natural direction of this point

$$\frac{d\eta}{d\vartheta} = 1 + w^{-2} exp(-m - v^{-1}\sigma^{-1}) \qquad (9.41)$$

and was continued until η vanished (Zhizhin, Poritskaya, 1994).

The value of ϑ for $\eta = 0$ corresponds to the wave-front temperature. From equations (9.39) and (9.41) specifying the value of the velocity w found by the semi-infinite reaction-zone method for definite values of m, ϑ, and v, one can calculate the wave front temperature from the differential equation (9.39) and compare it with the wave-front temperature obtained from the approximated analytical solution. Figure 6 gives calculated wave-front temperature (curve 3) for $v = 1$ and $\sigma = 1.6$ versus the parameter m. Curve 3 corresponds to the results of calculation using both the differential equation (9.39) and the algebraic equation (9.35), which are indistinguishable in the scale of the figure. We note that with increase in m and the corresponding increase in w, the natural direction (9.41) approaches 1, i.e., the zero isocline of equation (9.6) (the dashed straight line in Figure 1). Under these conditions, computer calculations of equation (9.41) become inconvenient because any finite variation in the phase coordinates can lead to a passage through the zero isocline and a change in the sign of the derivative $d\eta/d\vartheta$. However, in the exact solution, the sign of the derivative $d\eta/d\vartheta$ does not change if the solution passes through the singular point b. In this case, after exit from the point b in the natural direction, it is necessary to pass from numerical integration of (9.39) to calculation by the equation $\vartheta + 1 - \eta = w^{-2}\exp(-\nu^{-1}(\vartheta + \sigma)^{-1})$. It describes the solution on the left in the neighborhood of the zero isocline.

CONCLUSIONS

A mathematical model for an ideal solid-flame combustion with a variable chemical-reaction surface was constructed. The studies performed show that the solutions obtained adequately describe the effect of autoinhibition on the main characteristics of propagation of solid-flame combustion waves. It should be noted that a steady-state solid-flame combustion wave exists for limited values of the parameter m. Analysis of the data in Figure 9.3 leads to the approximate inequality necessary for the existence of steady-state wave: $m \leq (\sigma - 1)^{-1}\nu^{-1} = E(RT_0)^{-1}$. This is a fundamental feature that distinguishes the proposed model from the model of A.P. Aldushin et al. (Aldushin, Martem`yanova, Merzhanov, Khaykin, Shkadinsky, 1972), which ignores the variation in chemical reaction surface with time. It is noteworthy that in the article by these authors, an asymptotic expression for the wave velocity was obtained for large values of the parameter m. Such an opposite result indicates the importance of taking into account the length of the chemical reaction zone and the real variable surface area of chemical interaction in the analysis of processes in disperse systems. Ignoring these patterns can lead to fundamentally incorrect results of the analysis.

REFERENCES

Aldushin, A.P., & Martemyanova, T.M., Merzhanov, A.G., Khaikin, B.I., Shkadinsky, K.G. (1972). Propagation of the front of an exothermic reaction in condensed mixtures during the interaction of components through a layer of a refractory product. *Physics of combustion and explosion,* 2, 202 - 212.

Brown, M., Dollimore, D., & Galvey, A. (1980). *Reaction in the Solid State.* Elsevier.

Merzhanov, A. G. (2000). *Solid – Flame Cjmbustion* [in Russian]. Izd. ISMAN.

Merzhanov, A. G. (2005). *Better to be needed than free....* Chernogolovka: Territory. Zel'dovich, Ya. B., & Kamenetskii, D.A. Theory of thermal propagation of flames. (1938). *Zh. Fiz. Khim, 12*(1), 100–105.

Merzhanov, A.G., & Borovinskaya, I.P., Shkiro, V.M. (1967). *The phenomenon of wave localization of self-braking solid-phase reactions. Diploma No. 287.* USSR. Priority from 07/05/1967. Bulletin of Inventions, 1984, No. 32, 3. Bulletin of the Academy of Sciences of the USSR, 1984, 10, 141.

Merzhanov, A. G., Borovinskaya, I. P., & Shkiro, V. M. (1972). *France Patent N° 2088668.*

Merzhanov, A. G., Borovinskaya, I. P., & Shkiro, V. M. (1973). *US Patent N°3726643.*

Merzhanov, A.G., & Borovinskaya, I.P., Shkiro, V.M. (1974). *England Patent N°1321084ю*

Merzhanov, A. G., Borovinskaya, I. P., & Shkiro, V. M. (1982). *Japan Patent N°1098839.*

Zhizhin, G. V. (1992). *Macrokinetics in Frontal – Polymerization Reactors.* Politechnika.

Zhizhin, G. V. (1997). *Mathematical Models for Radical- Polymerization Waves.* Izd. CZPI.

Zhizhin, G. V. (2004). Model of an Ideal Solid-Flame Combustion Wave with a Variable Chemical -Reaction Surface. *Combustion, Explosion, and Shock Waves, 40*(1), 85–91. DOI: 10.1023/B:CESW.0000013671.95999.6c

Zhizhin, G. V., & Poritskaya, I. Y. (1994). Self-regulating waves for order n exothermic reactions in condensed media. *Combustion, Explosion, and Shock Waves, 30*(6), 781–787. DOI: 10.1007/BF00755250

Chapter 10
Gunpowder Burning Wave

ABSTRACT

An essentially non-stationary model of gunpowder combustion is constructed. The model uses empirical dependences of the propellant burning rate under stationary conditions on the initial temperature and the temperature of the hot end of the condensed phase. A non-stationary relaxation equation for the temperature gradient at the hot end of the condensed phase is introduced, based on the experimental laws of gunpowder combustion. A closed system of differential equations has been constructed that describes non-stationary modes of powder combustion, which has equilibrium positions corresponding to stationary modes of powder combustion. The system of differential equations has been qualitatively studied in the phase space. The conditions for the transition between equilibrium positions are determined. Regimes with a critical burning rate of gunpowder have been found. The conditions for the existence of the combustion of gunpowder with supercritical velocities are investigated.

INTRODUCTION

Gunpowder or solid rocket fuels belong to the class of condensed substances, the richest in chemical energy that can be converted into heat. After the initiation of combustion, a wave of chemical transformations propagates through the powder, in which it first warms up to a temperature that provides a sufficiently high rate of the gasification reaction of the solid, and then the gasification products react in the gas phase to form the final combustion products of the powder. To build an adequate theory of gunpowder combustion, it is necessary to have basic experimental information about the structure of the gunpowder combustion wave. Various issues of gunpowder combustion have been experimentally investigated in a number of works

DOI: 10.4018/978-1-6684-7745-8.ch010

(see, for example, Andreev, 1966; Zenin, 1966; Zenin, Nefedova, 1968; Belyaev et al., 1973; Zenin, 1973, 1980).

In a number of well-known works (Zeldovich, 1942; Zeldovich, Leipunsky, Librovich, 1975; Novozhilov, 1965, 1966, 1973, 2004), a theory of gunpowder combustion was constructed, in which the main assumptions were assumptions about the limiting effect on the combustion of gunpowder of thermal processes in the condensed phase and on the possibility of using a stationary (Mikhelson) expression for the temperature gradient in the condensed phase at its hot end when studying non-stationary modes of combustion of gunpowder. Using the analysis of the evolution of small perturbations of stationary combustion regimes in a linear approximation, the stability of the combustion of gunpowder was studied (Novozhilov, 1965, 1966, 1973) and, at the same time, it was noted that the linear approximation significantly limits the study, primarily due to the presence of a nonlinear convective term in the heat conduction equation. The method of moments with the help of a numerical solution of the heat equation was used to study non-stationary modes of gunpowder combustion at small deviations of the system from stationary states (Novozhilov, 2004). The non-stationary heat conduction equation was reduced to an infinite system of ordinary differential equations, and it was found that in a certain region of the problem parameters, there are regimes of chaotization of the gunpowder burning rate. In this chapter, a gunpowder combustion model is constructed that describes the combustion processes at any distance from stationary modes. The gunpowder combustion equations are reduced to an autonomous system of two first-order differential equations with an independent time variable, i.e. to the usual form in the study of dynamical systems. This allows one to represent various non-stationary modes of powder combustion by trajectories in the phase plane, to investigate the behavior of trajectories not only in the vicinity of singular points - stationary modes, but also far from them, and also to determine the nature of the change in the variety of equilibrium positions with a change in the parameters of the problem. Such a statement naturally required abandoning the use of the stationary Mikhelson expression for the temperature gradient at the hot end of the condensed phase. One of the equations of the system is a differential equation for the burning rate. It is obtained by integrating the non-stationary heat conduction equation. As the second equation of the system, a relaxation equation for the temperature gradient at the hot end of the condensed phase is proposed, generalizing previous studies on the dynamics of gunpowder combustion (Zeldovich, 1942; Zeldovich, Leipunsky, Librovich, 1975; Novozhilov, 1965, 1966, 1973, 2004). As a result of studying the obtained model, it was found that, in addition to two singular points corresponding to stationary combustion modes with a higher and lower burning rate and changing their stability depending on the values of the problem parameters, the system has a line of singular points in the phase plane, in which one of the phase velocities has discontinuity $\pm\infty$. The points of this line correspond to the limit states of the system with a critical burning rate. Limiting states arise either at the initial moment of combustion of the condensed phase, or at the moment of its complete combustion. It is essential that the line of singular points destroys possible modes of velocity chaotization and has a singular point through which there is a continuous transition when the sign of the difference between the heat flux entering the condensed phase from the reaction zone and the convective heat flux through the condensed phase changes.

10.1. MATHEMATICAL MODEL OF AN ESSENTIALLY UNSTEADY PROPELLANT COMBUSTION WAVE

Due to the assumption that the processes of heat transfer in the condensed phase have a limiting effect on the entire combustion process of gunpowder (Zeldovich, 1942), the main equation for describing the combustion of gunpowder is the non-stationary equation of heat conduction in the condensed phase, which, in a one-dimensional approximation, in coordinates associated with the

$$\rho c\frac{\partial T}{\partial t} = \lambda\frac{\partial^2 T}{\partial x^2} - \rho u(t)c\frac{\partial T}{\partial x}. \tag{10.1}$$

Here t is time; x is the spatial coordinate; c, ρ, λ are the assumed constant specific heat capacity, density and thermal conductivity of the condensed phase; u(t) is the velocity of the condensed phase in the direction of the x axis, which is equal in absolute value to the velocity of the hot end of the condensed phase due to burnout in the stationary system of coordinates (we will call it the

We assume that, in the general case, the condensed phase has an infinite extension to the region of negative values of the x coordinate, and the zero value of this coordinate is associated with the hot end of the condensed phase, and the x axis is directed from the end of the condensed phase to the outer region.

Solutions to equation (10.1) must satisfy the boundary conditions

$$T|_{x=0} = T_h(t), \frac{\partial T}{\partial x}\Big|_{x=0} = \varphi(t), T|_{x=-\infty} = T_0, \frac{\partial T}{\partial x}\Big|_{x=-\infty} = 0. \tag{10.2}$$

The peculiarities of the boundary conditions (10.2) are that the functions are unknown functions and they must be determined as a result of solving the problem. Moreover, these functions must be found from the solutions of equation (10.1) using experimental data on the combustion of gunpowder under stationary conditions (Novozhilov, 1965, 1966, 1973). In particular, these data can be presented as dependencies (Novozhilov, 2004)

$$u^s = B\exp(\beta_1 T_h^s), \tag{10.3}$$

$$u^s = A\exp(\beta T_0), \tag{10.4}$$

where A, B, β, β_1 are the experimentally determined parameters, and the index s indicates that the values of the combustion rate and the temperature of the hot end of the condensed phase correspond to the stationary combustion regime. According to the works of B.V. Novozhilov (Novozhilov, 1966, 1967, 1973), the empirical values of the coefficients included in (10.3), (10.4) best accumulate the complex kinetics of gasification processes in the boundary layer between the purely condensed and gas phases, as well as chemical reactions in the gas and condensed phases. We emphasize that when analyzing the combustion of gunpowder, it makes no sense to introduce ideas about a narrow reaction zone, since, in principle, chemical reactions can occur in both phases.

Equation (10.1) in stationary conditions has the form

$$\lambda\frac{d^2T^s}{dx^2} - \rho c u^s\frac{dT^s}{dx} = 0 \tag{10.5}$$

and its solution is well known (Michelson's solution)

$$T^s = T_0 + (T_h^s - T_0)\exp(x\rho c u^s/\lambda). \tag{10.6}$$

Under stationary conditions, the values of temperature and the derivative of temperature with respect to the spatial variable at the hot end of the condensed phase are related by the equality

$$\rho u^s c(T_h^s - T_0) = \lambda\varphi^s. \tag{10.7}$$

This means that the heat flux entering the condensed phase through the hot end is completely spent on increasing the temperature of the condensed phase from the initial T_0 to the final one T_h^s. According to Ya.B. Zel'dovich (Zel'dovich, 1942), this equality can also be used in non-stationary conditions. However, in non-stationary processes, strictly speaking, instead of equality (10.7), one should use the equality of heat fluxes from both sides of the hot end of the condensed phase, which is performed at any time, i. e. the equality of heat fluxes leaving the zone of kinetic processes and entering the condensed phase. Let us show that the use of equality (10.7) in non-stationary processes leads to a contradiction.

Let's introduce a variable

$$y(t) = \int_{-\infty}^{0} (T(t,x) - T_0)\,dx. \quad (10.8)$$

Integrating (10.1) over x, using (10.2), we get

$$\rho c \frac{dy}{dt} = -\rho u c(T_h - T_0) + \lambda\varphi. \quad (10.9)$$

If we use equality (10.7) under non-stationary conditions, then from (10.9) it follows that the variable *y* does not depend on time, but this contradicts the definition (10.8) *y* as a function of time in a non-stationary process. The value of a variable *y* in a stationary process can be easily found from (10.8) using (10.7)

$$y^s = (T_h^{\,s} - T_0)\lambda/(\rho u^s c). \quad (10.10)$$

The variable *y* by definition (10.8) at each moment of time *t* in the plane (*T*, *x*) is the area (Figure 1), bounded by the *y*-axis, the function *T(x)* and the straight - line $T = T_0$.

Figure 1. Graphical definition of variables y, y^s

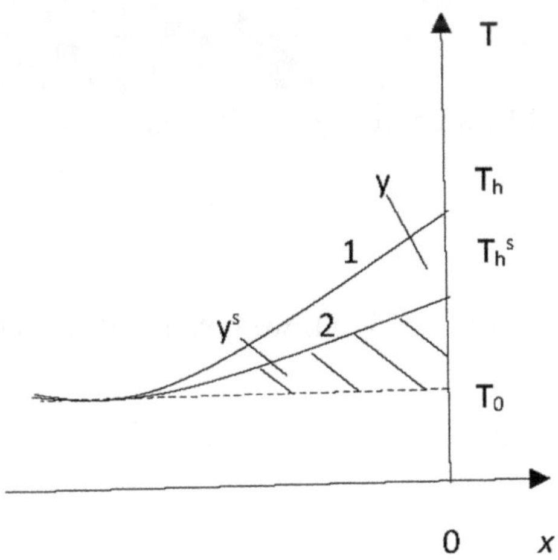

1 – function T(x); 2 - Michelson's solution (10.6)

Accordingly, the quantity y^s is the area limited by the Michelson solution (10.6), the y-axis and the straight line $T = T_0$. Because of this, taking into account that the difference between the temperatures T and T^s decreases with distance from the hot end of the condensed phase along the x coordinate exponentially with a characteristic length $\lambda/(\rho uc)$ (since the process is nonstationary), we represent y as a sum

$$y = y^s + \int_{-\infty}^{0} (T_h - T_h^s)\exp(x\rho uc/\lambda)\,dx. \tag{10.11}$$

Calculating the integral in (10.11), we have

$$y = y^s + (T_h - T_h^s)\lambda/(\rho uc). \tag{10.12}$$

Differentiating (10.12) with respect to t, we obtain

$$\frac{dy}{dt} = \lambda/(\rho uc)\frac{dT_h}{dt} - (T_h - T_h^s)\lambda(\rho u^2 c)^{-1}\frac{du}{dt}. \tag{10.13}$$

Since the burning rate is mainly determined by the maximum temperature of the condensed phase, i. e. temperature at the hot end of the condensed phase, it is reasonable to assume that dependence (10.3) can also be used for non-stationary processes

$$u = B\exp(\beta_1 T_h). \tag{10.14}$$

Differentiating (10.14), we have

$$\frac{dT_h}{dt} = (\beta_1 u)^{-1}\frac{du}{dt}. \tag{10.15}$$

Comparing (10.9), (10.13), (10.15), we obtain a differential equation for the burning rate

$$\frac{du}{udt} = \frac{\lambda(\rho c)^{-1}\varphi - u(T_h - T_0)}{\lambda(\rho uc)^{-1}(\beta_1^{-1} - T_h - T_h^s)}, \tag{10.16}$$

in which T_h, according to (10.14) is a function of the burning rate u.

Equation (10.16) contains the variable φ - the temperature gradient at the hot end of the condensed phase. By virtue of the rejection of the stationary condition (10.7), in order to close the system of equations, it is necessary to have an equation relating the non-stationary values of the temperature gradient at the hot end of the condensed phase and the combustion rate u. Before obtaining such an equation, we first turn again to the analysis of the stationary combustion regime.

Eliminate from equality (10.7) T_h^s, T_0 with the help of (10.3), (10.4)

$$\varphi^s = \rho c u^s(\lambda\beta)^{-1}(\ln m - (1-r)\ln(u^s/B)), m = A/B, r = \beta/\beta_1. \quad (10.17)$$

Since usually (Novozhilov, 1965, 1966, 1973) r < 1, then from (10.3), (10.4), taking into account that $T_h^s \geq T_0$, it follows that m > 1. The graph of equation (10.17) (Figure 2) is a line in the plane (ϕ^s, u^s) with a maximum passing through two points on the axis

u^s: $a_1(\rho cB\lambda^{-1}\beta^{-1}\ln m; B), a_2(\rho cB\lambda^{-1}\beta^{-1}\ln m; Bm^{1/(1-r)})$.

Figure 2. The graph of equation (10.17)

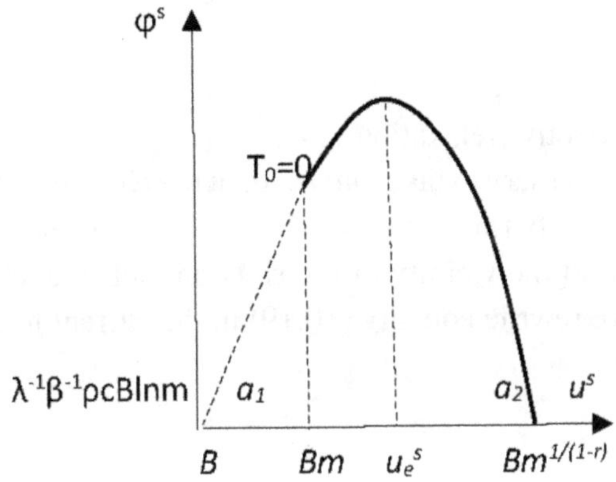

Thus, each value φ^s below the maximum corresponds to two speed values. Differentiating (10.17) you can find the coordinates of the maximum

$$u_e^s = (A/B)^{1/(1-r)}e^{-1}, \varphi_e^s = u_e^s(1-r)\rho c/(\lambda\beta).$$

As shown by linear analysis (Novozhilov, 1966, 1967, 1973), out of two stationary solutions corresponding to a certain value of the gradient φ^s, the solution with a lower burning rate is unstable, and the stationary solution with a high burning rate is stable, i.e. the temperature gradient φ returns to a stationary value φ^s when the right branch of the function $\varphi^s(u^s)$ deviates from it, and φ leaves the stationary value φ^s when it deviates from it on the left branch of the function $\varphi^s(u^s)$. However, it should be noted that when moving along the graph of the function $\varphi^s(u^s)$, in accordance with (10.3), (10.4), the temperatures $T_h^{\ s}, T_0$ change (they decrease with decreasing u^s). Moreover, with a value $u^s=Bm$ on the left branch of the function $\varphi^s(u^s)$ $T_0 = 0$. Thus, only the part of the dependence $\varphi^s(u^s)$ that lies at $u^s>Bm$ has physical significance (the solid part of the graph $\varphi^s(u^s)$ in Figure 2). The parameter $k = \beta(T_h^{\ s} - T_0)$ (10.18) used in the theory of gunpowder combustion (Zeldovich, 1942) also varies along the function $\varphi^s(u^s)$. The parameter k decreases with increasing u^s and is equal to 1- r at $u^s = u_e^s$, i.e., $k > 1 - r$, if $Bm < u^s < u_e^s$, and $k < 1 - r$ if it is $Bm^{1/(1-r)} > u^s > u_e^s$. The nature of the change in the temperature gradient at the hot end of the condensed phase φ upon deviation from stationary values can be reflected by the relaxation equation, which is often used in problems of physics and mechanics (see, for example, Zel'dovich and Raiser, 1966). In its simplest form, the relaxation equation for the temperature gradient at the hot end of the condensed phase can be written as

$$\frac{d\varphi}{dt} = \frac{\varphi^s - \varphi}{\tau_f}. \tag{10.19}$$

Here τ_f is some effective relaxation time.

In this case, it is necessary to take into account the change in the sign of the right side of (10.19) when the burning rate passes through the value u^s (the nature of the change φ is different in the vicinity of the right and left branches of the function $\varphi^s(u^s)$). Therefore, we rewrite equality (10.19) in the current form

$$\frac{d\varphi}{dt} = \frac{u - u_e^s}{u_e^s}\frac{\varphi^s - \varphi}{\tau_0}. \tag{10.20}$$

Here $\tau_0=\lambda/(\rho cB^2)$ is the simplest parameter with the dimension of time, which can be formed from the constant parameters of the problem.

The value on the right side of (10.20) can be expressed in terms of constant parameters using (10.18), (10.6), (10.3), (10.4)

$$\varphi^s = \rho ckB\lambda^{-1}\beta^{-1}e^{-k/(1-r)}m^{1/(1-r)}. \tag{10.21}$$

10.2. QUALITATIVE STUDY OF AN ESSENTIALLY NON-STATIONARY MODEL OF GUNPOWDER COMBUSTION IN THE PHASE PLANE

Equations (10.16), (10.20) represent the desired closed system of equations describing non-stationary modes of powder combustion. Assuming that this system qualitatively correctly describes non-stationary regimes of gunpowder combustion, we explore possible solutions to this system. Let us introduce dimensionless variables: dimensionless burning rate $w = \ln\left(\frac{u}{Bm^{\frac{1}{1-r}}}\right)$, dimensionless time $\tau = t\rho cB^2/\lambda$, dimensionless temperature gradient at the hot end of the condensed phase

$$F = \varphi\lambda\beta_1 r(\rho cBkm^{1/(1-r)})^{-1} e^{k/(1-r)}.$$

Then system (10.16), (10.20) takes the form

$$\frac{dw}{d\tau} = m^{2/(1-r)} e^{w} \frac{Fkr^{-1}e^{-k/(1-r)} - e^{w}(w + kr^{-1}(1-r)^{-1})}{1 - k(1-r)^{-1} - w}, \tag{10.22}$$

$$\frac{dF}{d\tau} = (1 - F)(e^{1+w} - 1). \tag{10.23}$$

System (10.22), (10.23) is autonomous and its solutions can be represented by trajectories in the phase plane (F,w). The location of the trajectories on the phase plane is determined by zero isoclines:

line

$$F = rk^{-1}e^{w+k/(1-r)}(w + kr^{-1}(1-r)^{-1})$$

, on which $dw/d\tau=0$ (10.24)

lines $w = -1$, $F = 1$, on which $dF/d\tau=0$

In addition, on the phase plane(F,w) there is, as follows from equation (10.22), a line of singular points

$$w=1 - k(1-r)^{-1}=w_*, \tag{10.25}$$

in which the derivative $dw/d\tau$ has a discontinuity $\pm\infty$.

The points of intersection of zero isoclines are the equilibrium positions of the system, corresponding to stationary regimes of gunpowder combustion. There are two positions of equilibrium a)

$$w_a = -1, F_a = e^{-w_*}((1-r)^{-1} - rk^{-1}), \tag{10.26}$$

b)

$$(1-r)^{-1} + rk^{-1}w_b = e^{w_*-1-w_b}, F_b = 1. \tag{10.27}$$

In addition, on the zero isocline (10.24) there is a point c through which the line of singular points (10.25) passes

c)

$$w_c = w_*, F_c = (rk^{-1} + 1)e. \tag{10.28}$$

At this point, the derivative dw/dτ has uncertainty 0/0, i.e. it is a singular point not only in the phase plane, but also in the integral space. At point c, the uniqueness of both trajectories and integral curves is violated, and it is continuously traversable in the independent variable τ. Singular points of this kind are characteristic of dynamical systems of gas dynamics. It is through such points that the gas velocity passes through the local value of the sound velocity, i.e. there is a continuous transition through the Mach number equal to one (Zhizhin, 1972; Vulis, Gusika, Zhizhin, 1972; Kulikovsky, Slobodkina, 1967). In this case, the analogue of the characteristic dimensionless gas velocity is the characteristic dimensionless propellant burning rate, which is determined by the experimental laws of propellant combustion (10.3), (10.4).

Let us determine the eigenvalues of the singular points of the equilibrium positions a) and b). From the discriminant of the system (10.22), (10.23) in the vicinity of the point a) we find the eigenvalues

$$\xi_{a1,2} = -\frac{m^{2/(1-r)}k}{2(1-r)re^2(1+w_*)} \pm\sqrt{\frac{m^{4/(1-r)}k^2}{4(1-r)^2r^2e^4(1+w_*)^2} + \frac{m^{2/(1-r)}k}{r(1+w_*)e^2}(e^{w_*} + rk^{-1} - (1-r)^{-1})}. \tag{10.29}$$

It follows from (10.29) that at k < 1 - r the eigenvalues $\xi_{a1,2}$ are real and have different signs, i.e. the singular point a) has saddle type. Moreover, if k < r (1 – r), then $F_a < 0$ and the equilibrium position a) is absent in the physical part of the phase plane, since, by definition, F_a is always greater than zero. In the range 1 – r < k < 2(1 – r), the equilibrium position a) is stable. Moreover, for

$$\frac{m^{2/(1-r)}k}{4r(1-r)^{2}e^{2}(1+w_{*})} > (1-r)^{-1} - rk^{-1} - e^{w_{*}},$$

the eigenvalues $\xi_{a1,2}$ are real and negative, i.e. the equilibrium position a) has the type of a stable knot. If

$$\frac{m^{2/(1-r)}k}{4r(1-r)^{2}e^{2}(1+w_{*})} < (1-r)^{-1} - rk^{-1} - e^{w_{*}}$$

then the eigenvalues $\xi_{a1,2}$ are complex with a negative real part, i.e. the equilibrium position a) has the type of stable focus. If k > 2(1 – r) then the eigenvalues $\xi_{a1,2}$ are again real and of different signs, i.e., the equilibrium position a) again has the saddle type.

From the discriminant of system (10.22), (10.23) in a neighborhood of point b) we have

$$\xi_{b1} = 1 - e^{1+w_b}, \xi_{b2} = -m^{2/(1-r)}e^{2w_b}\frac{kr(1-r)^{-1} + w_b + 1}{w_* - w_b}, \tag{10.30}$$

moreover, as follows from (10.27), $w_b < 0$. Note that the value of the dimensionless burning rate of gunpowder, according to the definition of this rate, can be negative if the value under the logarithm is less than one. However, the value of the dimensional burning rate of gunpowder is naturally always positive or zero.

From (10.30) it follows that three ranges of the absolute value of the dimensionless burning rate of gunpowder of the dimensionless $|w_b|$ can be distinguished with different signs of the eigenvalues:

1) $0 < |w_b| < w_*$, here $\xi_{b1} < 0$, $\xi_{b2} < 0$, and the equilibrium position b is of the stable knot type ;
2) $0 < |w_b| < 1$, here $\xi_{b1} < 0$, $\xi_{b2} > 0$, and the equilibrium position b has the saddle type;
3) $1 < |w_b| < kr^{-1}(1-r)^{-1}+1$, here $\xi_{b1} < 0$, $\xi_{b2} > 0$, and the equilibrium position b is of the unstable knot type.

It can be shown that for $|w_b| > kr^{-1}(1-r)^{-1}+1$ equation (10.27) has no roots.

Consider now a passing singular point c. In this case (Zhizhin, 1972) it is necessary to reveal the uncertainty 0/0 of the derivative dw/dτ at the point c. Revealing this uncertainty according to L'Hopital's rule (Luzin, 1960), we find that there can be two eigendirections at the point c

$$\left(\frac{dw}{d\tau}\right)_{1,2} = -0.5K \pm \sqrt{0.25K^2 - m^{2/(1-r)} k r^{-1} e^{-1-2w_*}(e^{1+w_*} - 1)(1 - rk^{-1}e^{-1} - e^{-1})},$$

$$K = m^{2/(1-r)} e^{2w_*}(2 - kr^{-1}) - w_*. \qquad (10.31)$$

It follows from (10.31) that for r < (e – 1)(e – 0.5)$^{-1}$ the point c has a saddle type by the nature of the location of the trajectories in its neighborhood. However, the direction of change of the independent variable on the trajectories does not change when passing through the point c; the point depicting the state of the system, moving along the trajectory, does not asymptotically approach or move away from the point c, but passes through the point c at a finite time value.

At r > (e – 1)(e – 0.5)$^{-1}$ point c can be of the node, focus, or center type (K = 0). However, the difference between the focus and the center is not significant in this case, since all trajectories in the vicinity of the point c are cut by a line of singular points passing through the point c. On this line, the direction of change of the independent variable changes to the opposite at a finite time value. So, there is no kind of oscillatory processes in the vicinity of the point c. The existence of a line of singular points means the existence of a limiting critical speed for the burning of gunpowder $w=1-k(1-r)^{-1}=w_*$, which can be achieved with complete combustion of the condensed phase or at the beginning of its combustion, similar to the existence of a limiting critical speed of gas movement in channels equal to the local value of the speed of sound achieved at the ends of a channel of limited length (Vulis, 1946, 1947, 1950; Zhizhin, 2004, 2021). Physically, the limiting critical burning rate of gunpowder corresponds to such a burning rate at which an increase in the temperature of the condensed phase (for each value of the spatial coordinate) caused by an increase in the temperature of the hot end of the condensed phase is compensated by a decrease in the temperature of the condensed phase caused by an increase in the burning rate (equation (10.13)). In gas dynamics, this corresponds to the zero sum of external influences on the gas flow (Vulis, 1946, 1947, 1950; Zhizhin, 1972, 2004, 2021; Vulis, Gusika, Zhizhin, 1972). If, simultaneously with reaching zero of the sum of influences on the gas flow, the flow rate becomes equal to the critical one, then it becomes possible to continuously pass the flow rate through the critical value of the speed.

Summing up the results of the analysis of singular points, zero isoclines and lines of singular points, we can distinguish four qualitatively different options for the phase portrait of the system (10.22), (10.23), shown in Figure 3 a, b, c, d.

Figure 3. Phase portraits of the system (10.22), (10.23) for different values of the parameters

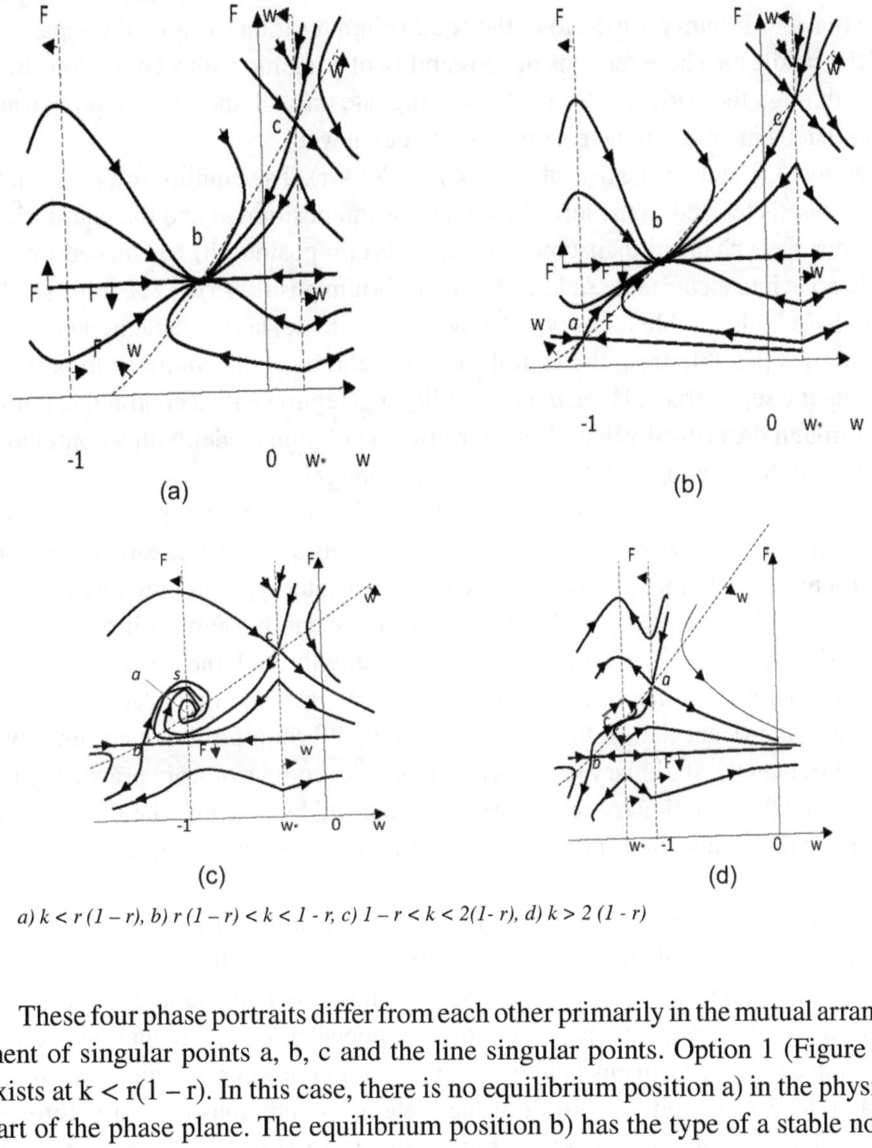

a) k < r (1 – r), b) r (1 – r) < k < 1 - r, c) 1 – r < k < 2(1- r), d) k > 2 (1 - r)

These four phase portraits differ from each other primarily in the mutual arrangement of singular points a, b, c and the line singular points. Option 1 (Figure 3a) exists at k < r(1 – r). In this case, there is no equilibrium position a) in the physical part of the phase plane. The equilibrium position b) has the type of a stable node, and the passing point c has the type of a saddle. One of the separatrices of the saddle ends at the equilibrium position b), passing continuously through the singular point c. The second separatrix of the saddle continuously passes through the value of the critical burning rate of gunpowder, continuing to move with an increase in the burning rate of gunpowder.

Option 2 (Figure 3b) exists at r (1 – r) < k < 1 - r. In this variant, in contrast to the previous variant, an equilibrium position a) has appeared, which has a saddle type located to the left of the equilibrium position b). Here, a transition is possible from the equilibrium position a) to the equilibrium position b) along the separatrix *s* of the saddle a). There remains the possibility of a regime with a continuous transition through the critical value of the burning rate through the passage point c and a continuous increase in the burning rate of gunpowder.

Option 3 (Figure 3c) exists at 1 – r < k < 2(1 – r). The equilibrium position b) in this case is located to the left of the equilibrium position a) and the equilibrium positions have changed their type. The equilibrium position b) has turned from a stable node into an unstable saddle. The equilibrium position a) has become stable. It can have both a stable focus type (shown in Figure 3c) and a stable node type. A transition is possible from the equilibrium position b) to the equilibrium position a) along the separatrix s. Here, the possibility of a regime with a continuous transition through the critical value of the burning rate of gunpowder with a continuous increase in this rate through the point c is also preserved.

Option 4 (Figure 3e) exists at k > 2(1 – r). In this case, the passing singular point c) is located between the equilibrium positions a) and b). It can have a knot type (depicted in Figure 3e) as well as a focus or center type (in previous versions it had a saddle type). It is possible to pass from the equilibrium position b) to the equilibrium position a) along the separatrix passing through the point c). There is no regime with a continuous increase in the burning rate of gunpowder in this case. It is also significant that in this variant, both equilibrium positions are unstable, and, consequently, stationary regimes of gunpowder combustion are practically not realized, and non-stationary processes are completed by reaching the limiting rate of gunpowder combustion at the moment of complete combustion of the condensed phase.

The transition from option 1 to option 4 of the phase portrait of the system is accompanied by a shift in the line of limiting states from the region of positive values of the dimensionless burning rate of gunpowder to the area of negative values of the dimensionless burning rate of gunpowder, i.e., a gradual increase in the parameter k. The dimensional burning rate of gunpowder remains positive. In variants 1 - 3 of the phase portrait of the system, a continuous transition through the limiting burning rate is possible both with an increase and with a decrease in the burning rate. In this case, a decrease in the temperature gradient at the hot end of the condensed phase necessarily occurs.

In variant 4, in which there are no stable stationary regimes of powder combustion, non-stationary processes are possible with a continuous transition from sub limiting combustion rates to super limiting combustion rates with a constant increase in the temperature gradient at the hot end of the condensed phase. The modes of chaotization of gunpowder burning rates found earlier in the quasi-stationary model (Novozhilov, 2004) correspond in terms of parameters to variant 4. However, as shown by this study of an essentially non-stationary model of gunpowder combustion, chaotization modes are impossible in this case. The main reason for the destruction of the modes of chaotization of the burning rate is the existence of the limiting critical burning rate of gunpowder established in this work. The variety of limiting states with a critical burning rate of gunpowder was not found in the work of B.V. Novozhilova. And it is of essential importance, since it breaks the solutions located between the stationary states.

CONCLUSION

Starting from the well-known work of Ya.B. Zeldovich, published in 1942 (Zeldovich, 1942), the idea of gunpowder combustion prevailed, according to which thermal processes in the condensed phase have a limiting effect on the combustion of gunpowder, and when studying non-stationary processes of gunpowder combustion, Michelson's stationary expression for the temperature gradient at the hot end of the condensed phase is applicable. In this case, the boundary conditions for the heat equation in the condensed phase must be specified from the results of the corresponding experiments. B.V. Novozhilov proposed to represent the results of these experiments in the form of empirical dependences (Novozhilov, 1966, 1967, 1973, 2004) of the stationary combustion rate of gunpowder on the initial temperature and the temperature of the hot end of the condensed phase. In this chapter, using the empirical dependences of the steady-state combustion rate of gunpowder, it is shown that the application of Michelson's stationary solution under non-stationary conditions of gunpowder combustion leads to a contradiction. To obtain a closed system of equations describing the non-stationary combustion of gunpowder, at an unknown gunpowder burning rate and an unknown value of the temperature gradient at the hot end of the condensed phase, a differential relaxation equation for the dimensionless temperature gradient at the hot end of the condensed phase is proposed, based on the qualitative nature of the empirical dependences of the gunpowder combustion process. The resulting system of differential equations is qualitatively investigated in the phase plane, in which there are two singular points - two equilibrium positions. The equilibrium positions of this phase plane correspond to the stationary regimes of gunpowder combustion. The study was carried out in the phase plane as a whole, i.e. not only in the vicinity of the equilibrium positions, but also at any distance from the equilibrium positions. This significantly distinguishes the study from previous studies limited to the neighborhood of equilibrium positions. This made it possible to determine the conditions for the transition from one stationary regime of gunpowder combustion to another and vice versa, as well as the conditions for changing the type of equilibrium positions and their stability. It is important that the study for the first time made it possible to establish the existence of a critical speed of burning gunpowder, corresponding to the critical speed of gas movement in the channels, corresponding to the achievement of the gas movement speed by the value of the local speed of sound in the flow. It has been established that during the combustion of gunpowder, a continuous transition through the critical value of the combustion rate is possible if the heat flux entering the condensed phase through its hot end and the convective heat flux passing through the condensed phase are simultaneously equal. The point of transition through the critical value of the burning rate of gunpowder is a singular point, at which there is an uncertainty zero to zero both in the phase and in the integral space. The possibility of passing through the critical speed of burning of gunpowder makes it possible to carry out combustion modes with a continuous increase in the speed of burning of gunpowder, which makes it possible to significantly increase the momentum created by the burning of gunpowder. The existence of the line of the critical burning rate of gunpowder in the phase plane cuts the trajectories of the process that do not take into account the existence of the critical burning rate. This proves that the chaotization regimes of the burning rate of gunpowder, established earlier, do not actually exist, since when the critical burning rate is reached, the movement along the trajectory stops, unless the state of the system corresponds to passable (singular) points.

REFERENCES

Andreev, K. K. (1966). *Thermal decomposition and combustion of explosives*. Nauka.

Belyaev, A. F., Bobolev, V. K., Korotkov, A. I., Sulimov, A. A., & Chuiko, S. V. (1973). *Transition of combustion of condensed systems into explosion*. Nauka Publishing House.

Kulikovsky, A. G., & Slobodkina, F. A. (1967). On the stability of arbitrary stationary flows in the vicinity of transition points through the speed of sound. *Applied Mathematics and Mechanics*, *31*(4), 5–21.

Luzin, N. N. (1966). *Differential Calculus*. Higher School.

Novozhilov, B. V. (1965). Criterion of stability of the stationary mode of combustion of gunpowder. *Journal of Applied Mechanics and Technical Physics*, *4*, 157–160.

Novozhilov, B. V. (1966). Nonlinear fluctuations in the burning rate of gunpowder. *Journal of Applied Mechanics and Technical Physics*, *5*, 31–41.

Novozhilov, B. V. (1973). *Unsteady combustion of solid rocket propellants*. Nauka.

Novozhilov, B. V. (2004). Chaotization of the non-stationary burning rate of gunpowder. *Chemical Physics*, *5*, 68–74.

Vulis, L.A. (1946). On the transition through the speed of sound in a gas flow. *Reports of the Academy of Sciences of the USSR*, LIV, 8, 669 - 672.

Vulis, L.A. (1947). On the law of reversal of influences in a real gas flow. *Reports of the Academy of Sciences of the USSR*, LVI, 8, 799 - 801.

Vulis, L. A. (1950). *Thermodynamics of gas flows*. Gosenergoizdat.

Vulis, L. A., Gusika, P. L., & Zhizhin, G. V. (1972). Two-phase flows in a channel of constant cross section (qualitative study). *Journal of Applied Mechanics and Technical Physics*, *5*, 143–156.

Zeldovich, Ya. B. (1942). On the theory of combustion of gunpowder and explosives. *Journal of Experimental and Theoretical Physics*, *12*(11-12), 498–524.

Zeldovich, Ya. B., Leipunsky, O. I., & Librovich, V. B. (1975). *The theory of non-stationary combustion of gunpowder*. Nauka.

Zeldovich, Ya.B., & Raiser, Yu.P. *Physics of shock waves and high-temperature hydrodynamic phenomena*. Moscow: Nauka.

Zenin, A.A. (1966). Formal kinetic characteristics of reactions occurring during the combustion of gunpowder. *Physics of combustion and explosion*. 2, 28 - 32.

Zenin, A. A. (1973). On one model of the reaction layer of the condensed phase of ballistic powder. *Reports of the Academy of Sciences of the USSR, 213*(6), 1357–1360.

Zenin, A. A. (1980). *Processes in the combustion zones of ballistic powders. Physical processes during combustion and explosion*. Atomizdat.

Zenin, A.A., & Nefedova, O.I. (1968). On the combustion of ballistic powder in a wide range of initial temperatures. *Physics of combustion and explosion.* 9(4), 343-349.

Zhizhin, G. V. (1972). *Qualitative study of one-dimensional steady flows*. Dissertation of the Candidate of Physical and Mathematical Sciences. Leningrad: Leningrad Polytechnic Institute.

Zhizhin, G.V. (2004). Synergetics of flows. *Life and safety. Ecology*, 2–3 a, 366–385.

Zhizhin, G. V. (2021). Synergetics in Gas Streams. *Journal of Nature Science and Technology, 15*(3), 127–144.

Chapter 11
Waves in Biological Populations

ABSTRACT

Self-regulating nonlinear waves in various biological populations are considered as moving attractors in excitable media. Mathematically, waves in populations are solutions of nonstationary parabolic systems of differential diffusion equations with source terms and the velocity of the wave is an eigenvalue of the problem, and its profile is an eigenvalue function of the problem. There is no general exact method for solving such a problem. An approximate method for its solution is proposed (the semi - infinite reaction zone method), which essentially reduces to solving an algebraic system of equations. This method is used to calculate the waves in various biological populations. It is shown that there are two types of waves: a wave of conquest and a solitary wave. In all cases considered, formulas for calculating the velocity of the wave and its profile were obtained. One of the important examples considered is the analysis of solitary waves in populations of the herd locust.

WAVES IN THE LOGISTIC BIOLOGICAL POPULATIONS

To study the waves of biological populations, one will use the semi - infinite reaction zone method developed by the author in the study of the polymerization waves and combustion waves (Zhizhin, 1982, 1988, 1992, 1997 a, b, 2004a, b, 2008; Zhizhin, & Poritskaya, 1994; Zhizhin, & Larina, 1994).

The nonstationary equation for the change in the concentration N of individuals in the logistic population in a one - dimensional range, taking into account the chaotic mobility μ of individuals, has the form (Svirezhev, 1987)

$$\frac{\partial N}{\partial t} = \mu \frac{\partial^2 N}{\partial x^2} + F(N). \tag{11.1}$$

DOI: 10.4018/978-1-6684-7745-8.ch011

Here $F(N)$ is the function describing the local law of the population growth, and

$$F(N) = (B - D)N, \quad (11.2)$$

where B and D are the fertility and mortality functions.

Equation (11.2) describes the fact that the local law of the population growth is determined by two processes — birth and death. In the model of the logistic population it is assumed that mortality D is a linearly increasing function of the concentration of individuals

$$D(N) = \alpha_1 + \alpha_2 N, \alpha_1 > 0, \alpha_2 > 0.$$

Here α_1 is natural mortality. Increase in mortality with increasing concentration N is due to increased competition with limited resources (food, space, etc.). In addition, it is assumed that the fertility function B it is determined only by the physiological limits of fertility and is independent of N, i.e. $B = m =$ constant, where m is the so - called natural fertility or fecundity. Given these assumptions and equation (11.2), equation (11.1) takes the form

$$\frac{\partial N}{\partial t} = \mu\frac{\partial^2 N}{\partial x^2} + \alpha N(1 - N/N_+). \quad (11.3)$$

Here $\alpha = m - \alpha_1$ is Malthusian parameter, and $N_+ = (m - \alpha_1)/\alpha_2$ is the capacity of the medium, i.e. maximum possible concentration of individuals in the environment.

If the competition between individuals of the population is completely absent ($\alpha_2 = 0$), $N_+ = \infty$, then the law of the local population growth takes a simple form $F(N) = \alpha N$. In this case, the population density increases indefinitely (the Malthus model of exponential growth) and the wave solution, as a transition from one equilibrium position to another, does not exist ($\alpha > 0$).

To find the stationary wave solution of equation (11.3), one can introduce the wave coordinate $z = ut - x$, here u is the wave velocity. One also introduces dimensionless variables

$$n = N/N_+, \varsigma = z(\alpha/\mu)^{1/2}, w = u(\alpha\mu)^{-1/2}.$$

Then equation (11.3) can be written as a system

$$\frac{dp}{d\varsigma} = wp - F(n), F(n) = (1 - n)n, \quad (11.4)$$

$$\frac{dn}{d\varsigma} = p. \quad (11.5)$$

The solutions of the autonomous system (11.4), (11.5) can be represented by trajectories in the phase plane (p,n). Their locations are determined by the zero isocline $p=n(1-n)/w$, on which $dp/d\varsigma=0$, the zero isocline $p = 0$, on which $dn/d\varsigma=0$, as well as by special points $a(0,0)$, $b(0,1)$. The eigenvalues in the special point a (the initial equilibrium position) are equal to the eigendirections at this point

$$\lambda_{1,2} = w/2 \pm \sqrt{w^2/4 - 1} = (dp/dn)_{1,2}, \tag{11.6}$$

and the eigenvalues at the special point b (the final equilibrium position) are equal to the eigendirections at the special point b

$$\lambda_{3,4} = w/2 \pm \sqrt{w^2/4 + 1} = (dp/dn)_{3,4}. \tag{11.7}$$

Let us consider in more detail than in the well-known work of A.N. Kolmogorov, I.G. Petrovsky, N.S. Piskunov (Kolmogorov, Petrovsky, Piskunov, 1937), the behavior of trajectories in the phase plane, omitting the already known details. From (11.7) it follows that the final equilibrium position is of a saddle type, since $\lambda_1>0$, $\lambda_2<0$

From (11.6) it follows that the initial equilibrium position can be of the type of an unstable node with $w\geq 2$ or a focus with $w < 2$. Since at $w < 2$ trajectories will fall into the region of negative concentrations, this area of parameters is not of interest. One confines ourselves to the values of speed $w\geq 2$. Then the point a has the type of an unstable node, since $\lambda_{1,2}>0$ and has two proper directions, and both are located in the quadrant $p>0$, $n>0$. The wave solution leaves the point a and enters point b, i.e. boundary conditions must be satisfied $p=n=0$ at $\varsigma\to-\infty$ and $p=0$, $n=1$ at $\varsigma\to+\infty$. Obviously, for any value of speed, there is a trajectory emanating from point a, and entering point b (Figure 1).

Figure 1. Phase diagram of waves in the logistic population

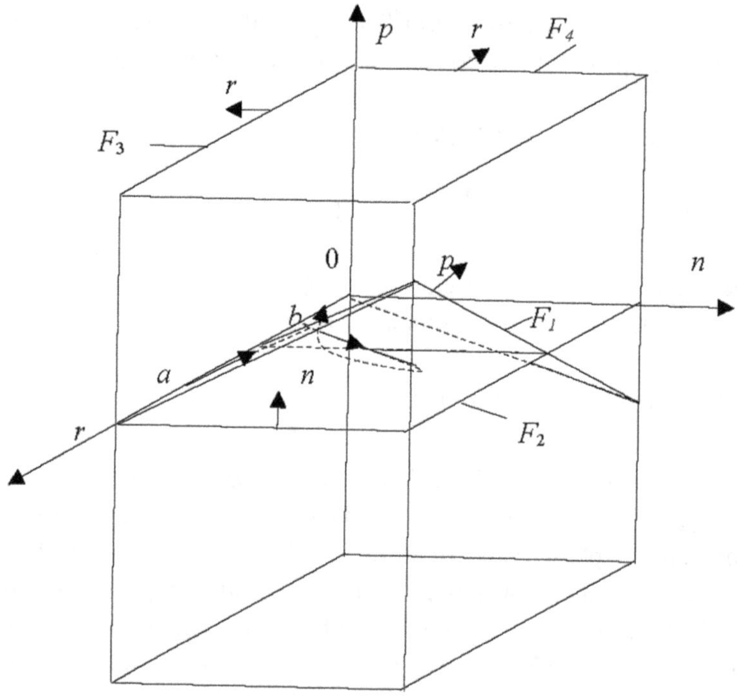

However, with which eigenvalues λ_1 or λ_2 the trajectory leaves point *a,* if it enters point *b*, is not known in advance, since both options may be. In the first case, the trajectories emanating from point *a* along its own direction 2 intersect the zero isocline $p=n(1-n)/w$ and, without entering point *b*, intersect the axis $p = 0$. In this case, one of the trajectories emanating from point *a* along its own direction 1 enters point *b*. In the second case, the trajectories emanating from point *a* in its own direction 1, not reaching point *b*, leave in their own direction3 points *b*. For $w=2$ the eigenvalues of the point *a* coincide; their own directions coincide, and the point *a* has the type of a degenerate unstable node. Figure 1 shows the case when the trajectory emanating from point *a* in its own direction 1 enters the special point *b* along the own direction 4. The corresponding integral curve is shown in Figure 2.

Figure 2. Wave profile of conquering the logistic population: n_f is the concentration of individuals at the wave front (ς=0), n_Δ is the concentration of individuals at the bound of zone 2 ($\varsigma=\Delta$)

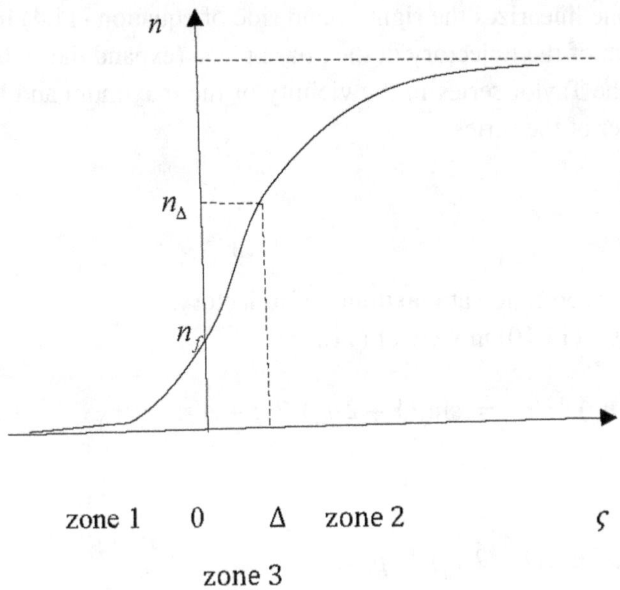

It represents the conquest wave profile from the wave, moving along the spatial coordinate x, with a speed u, fills the space with the maximum concentration of individuals N_+ (according to the introduced wave coordinate, the directions x and z are opposite). One divides the wave into three zones (Figure 2). In zone 1 (-∞<ς<0), one considers the solution of the system linearized in the vicinity of the initial equilibrium state a to be fair

$$n = n_f \exp(\lambda_{1,2}\varsigma). \tag{11.8}$$

Here n_f is the concentration of individuals at the wave front ς=0, $\lambda_{1,2}$ is the eigenvalue at the point a (which of them will be established later).

In zone 2 (Δ≤ς<+∞), one considers the solution of the system linearized in the vicinity of the final equilibrium state b to be valid

$$n = 1 - (1 - n_\Delta)\exp(\lambda_4(\varsigma - \Delta)). \tag{11.9}$$

The conditions in zone 1 and zone 2 ensure that the approximate solution of the system emerges from the initial equilibrium position and enters the final equilibrium position with almost the same eigenvalues and proper directions as the exact solution of system (11.4), (11.5).

In zone 3, one linearizes the right - hand side of equation (11.4) in the vicinity of the maximum of the trajectory in the phase plane (expand the right - hand side of (11.4) into the Taylor series in the vicinity of the maximum and limit the first nonzero member of the series)

$$\frac{dp}{d\varsigma} = (2n_m - 1)(n - n_m). \tag{11.10}$$

Here n_m is concentration at maximum of trajectory.

One integrates (11.10) in view of (11.5)

$$(n - n_m)(1 - 2n_m)^{1/2}/p_m = \sin((1 - 2n_m)^{1/2}\varsigma + A). \tag{11.11}$$

Here

$$A = \arcsin((n_f - n_m)(1 - 2n_m)^{1/2}/p_m), \tag{11.12}$$

$$p_m = n_m(1 - n_m)/w$$

is sense of p at maximum of trajectory. (11.13)

Put in (11.11) $\varsigma=\Delta$, then one has

$$(n_\Delta - n_m)(1 - 2n_m)^{1/2}/p_m = \sin((1 - 2n_m)^{1/2}\Delta + A). \tag{11.14}$$

Here n_Δ is sense of n at $\varsigma=\Delta$.

Differentiating (11.8), (11.9) once, one equate the first derivatives on the border of zones 1 - 3 and 2 – 3

$$n_f^2\lambda_{1,2}^2 = p_m^2 - (1 - 2n_m)(n_f - n_m)^2, \tag{11.15}$$

$$\lambda_4^2(1 - n_\Delta)^2 = p_m^2 - (1 - 2n_m)(n_\Delta - n_m)^2. \tag{11.16}$$

Differentiating (11.8), (11.9) twice, one equate the second derivatives on the border of zones 1 - 3 and 2 – 3

$$n_f\lambda_{1,2}^2 = (2n_m - 1)(n_f - n_m), \tag{11.17}$$

$$\lambda_4^2(1 - n_\Delta) = (1 - 2n_m)(n_\Delta - n_m). \qquad (11.18)$$

So, one has a system of nine algebraic equations (11.6), (11.7), (11.12) - (11.18) with respect to nine unknowns $n_f, n_m, n_\Delta, p_m, A, \lambda_{1,2}, \lambda_4, w, \Delta$, i.e. system is closed. Solving it, can get the equation to determine the speed of the wave

$$\lambda_4^2(4w^2(-K_{1,2} \pm T_{1,2} - 1)^{2} - (-K_{1,2} \pm T_{1,2})^{-1}) = 1. \qquad (11.19)$$

Here

$$K_{1,2} = 0.5(\lambda_4^2 + \lambda_{1,2}^2 - w^2\lambda_{1,2}^2 + w\lambda_4^2), T_{1,2} = (K_{1,2}^2 + (w^2 - 1)\lambda_{1,2}^2\lambda_4^2).$$

After finding the velocity in (11.19), the remaining parameters are determined by the equations already obtained.

The results of the calculation of the left side of equation (11.19) (*Y*) as functions of speed are shown in Figure 3.

Figure 3. Decision of equation (11.19)

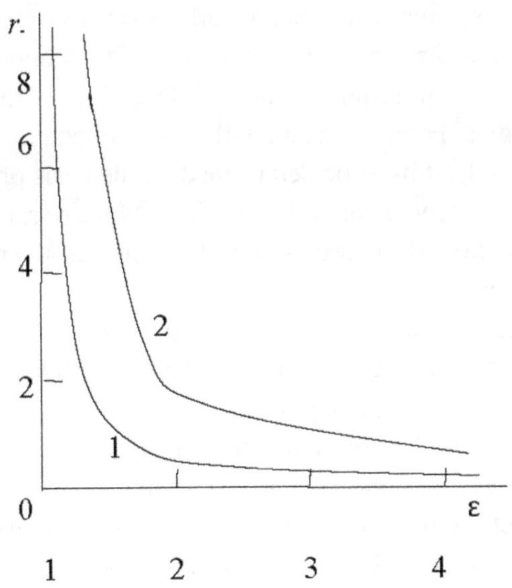

1 – Y(w,λ₁), 2 - Y (w,λ₂)

The branches of the function $Y(w)$, corresponding to the values of λ_1, are designated 1. The branches of function $Y(w)$, corresponding to the values of λ_2, are designated 2. The signs "+" and "-"at 1 and 2 correspond to the signs $T_{1,2}$. From

Figure 11.3 it follows that equation (11.19) satisfies only the branch $Y(w)$, corresponding to the eigenvalue λ_1 at the sign "+" at T_1. The value of speed satisfying equation (11.19) is 2.0015. It is very close to the speed value of 2, found by A.N. Kolmogorov, I.G. Petrovsky, & N.S. Piskunov as a result of a complex analysis of the nonstationary equation (11.3).

The main attention in this work was paid to determining the shape and velocity of a population wave in a stationary mode. With the help of a mathematical study of the nonstationary equation (11.1), they shown that over time the initial perturbation of the concentration of a population in the form of a step transforms and the population wave takes a stationary form of a hollowed out step. For it the wave propagation velocity tends to a constant value — the minimum of all possibility the values of the diffusion coefficient and the coefficient of proportionality in the law of reproduction of the population. Unfortunately, the possibility of applying the results of A.N. Kolmogorov, I.G. Petrovsky, N. S. Piskunov is limited by the framework of a very simple model of biological population and accepted form of the initial perturbation of the concentration. In the monograph by A.A. Samara et al. (Samara, Galaktionov, Kurdyumov, & Mikhailov, 1987) showed that if the initial concentration perturbation is different from the step, then the propagation velocity of a stationary wave may differ from the minimum. Therefore, it should be considered incorrect the assumptions of some authors (for example, Svirezhev, 1987; Ivanov, 1984), who consider more modern models of the population, that the wave propagation velocity has a minimum value of 2. In each case, the wave propagation velocity of the biological population and in the study of equations of mathematical physics (Polozhiy, 1964), it must be determined on the basis of this particular system of equations, as an eigenvalue of the problem. Moreover, this eigenvalue must correspond to the solution of the problem of the profile of the population wave, as an eigenvalue function.

Many authors generally do not set the task of determining the wave velocity of a biological population (see, for example, Keller, & Segel, 1971; Ivanitsky, Medvedinsky, & Tsyganov, 1991; Kerner, & Osipov, 1991), considering it to be an unknown parameter, and classify possible solutions (Berezovsky, & Karev, 1999) depending on wave velocity values. At the same time, it is as if they forget that the speed of wave propagation is a function of the conditions in which the population is located and the characteristics of the population itself. All these conditions and characteristics should be included in the equations of the population under consideration, and thus these equations themselves should determine the value of the wave velocity. Only those profiles can be considered in the wave, which correspond to the found speed values. The reason for this situation is the lack of methods for determining the speed of stationary propagation of a wave of a biological population.

The semi - infinite reaction zone method (Zhizhin, 2004), given in this section, allows finding the agreed values of the wave velocity of a biological population and its structure. This makes it possible to solve problems on the speed and structure of waves of biological populations of various models, in essence, by algebraic methods. It should be noted that the solution of the question of the velocity and structure of the wave of a biological population by the method proposed by A.N. Kolmogorov, I.G. Petrovsky, & N.S. Piskunov. However, this method is so complex that so far no one has been able to extend it to solve the problems of the population wave with more complex models. The theory of autosolitons (Kerner, & Osipov, 1991), which proposed methods for finding the speed and structure of autosolitons, does not describe a wave of populations of living organisms (Zhizhin, 2004).

MODEL OF A SOLITARY WAVE OF UNICELLULAR ORGANISMS TAKING INTO ACCOUNT THEIR NUTRITION, REPRODUCTION AND DEATH

In the mid - 60s of the last century, Adler and his students conducted a series of experimental studies of the propagation of waves of a population of bacteria in capillary tubes containing a nutrient medium (Adler, 1966; Adler, & Templeton, 1967; Adler, & Dahl, 1967). The experiments were carried out on bacteria *Escherichia coli* in a nutrient medium containing various salts. When bacteria were launched on one side of the tube, a wave of population (wave of conquest) was observed as it ran through the tube. When methionine was added to the nutrient medium in experiments, a different mode of wave propagation was observed — an antinode of bacterial concentration was formed, which moved along the tube. This means that a solitary wave of a population of bacteria was observed. The mechanism that leads to the formation of a solitary wave of a population in experiments has remained unclear. Keller and Segel (Keller, & Segel, 1971) suggested that the cause of the formation of a solitary population wave is the chemotaxis of bacteria, i.e. directional movement of bacteria towards an increase in the gradient of the nutrient resource, which naturally occurs due to the absorption of the resource by bacteria in the place of their appearance. The mathematical model of a solitary wave constructed by them did not take into account the multiplication of bacteria and their death, which contradicts the possibility of a wave being formed. An analysis of the possible effects of chemotaxis on the formation of a solitary wave conducted by the author (Zhizhin, 2004) showed that the established opinion in the scientific literature on the leading role of chemotaxis in the formation of a solitary population wave (Ivanitsky, Medvedinsky, & Tsyganov, 1991, 1994) is hyperbolized. Kennedy and Aris (Kennedy, & Aris, 1980) for the first time showed that a solitary population wave can exist without chemotaxis, if one take into account the chaotic mobility of individuals, the reproduction of the population, the death of individuals, the change in the concentration of the resource due to the feeding of individuals. The velocity of the wave, however, in this work and in its continuation (Lauffenburger, Kennedy, & Aris, 1984) was assumed minimal and the dependence of the velocity on the parameters was not established. In this regard, in the book consider the population model proposed by Kennedy and Aris, disregarding chemotaxis, and apply the semi - infinite reaction zone method to determine the wave velocity and its dependence on parameters.

The system of non-stationary equations describing in a one - dimensional formulation the change in the concentrations of unicellular organisms N and the resource R will be written as

$$\frac{\partial N}{\partial t} = \mu \frac{\partial^2 N}{\partial x^2} + N(\eta \tau_f^{-1} + \tau_e^{-1}), \tag{11.20}$$

$$\frac{\partial R}{\partial t} = -N/\tau_f. \tag{11.21}$$

Here τ_e is the effective lifetime of unicellular organisms, which will be considered permanent ($\tau_e=\tau_{e0}$); η is coefficient of resource processing into biomass of single - cell organisms; $\tau_f=\tau_{f0}(K_p+R)R^{-1}$ is effective time for feeding unicellular organisms (equation of Manod (Manod, 1950)); τ_{f0} is characteristic time of food with an excess of resource; resource concentration at which the food process time is twice as long τ_{f0}.

In equations (11.20), (11.21), the increase in the amount of biomass of a population due to reproduction is proportional to the amount of the resource consumed by it (proportionality factor η). Thus, it is possible to enter the characteristic time of reproduction of the population $\tau_p=\tau_f/\eta$. One also introduces the wave coordinate $z = u\,t - x$, dimensionless variables and parameters

$$r = R/K_p, \varepsilon = \tau_{e0}/\tau_{p0}, n = \varepsilon N(\eta K_p)^{-1}, \varsigma = (\tau_{e0}\mu)^{-1/2}, w = u(\tau_{e0}/\mu)^{1/2}.$$

Then the system (11.20), (11.21) is reduced to a system of three ordinary differential equations of the first order

$$\frac{dp}{d\varsigma} = wp + n(1 - \varepsilon r(1 + r)^{-1}), \tag{11.22}$$

$$\frac{dr}{d\varsigma} = -nrw^{-1}(1 + r)^{-1}, \tag{11.23}$$

$$\frac{dn}{d\varsigma} = p. \tag{11.24}$$

Solutions of the autonomous system (11.22) - (11.24) can be represented by trajectories in a three -dimensional phase space. It has the following four zero surfaces (Figure 4)

$F_1 : p = nw^{-1}(\varepsilon r(1+r)^{-1} - 1)$,

in its points $dp/d\varsigma = 0$;
F_2: $p=0$, in its points $dn/d\varsigma=0$;
F_3 and F_4: $n=0$ and $r=0$, in its points $dr/d\varsigma=0$.

Figure 4. The phase space of the system (11.22) – (11.24)

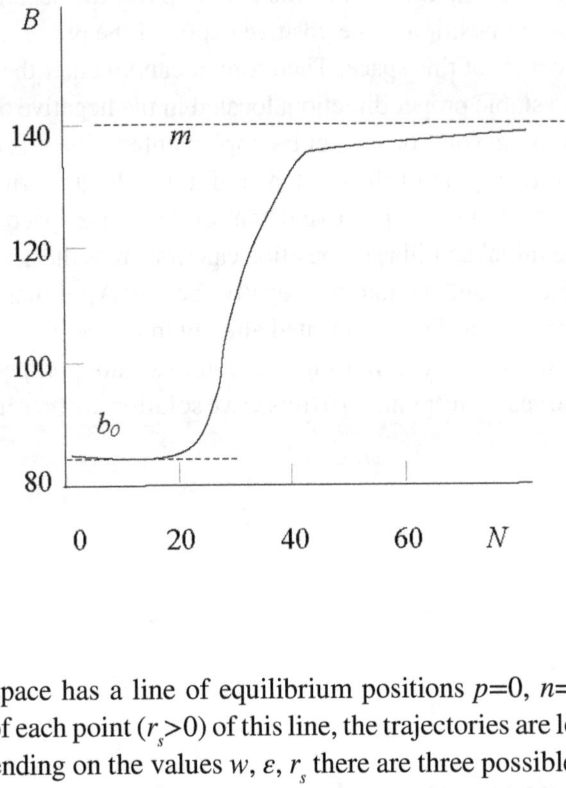

The phase space has a line of equilibrium positions $p=0$, $n=0$, axis r. In the neighborhood of each point ($r_s>0$) of this line, the trajectories are located in a plane. Moreover, depending on the values w, ε, r_s there are three possible locations of the trajectories. From the discriminant of equations (11.22), (11.23) can find that the eigenvalues are equal to their own directions

$$\lambda_{1,2} = (dp/dn)_{1,2} = w/2 \pm \sqrt{w^2/4 + 1 - \varepsilon r_s(1+r_s)^{-1}}. \quad (11.25)$$

From (11.25) it follows that with $\varepsilon<(1+r_s)/r_s$ special points of the axis are of the saddle type ($\lambda_1>0$, $\lambda_2<0$), with $(1+r_s)/r_s$ the special points have the type of unstable knot($\lambda_1>0$, $\lambda_2>0$), and with $\varepsilon>(w^2/4+1)(1+r_s)/r_s$ - the focus. The desired wave solution can leave the initial equilibrium position only in an unstable proper direction ($\lambda_i>0$, i=1,2). From the above it is clear, that two variants of the wave solution are provided. 1) A trajectory emerging from the initial equilibrium position of the saddle type along the separatrix S_1, and entering into the final equilibrium position of the saddle type along the separatrix S_2. 2) Trajectories emerging from the initial equilibrium position of a knot type along one of its own directions and entering the final equilibrium position of a saddle type along the separatrix S_2. Since at $r_s<(\varepsilon-1)^{-1}$ the equilibrium positions are of saddle type, but the zero surface F_1 is located in the negative octant (on p) of the phase space, the separatrix S_1, leaving the saddle equilibrium position in the positive octant of the phase space, cannot fall into the negative octant of this space. Therefore, it cannot enter the saddle position of equilibrium in a stable proper direction, located in the negative octant. Thus, the first variant of the wave solution cannot be implemented. There is only the second option. Indeed, in the region of the existence of a nodal equilibrium position, the zero surface F_1 is located in the positive octant of the phase space and trajectories emerging from the initial equilibrium position can first cross the zero surface F_1 and then the zero surface F_2 and get into the negative octant. After that, the trajectories can again cross the zero surface F_1, located already in the negative octant, and one of them can enter the saddle equilibrium position in a stable proper direction.

The integral curves corresponding to this wave solution are presented in Figure 5.

Figure 5. The structure of the wave solution in the population with regard to nutrition, reproduction, death and the absence of chemotaxis

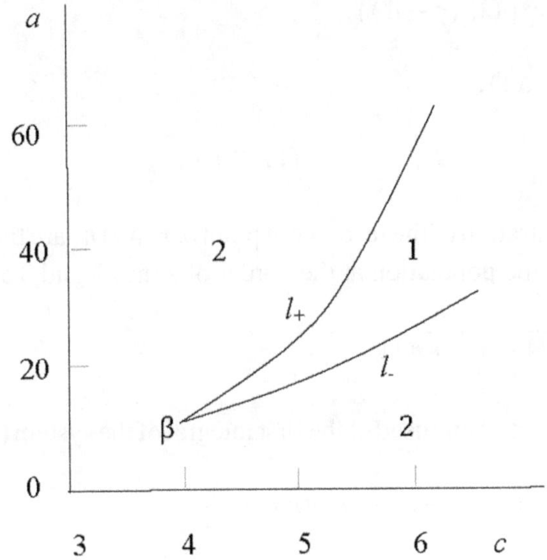

As follows from the graphs, the concentration of the population first grows along the wave coordinate and then one decreases. Thus, the wave solution has the form of a solitary population wave. It should be noted that in this case there were no chemotaxis.

To obtain an approximate analytical solution, one applies the semi-infinite reaction zone method. Divide the wavelength into three zones (Figure 11.5). In the first zone ($-\infty<\varsigma<0$), one considers the solution of the system (11.22) – (11.24) linearized in the vicinity of the initial equilibrium state a to be fair

$$r = r_- - (r_- - r_f)\exp(\lambda_{1,2-}\varsigma), \tag{11.26}$$

$$n = n_f \exp(\lambda_{1,2-}\varsigma). \tag{11.27}$$

Here r_- is the initial concentration of the resource, r_f and n_f are the concentration of the resource and population at the wave front at $\varsigma=0$,

$$\lambda_{1,2-} = w/2 \pm \sqrt{w^2/4 + 1 - \varepsilon r_-(1 + r_-)^{-1}}. \tag{11.28}$$

In the second zone ($\Delta<\varsigma<+\infty$), one considers the solution of the system (11.22) – (11.24) linearized in the vicinity of finite equilibrium state b to be fair

$$r = r_+ - (r_\Delta - r_+)\exp(\lambda_{2+}(\varsigma - \Delta)), \tag{11.29}$$

$$n = n_\Delta \exp(\lambda_{2+}(\varsigma - \Delta)).$$

(11.30)

Here r_+ is concentration of the resource at point b, r_Δ and n_Δ are the concentrations of the resource and the population at the border of zones 2 and 3 at $\varsigma=\Delta$,

$$\lambda_{2+} = w/2 - \sqrt{w^2/4 + 1 - \varepsilon r_+(1 + r_+)^{-1}}. \tag{11.31}$$

In the zone 3 ($0<\varsigma<\Delta$) one used of the first integral of the system (11.22) – (11.24)

$$dn/d\varsigma = wn - (\varepsilon - 1)w(r_- - r) + w\ln(r_-/r). \tag{11.32}$$

One linearizes the right - hand side of (11.32) in the neighborhood of the maximum of the trajectory $n(r)$. Expanding the right - hand side of (11.32) into a Taylor series in the vicinity of the maximum ($n=n_m$, $r=r_m$), one confines us to the first non - zero term

$$dn/d\varsigma = \beta_m(r - r_m). \tag{11.33}$$

Here

$$\beta_m = (\varepsilon - 1)w - w/r_m > 0, \tag{11.34}$$

$$n_m = (\varepsilon - 1)(r_- - r_m) - \ln(r_-/r_m). \tag{11.35}$$

In view of the finiteness of zone 3, one takes the linear law of variation of r in this zone

$$r = r_f - (r_f - r_\Delta)\varsigma/\Delta. \tag{11.36}$$

Then, one integrated (11.33) with account (11.36)

$$n = -0.5\Delta\beta_m(r - r_m)^2/(r_f - r_\Delta) + n_m. \tag{11.37}$$

Using the condition of continuity of concentrations at the zone boundary from (11.37) can to obtain

$$n_f = -0.5\Delta\beta_m(r_f - r_m)^2/(r_f - r_\Delta) + n_m, \tag{11.38}$$

$$n_\Delta = -0.5\Delta\beta_m(r_\Delta - r_m)^2/(r_f - r_\Delta) + n_m. \tag{11.39}$$

Using the condition of continuity of the first derivatives on the zone boundary from (11.26), (11.27), (11.29), (11.30), (11.36), (11.37) one get

$$\Delta(r_- - r_f)\lambda_{1,2-} = r_f - r_\Delta, \tag{11.40}$$

$$n_f\lambda_{1,2-} = (r_f - r_m)\beta_m, \tag{11.41}$$

$$\Delta(r_\Delta - r_+)\lambda_{2+} = r_\Delta - r_f, \tag{11.42}$$

$$n_\Delta\lambda_{2+} = (r_\Delta - r_m)\beta_m. \tag{11.43}$$

One assume that the linear section in the profile $r(\varsigma)$ is chosen so that the sum of the second derivatives $d^2r/d\varsigma^2$ at $\varsigma=0$ on the left and at $\varsigma=\Delta$ on the right is zero. Differentiating (11.26), (11.29), one has

$$(r_\Delta - r_+)\lambda_{2+}^2 = (r_- - r_f)\lambda_{1,2-}^2. \tag{11.44}$$

Equate the right side of equation (11.22) to zero on the border at $\varsigma=\Delta$

$$w\lambda_{2+} + 1 - \varepsilon r_\Delta/(1 + r_\Delta) = 0. \tag{11.45}$$

So, one has 12 equations (11.28), (11.31), (11.34), (11.35), (11.38) - (11.45) and 12 unknowns

$$\lambda_{1,2-}, \lambda_{2+}, w, n_m, r_m, \Delta, r_f, r_\Delta, r_+, n_f, n_\Delta, \beta_m.$$

The resulting system of equations is resolved (Zhizhin, 2004) and leads to one equation for the speed

$$(r_f - r_m)(2r_- - r_f - r_m)\beta_m/\lambda_{1,2-} = 2n_m(r_- - r_f). \tag{11.46}$$

All quantities in the last equation (11.46) are functions of speed. After determining the velocity value by (11.46), the remaining parameters are determined by the already obtained equations.

As rally from (11.33) with $r_{-} = r_{+} = r_{f} = r_{\Delta} = (\varepsilon - 1)^{-1}, w = 0$, this equation holds. The ratio $r_{-} = (\varepsilon - 1)^{-1}$ defines a line in the plane of control parameters (r_{-}, ε) (Figure 6).

Figure 6. Area of existence of a solitary wave of unicellular organisms: 1- the lower boundary of the domain of existence of solutions (complete degeneration of the wave), 2 - complete consumption of the resource in the wave

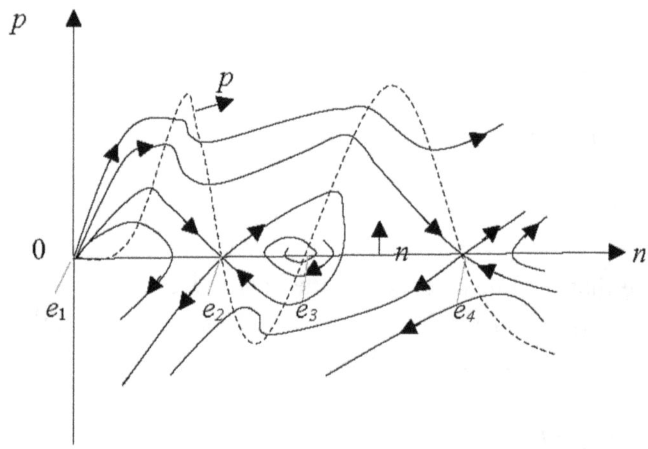

This line (1) is the lower boundary of the domain of existence of solutions. It corresponds to the complete degeneration of the wave. The upper limit of the domain of existence of solutions is line 2 (Figure 11.6), on which $r_{+}=0$. The equation of this line is obtained after substituting $r_{+}=0$ into the analytical solution. It responds to the mode of complete consumption of the resource in the wave. On this line, the wave velocity at the given ε is maximum. The value of the wave velocity decreases to zero, if one move in a straight line $\varepsilon =$ *constant* across the region of existence of solutions, from line 2 to line 1. The wave velocity along line 2 smoothly increases with increasing ε. Calculations of concentration profiles of a population and a resource with specific values of control parameters (Zhizhin, 2004) confirmed the appearance of the graphs in Figure 5, obtained as a result of a qualitative study of the original system of differential equations.

THE WAVES IN POPULATIONS ALLEE

When considering the logistic population, it was assumed that the fertility function B in the local law of growth is determined only by the physiological limits of fertility and does not depend on the concentration of the population. For many species of multicellular organisms that migrate freely across their range and reproduce sexually, this is not entirely true. At low population densities, reproduction is determined rather by the probability of meeting the mating partners, and not by physical fertility. In these cases, the $B(N)$ relationship has the form shown in Figure 7 as an example of the butterfly population.

Figure 7. Dependence of the fertility function on the population concentration in the Allee populations

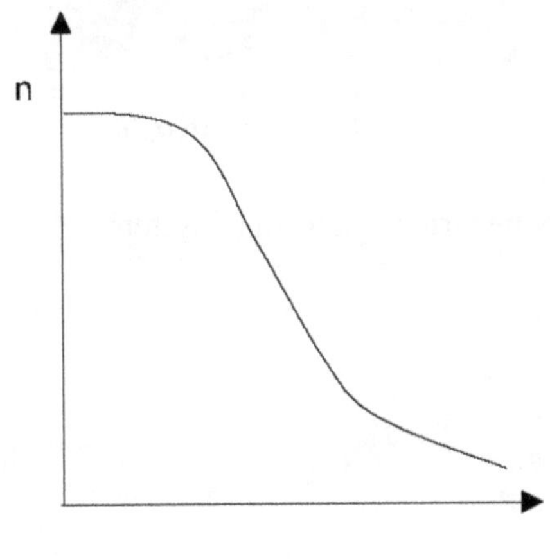

In this figure, the number of eggs laid by one butterfly *Anagasta kuhniella* (Svirezhev, 1987; Ullyet, 1945) is plotted on the ordinate axis, depending on the concentration of butterflies. Populations with this type of dependence $B(N)$ are called Allee populations (Allee, 1951).

The peculiarity of this dependence for such populations is the saturability of B with increasing N and the finiteness of B at low concentrations of N. The dependence that asymptotically tends to some limiting values for large and small values of the argument can be expressed, as in the theory of combustion, by an exponential function

$$B(N) = b_0 + b_1 \exp(-\alpha_0/N). \tag{11.47}$$

Here $b_1+b_0=m$ is physiological birth rate.

Assuming that mortality linearly depends on the concentration of the pupil, the function describing the local law of population growth in the nonstationary diffusion equation (11.1) has the form

$$F(N) = b + b_1 \exp(-\alpha_0/N) - \alpha_2 N. \tag{11.48}$$

Here $b=b_0-\alpha_1>0$.

Introducing the scaled concentration of the population $n=N/\alpha_0$, related to the parameter α_0, characterizing the rate of increase of B with increasing N, as well as the wave coordinate $z=ut-x$, one write the differential equation of diffusion in total derivatives

$$wdn/d\varsigma = d^2n/d\varsigma^2 + (1 + a\exp(-1/n) - cn. \tag{11.49}$$

Here

$$\varsigma = z(b/\mu)^{1/2}, w = u(\mu b)^{1/2}, a = b_1/b, c = \alpha_2\alpha_0/b.$$

Equation (11.49) may be written in form of system

$$dp/d\varsigma = wp - F(n), \tag{11.50}$$

$$dn/d\varsigma = p. \tag{11.51}$$

Here

$$F(n) = n(1 + a\exp(-1/n) - cn). \tag{11.52}$$

It should be noted that the representation of the local law of population growth $F(n)$ in the form of a polynomial of n, proposed in some works (Kerner, & Osipov, 1991), does not reflect the main feature of the Allee population - the saturability of the birth function at large n. Accounting for this feature leads to a function (11.52) that is not a polynomial.

The solution of system (11.50), (11.51) can be represented by trajectories in the phase plane (p,n). Their location is determined by zero isoclines $p = 0$, on which $dn/d\varsigma=0$, and

$$p = (1 + a\exp(-1/n) - cn)n/w, \tag{11.53}$$

on which $dp/d\varsigma=0$

The equilibrium positions of the system (11.50) – (11.52) are the intersections of the zero isoclines, i.e. they are the roots of the equation $0=(1+a\exp(-1/n)-cn)n$. Obviously, one root of this equation is $n=0$. Other roots are determined by the equality

$$a\exp(-1/n)=cn-1. \qquad (11.54)$$

The number of roots of equality (11.54) depends on the values of the parameters a, c. It can be equal to 1 or 3 according to the number of intersections of the right and left side of (11.54). The boundary value of the roots (11.54) is equal to two, one of which corresponds to the touch of the right and left parts of (11.54). Equating the derivatives of the right and left parts of (11.54) with the simultaneous fulfillment of equality (11.54) itself, one can find an expression for the boundary line in the plane (a, c) separating the parameter regions with two and four equilibrium positions

$$a = (c(0.5 \pm \sqrt{0.25-c^{-1}})-1\exp(0.5 \pm \sqrt{0.25-c^{-1}})^{-1}. \qquad (11.55)$$

The graph of this line (l) is presented in Figure 8.

Figure 8. Bifurcation system diagram (11.50)-(11.52): 1 – region of four equilibrium positions, 2 – region of two equilibrium positions

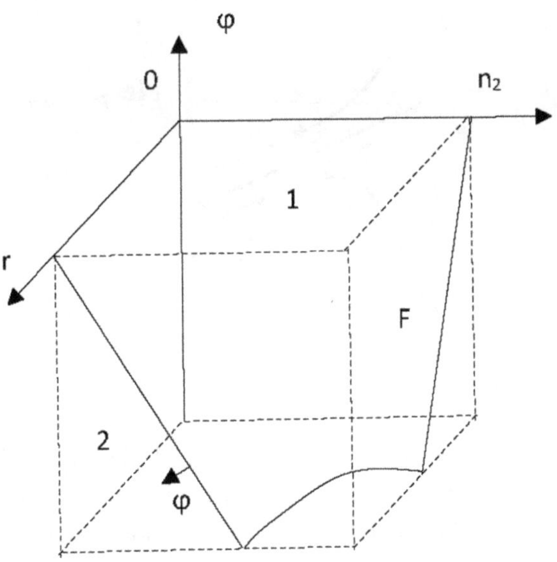

It consists of two branches (l_+ and l-) corresponding to the signs in front of the root in (11.55). In region 1 between branches of the line l, there are four equilibrium positions. In area 2, including the c axis, the system has two equilibrium positions. On the line (11.55) itself, the system has three equilibrium positions, one of which is not coarse and disappears with a small wiggling of the system. On the axis a, the system has one equilibrium position when n=0. The β point with the coordinates $a=e^2=7.389$, c=4 in which both branches of the line are connected is a catastrophe point of the system (11.50) – (11.52). Since $F(n)$ is not a polynomial, this point cannot be attributed to any of the elemental catastrophes (Breker, & Lander, 1977), but its neighborhood is close to a neighborhood of a hyperbolic ombelic catastrophe at a constant value of one of the deformation parameters. Of greatest interest is the case when the system (11.50) – (11.52) has four equilibrium positions, as in this case there is the greatest number of solutions (Zhizhin, 2004).

Here there are topologically distinct phase diagrams of the system (11.50) – (11.52), having as a solution conquest waves, solitary waves and switching waves (transitions from one equilibrium position to another, each of which is different from the origin) (Zhizhin, 2004). Choose one of them, containing waves of conquest and solitary waves (Figure 9).

Figure 9. Phase diagram of system (11.50) - (11.52) with four equilibrium positions

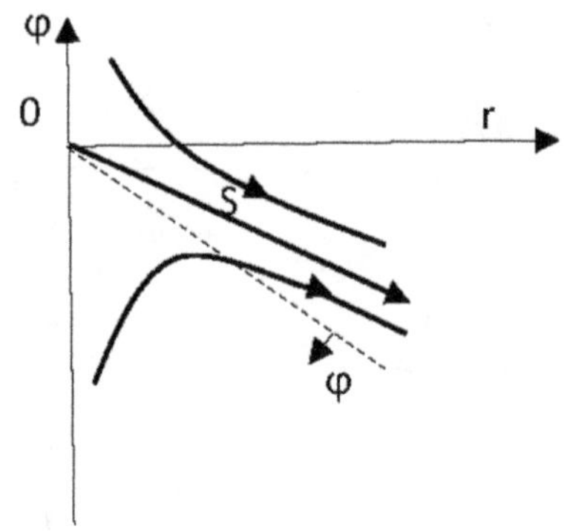

It contains the equilibrium positions e_i, $i=1\div4$. The types of special points (equilibrium positions) are determined from the discriminant of the system in the neighborhood of each of the special points. In the vicinity of the point $e_1(0,0)$ one has

$$\lambda_{1,2e_1} = 0.5w \pm \sqrt{0.25w^2 - 1}, \tag{11.56}$$

for $w>2$ the point e_1 has the type of an unstable node. In the vicinity of points e_i, $i=2,3,4$ one has

$$\lambda_{1,2e_i} = 0.5w \pm \sqrt{0.25w^2 - F'(n)}, i = 2,3,4. \tag{11.57}$$

Here

$$F'(n) = 1 - 2cn + (1 + 1/n)\exp(-1/n). \tag{11.58}$$

Since $F'(n) < 0$ in point e_3, so this point has the type of an unstable focus or an unstable node.

However, in topological terms, this difference is insignificant, which determines the topological equivalence of phase diagrams in these cases. Since at points e_2, e_4 $F'(n) > 0$ these points are of saddle type. To solve the problem of the existence of wave solutions, it is necessary to construct the entire system of trajectories on the phase plane so that the assumed location in the neighborhood of a particular point does not contradict the location of the trajectories in the neighborhood of other special points. This excludes some of the solutions found (Svirezhev, 1987) without considering this condition. From the phase diagram in Figure 9 it follows that the trajectory that goes from the equilibrium position e_1 to the equilibrium position e_2 corresponds to the wave of conquest. The solitary wave corresponds to the trajectory out of the equilibrium position e_2 and entering the equilibrium position e_2.

Consider the waves of conquest. Using the semi - infinite reaction zone method, one divides the wave into three zones. In the first zone ($-\infty<\varsigma<0$), one considers the solution of the system in the vicinity of the equilibrium state e_1 to be fair

$$n = n_f \exp(\lambda_{1,2e_1} \varsigma). \tag{11.59}$$

Here n_f is the concentration value of the population at the wave front at $\varsigma=0$.

In the second zone $\Delta<\varsigma<+\infty$), one considers the solution of the system in the vicinity of the e_2 point to be fair

$$n = n_2 - (n_2 - n_\Delta)\exp(\lambda_{2e_2}(\varsigma - \Delta)). \tag{11.60}$$

Here n_Δ is the concentration value of the population for $\varsigma=\Delta$.

In the third zone, one linearizes the right - hand side of (11.50) in the vicinity of the maximum of the trajectory of the phase plane (p,n)

$$dp/d\varsigma = -F'_m(n - n_m), F'_m = F'_n(n_m). \tag{11.61}$$

One integrates (11.61) with account (11.51)

$$(n - n_m)p_m^{-1}(-F_m^{'})^{1/2} = \sin((-F_m^{'})^{1/2}\varsigma + A), \tag{11.62}$$

$$A = \arcsin((n_f - n_m)p_m^{-1}(-F_m^{'})^{1/2}), \tag{11.63}$$

$$p_m = F_m/w \tag{11.64}$$

One put in (11.62) $\varsigma=\Delta$

$$(n_\Delta - n_m)p_m^{-1}(-F_m^{'})^{1/2} = \sin((-F_m^{'})^{1/2}\Delta + A). \tag{11.65}$$

Differentiating (11.59), (11.60), (11.65) once, one equates the first derivatives on the border of zones 1 - 3, 2 – 3

$$n_f^2\lambda_{1,2e_1}^2 = p_m^2 - F_m^{'}(n_f - n_m)^2, \tag{11.66}$$

$$\lambda_{2e_2}^2(n_2 - n_\Delta)^2 = p_m^2 - F_m^{'}(n_\Delta - n_m)^2. \tag{11.67}$$

Differentiating (11.59), (11.60) twice taking into account (11.61), one equate the second derivatives on the border of zones 1 - 3, 2 - 3

$$n_f^2\lambda_{1,2e_1}^2 = -F_m^{'}(n_f - n_m), \tag{11.68}$$

$$\lambda_{2e_2}^2(n_2 - n_\Delta) = -F_m^{'}(n_\Delta - n_m). \tag{11.69}$$

Solving the resulting system of algebraic equations, one obtains two equations for determining the velocity of the wave w and the concentration of the population at the maximum of the trajectory n_m

$$1 = \lambda_{1,2e_1}^2(w^2 n_m^2 F_m^{-2} - 1/F_m^{'}), \tag{11.70}$$

$$F_m^{'}\lambda_{2e_2}^{-2} = F_m^{-2}(F_m^{'})^{-1}w^2(\lambda_{2e_2}^2(n_2 - 1) + (n_2 - n_m)F_m^{'})^2 - 1. \tag{11.71}$$

The solution obtained can be considered as a general solution of the problem of the conquest wave for an arbitrary function $F(n)$ in equation (11.50). Similarly, using the semi - infinite reaction zone method, analytical solutions can also be obtained for other wave solutions in the Allee populations.

MATHEMATICAL MODEL OF FLIGHT OF THE GREGARIOUS LOCUST

The propagation of wave in populations of animal organisms complicated by many accompanying processes. Such processes include changes in the concentration of the nutritional resource for the organisms of the population, environmental pollution with waste products of organisms, the desire for the herd of animal organisms, the interaction of animal populations and populations of plant organisms. Mathematical models of waves in populations of multicellular organisms taking into account changes in the concentration of nutrient resources and environmental pollution by products of vital activity of organisms are constructed and investigated in the already mentioned monograph (Zhizhin, 2004) and in work (Zhizhin, Bolshakova, 2000). The influence of the bark beetle herd in the process of attacking the forests in the form of a solitary wave has been studied in works (Zhizhin, 2011; Zhizhin, & Selikhovkin, 2012). In this section, the flight pattern of the gregarious locust will be investigated, since the locust still causes enormous damage to vegetation in many parts of the earth.

Many works have been devoted to the description of mass locust breeding and migration of its flocks (Bei - Bienko, & Mishchenko, 1951; Kopaneva, & Stebaev, 1985; Chetverikov, 1905; Shcherbinovsky, 1952, 1958). Well-known entomologist B.P. Uvarov in the 20s of the last centuries showed (Bey - Bienko, 1970) that the phenomenon of so-called phase variation is characteristic of the locust: depending on the density of the group and the concentration of food, either single forms or herd larvae form. A single form of larvae is formed when there is enough food. This form of the locust is characterized by inconspicuous coloring and calm behavior. With a lack of food and large crowding, a gregarious locust form is formed. This form is characterized by bright color, great mobility and aggressiveness. Processes, causing the dynamics of the number of insects, associated with the interaction of the two kingdoms living organisms: plants and animals. External influences — abiotic, biotic, and man - made — are multifaceted, and the response of bio – geo – cenosis is difficult to foresee, which requires the use of mathematical methods to describe and analyze them.

Existing work on the mathematical description of the behavior of insects includes questions of mathematical modeling of the distribution of locust populations over the territory. Topaz et al. (2012) consider a model of the locust transition to the active phase. Taylor (1979) assumes that the locust migration is completely determined by the flow of air. Topaz et al. (2008) raise of a model creeping locust migration.

In this section, a mathematical model of the flight of the gregarious locust is built over long distances when the locust is removed from the earth's surface at distances significantly exceeding the size of insects, as this is probably the most dangerous option for the environment to spread locusts in space.

This regime is not consistent with the locust migration crawling model. A model that considers locusts at the flight stage as inert dust particles in an air stream contradicts the established notions of aggressiveness and high mobility of the herd locust. Following the concept of S.S. Chetverikov (Chetverikov, 1905) one will consider the process of locust propagation as the propagation of living matter in space in the form of waves of life. Then, to describe the flight of the locust flock, parabolic systems of equations can be used, as soon as they describe the propagation of nonlinear waves in space in various excitable media (Kolmogorov et al., 1937; Svirezhev, 1987; Zhizhin, 2004, 2005, 2008).

The system of non - stationary equations describing in a one - dimensional formulation the change in the concentration of the biological population $N(x,t)$ and the food supply $R(x,t)$, averaging the discrete flight process of the herd locust in space and time, can be written as

$$\frac{\partial N}{\partial t} = \frac{\partial}{\partial x}\mu\frac{\partial N}{\partial x} + F(N), \tag{11.72}$$

$$\frac{\partial R}{\partial t} = -\tau_f^{-1}N. \tag{11.73}$$

Here t is a time, x is a spatial coordinate, μ is locust mobility coefficient, τ_f is characteristic locust feeding time, $F(N)$ is a function describing a local change in locust population concentration.

A local decrease in the concentration of locust individuals in flight occurs for various reasons (diseases, their swallowing up by birds and even other individuals of the flock). Locusts do not breed during the flight, but locusts have another way to combine the natural decline of the flock - this is attracting the locust to the flock from the environment through special chemical products released by insects.

In connection with this, equation (11.72) can be rewritten as

$$\frac{\partial N}{\partial t} = \frac{\partial}{\partial x}\mu\frac{\partial N}{\partial x} + (\tau_c^{-1} - \tau_e^{-1})N. \tag{11.74}$$

Here τ_e is the characteristic time of life of individuals of the population, τ_c is the characteristic time of replenishment of the population due to the desire of individuals of the locust to unite in a flock. One assumes τ_e, τ_c, τ_f as constants. The mobility coefficient of locust individuals increases with a decrease in the concentration of food supply (Kopaneva, & Stebaev, 1985). This behavior of locusts can be described as a first approximation by an inversely proportional relationship, considering the degree of the concentration of the food supply to be equal to one, μ~*A*/*R* . The coefficient of proportionality *A* in this relationship must have the dimension of the product [μ] [*R*]. Therefore, *A* can be represented as a product of two constants, one of which has the dimension of the concentration of food supply, and the other has the dimension of mobility of locust individuals, that is,

$$\mu = \frac{\mu_0 K_R}{R}. \tag{11.75}$$

Let us call K_R the unit concentration of the food base and μ_0 the mobility coefficient of the individuals at a single concentration of the food base.

The constants in formulas (11.73) - (11.75) obviously depend on the locust species, as well as on temperature as a parameter.

One will consider the stationary mode of locust population spread by the wave. If to introduce the wave coordinate *z*=*ut*–*x*, where *u* is the unknown wave propagation velocity (determined during the solution of the problem), as well as dimensionless variables and parameters

$$r = \frac{R}{K_R}, n = \frac{N}{K_R}, \varsigma = z\tau_0\mu_0/2, w = u(\tau_0/\mu_0)^{1/2}, \varepsilon = \tau_0/\tau_f, \tau_0 = \frac{\tau_c\tau_e}{\tau_e - \tau_c},$$

then equations (11.74), (11.73) take the following form

$$w\frac{dn}{d\varsigma} = \frac{d}{d\varsigma}\left(\frac{1}{r}\frac{dn}{d\varsigma}\right)+n, \tag{11.76}$$

$$w\frac{dr}{d\varsigma} = -\varepsilon n. \tag{11.77}$$

Equation (11.76) with regard to (11.77) and the boundary condition $n=0, \frac{dn}{d\varsigma} = 0$ for $r=r_-$ (the initial concentration of the food supply) has the first integral

$$\frac{dn}{d\varsigma} = rw\left(n - \frac{r_- - r}{\varepsilon}\right). \tag{11.78}$$

The system (11.78), (11.77) is autonomous and has in the phase plane (n, r) zero isoclines $r=0$, $n = \frac{r_- - r}{\varepsilon}$, on which $\frac{dn}{d\varsigma} = 0$, and $n=0$, on which $\frac{dr}{d\varsigma} = 0$, as well as two special points $\alpha(0,0)$, $\beta(0, r_-)$. From the discriminant of the system in the neighborhood of a point β one find the eigenvalues of this special point

$$\lambda_{1,2} = 0.5 r_- w \pm \sqrt{(0.5 r_- w)^2 - r_-}. \tag{11.79}$$

From (11.79) it follows that with $(0.5r_-w)^2 - r_- > 0$ the point β has the type of an unstable node, and with $(0.5r_-w)^2 - r_- < 0$ an unstable focus. Since we are looking for a solution that describes the change in the concentration of locust glands along the wave coordinate from zero to zero, one assumes that the radical expression in (11.79) is positive. Then the point β will have the type of an unstable node, and from it the trajectories emanate from its own directions $(\frac{dn}{dr})_{1,2} = -\lambda_{1,2} w/\varepsilon$, one of which corresponds to the desired wave solution.

From the discriminant of system (11.77), (11.78) in the vicinity of a point α, it follows that a point α has a saddle type with eigenvalues $\lambda_{3,4} = -r_-^{1/2}$, and the proper direction $\lambda_4 = -r_-^{1/2}$ in which the trajectory enters the point α is located in the positive quadrant of the plane. The sought solution leaves the point β along one of the separatrices of the node and enters the point α along the separatrix of the saddle (Figure 10), crossing the zero isocline $n = \frac{r_- - r}{\varepsilon}$.

Figure 10. The phase plane of solitary wave of the locust population

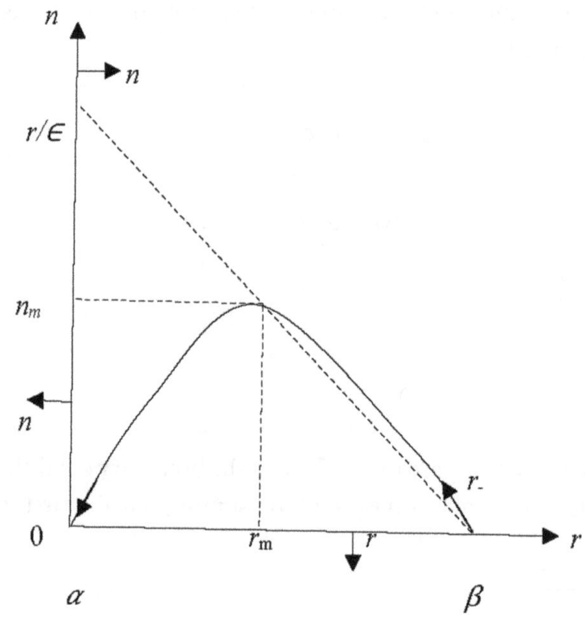

The arrows at the zero isoclines indicate the regions of positive values of the corresponding derivatives (Figure 10). The integral curves corresponding to this wave solution are shown in Figure 11.

Figure 11. Qualitative view of the integral curves of the system (11.77), (11.78)

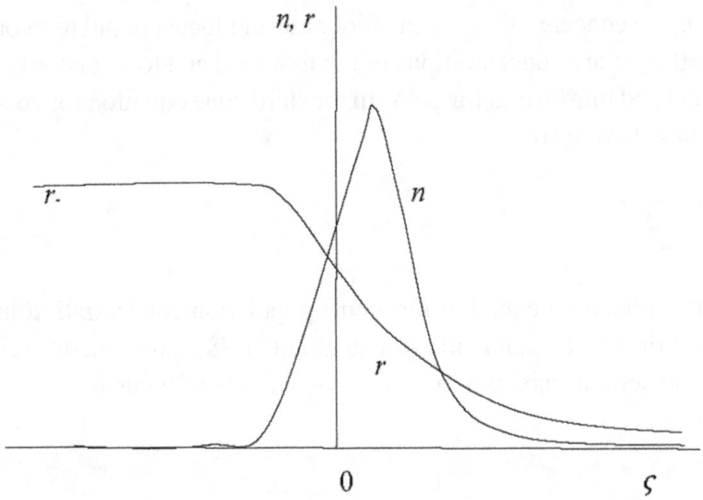

It can be seen that the concentration profile of the locust population has the form of a solitary wave (soliton), and the concentration profile of the food supply has the form of a nonlinear wave. Both proper directions in a point β are located above the zero isocline $n = \frac{r_- - r}{\varepsilon}$ and it is not possible to decide from qualitative analysis in which proper direction the wave solution leaves the special point β.

To solve this question, one applies the semi - infinite reaction zone method. Imagine that a wave consists of five zones. In the first zone ($-\infty<\varsigma<0$), one considers the solution of the system (11.77), (11.78) in the neighborhood of the point β to be fair. In the second zone ($\Delta<\varsigma<+\infty$), one considers the solution of the system in the neighborhood of the point α to be fair. In the third zone ($0<\varsigma<\Delta$), due to its limited nature, one will take a linear dependence of the concentration of the food base on the spatial coordinate. Then the equations for the concentrations in the first zone have the form

$$r = r_- - (r_- - r_f)\exp(\lambda_{1,2}\varsigma), \tag{11.80}$$

$$n = n_f \exp(\lambda_{1,2}\varsigma), \tag{11.81}$$

and in the second zone this form

$$r = r_\Delta \exp(-r_-^{1/2}(\varsigma - \Delta)), \tag{11.82}$$

$$n = n_\Delta \exp(-r_-^{1/2}(\varsigma - \Delta)). \qquad (11.83)$$

Here r_f, n_f are concentrations of the foot base and locust population on the front wave for ς=0; r_f, n_f are concentrations of the foot base and locust population on the bound second and third zones for $\varsigma=\Delta$. In the third zone equation for concentration of the foot base have form

$$r = r_f - \frac{(r_f - r_\Delta)\varsigma}{\Delta}. \qquad (11.84)$$

In order to obtain an equation for locust population concentration in the third zone, expand the right - hand side of equation (11.78) into a Taylor series in the vicinity of the soliton maximum ($n=n_m$, $r=r_m$) and confine us to the first term of the expansion

$$\frac{dn}{d\varsigma} = r_m n_m(\varsigma - \varsigma_m). \qquad (11.85)$$

With (11.78) one has

$$n_m = \frac{r_- - r_m}{\varepsilon}. \qquad (11.86)$$

Integrating (11.84), can get

$$n = n_m - 0.5\, r_m n_m(\varsigma - \varsigma_m)^2. \qquad (11.87)$$

The set of boundary conditions includes:

1) the equalities of concentrations of the locust population on both sides of the boundaries of the first and third zones at ς=0, as well as the boundaries of the second and second and third zones at $\varsigma=\Delta$

$$n_f = n_m - 0.5\, r_m n_m \varsigma_m^2, \qquad (11.88)$$

$$n_\Delta = n_m - 0.5\, r_m n_m(\Delta - \varsigma_m)^2 ; \qquad (11.89)$$

2) the equalities of the first derivatives of the concentrations of the components on ς with both sides of the boundaries of the first and third zones, as well as the boundaries of the second and third zones

$$(r_- - r_f)\lambda_{1,2} = \frac{r_f - r_\Delta}{\Delta}, \tag{11.90}$$

$$n_f\lambda_{1,2} = r_m n_m \varsigma_m, \tag{11.91}$$

$$r_\Delta r_-^{1/2} = \frac{r_f - r_\Delta}{\Delta}, \tag{11.92}$$

$$n_\Delta r_-^{1/2} = r_m n_m(\Delta - \varsigma_m); \tag{11.93}$$

3) the equality of the second derivative to zero of the n on ς in equation (11.76) in point ς=0, which leads to equation

$$w = \frac{r_f - r_\Delta}{\Delta r_f^2} + \frac{n_f}{r_m n_m \varsigma_m}. \tag{11.94}$$

In addition, from the equation (11.84) follows

$$r_m = r_f - \frac{(r_f - r_\Delta)\varsigma_m}{\Delta}. \tag{11.95}$$

The system of ten algebraic equations (11.79), (11.86), (11.88)–(11.95) is closed with respect to ten unknown parameters $\lambda_{1,2}, w, \Delta, r_m, r_\Delta, n_m, n_\Delta, r_f, n_f, \varsigma_m$.

Solving this system (Zhizhin, 2014), can to obtain an equation for calculating the wave propagation velocity

$$[r_f - (r_- - r_f)\lambda_{1,2}\varsigma_m](\varsigma_m/\lambda_{1,2} + 0.5\varsigma_m^2) = 1. \tag{11.96}$$

Here

$$\varsigma_m = b\frac{r_-^{-1/2} + 0.5b}{b + r_-^{-1/2} + \lambda_{1,2}^{-1}}, \tag{11.97}$$

$$b = \frac{r_f - (r_- - r_f)\lambda_{1,2} r_-^{-1/2}}{(r_- - r_f)\lambda_{1,2}}, \tag{11.98}$$

$$r_f = 0.5\lambda_{1,2}\frac{-\lambda_{1,2} + \sqrt{\lambda_{1,2}^2 + 4r_-(w\lambda_{1,2} - 1)}}{w\lambda_{1,2} - 1}. \tag{11.99}$$

To calculate the system, it is necessary to set the values of two parameters characterizing the initial concentration of the feed base and the ratio of the characteristic process times. A well-studied study showed that equation (11.96) has a solution if one of two eigenvalues $\lambda_{1,2}$ is chosen λ_2. Thus, the desired wave solution has its own number λ_2 and goes in its own direction $n = \frac{r_- - r}{\varepsilon}$, closest to the zero isocline. Figure 12 shows as an example the graphs of the functions calculated by the solution obtained.

Figure 12. Dependencies of the characteristics of the solitary locust wave on the initial concentration of the food supply (ε = 0.1): a) speed, b) own number of the initial equilibrium position, c) locust concentration at the maximum soliton

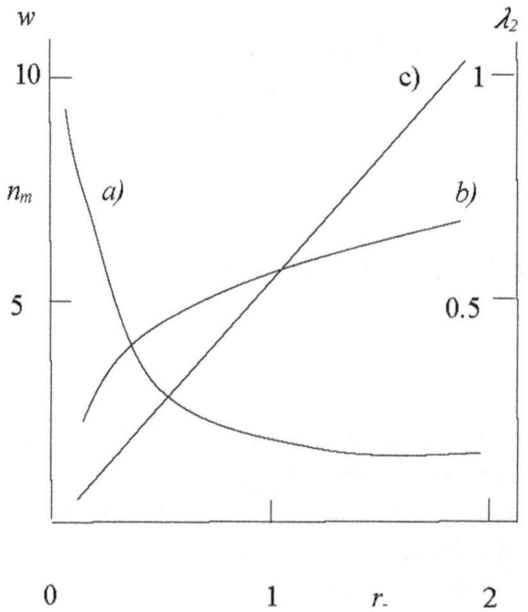

Figure 13 presents the results of the calculation of locust concentration profiles and the food resource in the wave.

Figure 13. Concentration profiles in the secluded locust population wave

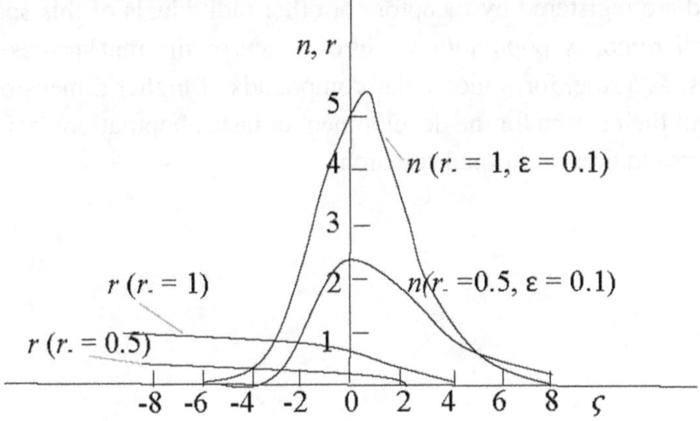

It can be seen that the locust concentration profile has the form of a soliton, and the food resource concentration profile has the form of a nonlinear switching wave from the initial final value to zero. Thus, the resulting solution is a solitary wave of the biological population.

Numerical studies have shown that the rate of a solitary population wave of locusts increases dramatically with a decrease in the initial concentration of the food supply. While the wave profile is stretched. This effect was also observed in other solitary population waves (Zhizhin, 2005).

CHEMORECEPTION OF BIOLOGICAL ORGANISMS

Biological animals, for their successful existence, must be able to determine the location of their food objects, promptly learn about the appearance of predators, scare off individuals of other species and find other individuals of their own species for mating. In the course of evolution, special organs were formed for transmission (generation) and for its perception (reception). Information is received by animals using sensory systems sensitive to light and sound signals, electromagnetic waves, temperature, etc. One of the “tools” for transmitting information is chemoreception - the ability of animals and plants to perceive specific chemical compounds and react to them in a certain way. Among animals, chemical communication is most common in insects, which use the language of smells to solve their basic life problems: nutrition, reproduction, and dispersal. Insects carry out chemical communication through special volatile low-molecular compounds - pheromones, which cause a specific response in individuals of the same biological species. Insects secrete one or more species-specific pheromones that spread from individual sources with air currents and are registered by receptors of other individuals of this species. With the help of pheromones, population waves of insects of different species are formed. Pheromones, as a rule, form molecular compounds of higher dimensions. The effectiveness of their action for the development of insect populations or for reducing their numbers requires additional research.

REFERENCES

Adler, J. (1966). Chemotaxis in bacteria. *Science*, *153*(3737), 708–716. DOI: 10.1126/science.153.3737.708 PMID: 4957395

Adler, J., & Dahl, M. M. (1967). Method for Measuring the Motility of Bacteria and for Comparing Random and Non-random Motility. *Journal of General Microbiology*, *46*(2), 161–173. DOI: 10.1099/00221287-46-2-161 PMID: 5339791

Adler, J., & Templeton, B. (1967). The Effect of Environmental Conditions on the Motility of Escherichia coli. *Journal of General Microbiology*, *46*(2), 175–184. DOI: 10.1099/00221287-46-2-175 PMID: 4961758

Allee, W. C. (1951). *Cooperation among animals with human implications*. N.Y. Shuman.

Berezovskaya, E. S., & Karev, G. P. (1999). Bifurcations of traveling waves in population taxis models. *Успехи физических наук*, *169*(9), 1011–1024. DOI: 10.3367/UFNr.0169.199909d.1011

Bey - Bienko, G.Ya. (1970). B.P. Uvarov (1889 - 1970) and his contribution to science and practice. *Entomological Review*, *49*(4), 915–922.

Chetverikov, S. S. (1905). Waves of life. From lepidopetrological observations of the summer of 1903. *Diary of the Zoological Department of the Imperial Society of Lovers of Natural Science. Anthropology and Ethnography.*, *3*(6), 106–111.

Isaev, A. S. (2001). *Population dynamics of forest insects*. Science.

Isaev, A. S., Nedorezov, L. V., & Khlebopros, R. G. (1979). Mathematical model of escape in the interaction of a predator and a prey. In *Mathematical analysis of the components of forest biocenoses* (pp. 74–82). Science.

Ivanitsky, G. R., Medvedinsky, A. B., & Tsyganov, M. A. (1991). From disorder to orderliness - on the example of the movement of microorganisms. *Успехи физических наук*, *161*(3), 13–71. DOI: 10.3367/UFNr.0161.199104b.0013

Ivanov, N.N. (1984). Solitary waves in the theory of populations. *News of the USSR Academy of Sciences. Biological series*, 5, 656 – 663.

Keller, E. F., & Segel, L. A. (1971). Traveling bands of chemotaxis bacteria: A theoretical analysis. *Journal of Theoretical Biology*, *30*(2), 238–248. DOI: 10.1016/0022-5193(71)90051-8

Kennedy, C. R., & Aris, R. (1980). Traveling waves in a simple population model involving growth and death. *Bulletin of Mathematical Biology*, *42*(3), 397–429. DOI: 10.1016/S0092-8240(80)80057-7 PMID: 7378612

Kolmogorov, A. N., Petrovsky, I. G., & Piskunov, N. S. (1937). The study of the diffusion equation, combined with an increase in the amount of substance, and its application to a single biological problem. *MSU Bulletin, series A. Mathematics and Mechanics*, *1*(1), 1–26.

Kopaneva, L. M., & Stebaev, I. V. (1985). *Locust life*. Agropromizdat.

Lauffenburger, D., Kennedy, R., & Aris, R. (1984). Traveling bands of Chemotaxis Bacteria in the Context of Population Growth. *Bulletin of Mathematical Biology*, *46*(1), 19–40. DOI: 10.1016/S0092-8240(84)80033-6

Manod, J. (1950). La technique de culture continue, theorie et applicationes. *Ann. Inst. Paster.*, 79.

Polozhiy, G. N. (1964). *Equations of mathematical physics*. High School.

Samara, A. A., Galaktionov, V. A., Kurdyumov, S. P., & Mikhailov, L. P. (1987). *Modes with aggravation in problems for quasilinear parabolic equations*. Science.

Scherbinovsky, N.S. (1952). *Desert Locust Schistocerka*. Moscow: State ed. rural - host lit.

Scherbinovsky, N. S. (1958). The main laws of mass breeding of the desert locust and the migration of its flocks. Moscow: *9 th International Conference on Plant Quarantine and Plant Protection from Pests and Diseases.*

Svirezhev, Yu. M. (1987). *Nonlinear waves, dissipative structures and disasters in ecology*. Science.

Taylor, R. A. J. (1979). A simulation model of locust migratory behavior. *Journal of Animal Ecology*, *48*(2), 577–602. DOI: 10.2307/4181

Topaz, C. M., Bernoff, A. J., Logan, S., & Toolson, W. (2008). A model for rolling swarms of locusts. *The European Physical Journal. Special Topics*, *157*(1), 92–109. DOI: 10.1140/epjst/e2008-00633-y

Topaz, C.M., D`Orsogua, M.R., Edelstein – Keshet, L., & Bernoff, J. (2012). Locust dynamics: Behavioral phase change and swarming. *PLOS Comput. Biol.* 8: e 1002642.

Ullyet, G. C. (1945). Oviposition by Ephestia kukniella. *Zell. J. Ent. Soc. S. Africa*, *8*, 53–59.

Volterra, V. (1931). *Lecons sur la Theorie Mathematique de La Lutte Pour la Vie*. Gautheer – Villars.

Zhizhin, G. V. (1982). The structure of the wave front of the polymerization. *Reports of the USSR Academy of Sciences*, *263*(6), 1399–1402.

Zhizhin, G. V. (1988). Autowave processes of distribution of chemical reactions in a dispersion medium. *Journal of Applied Mechanics and Technical Physics*, *6*, 35–43.

Zhizhin, G. V. (1992). *Macrokinetics of frontal polymerization reactors*. Polytechnica.

Zhizhin, G. V. (1997a). *Mathematical models of waves radical polymerization*. Publishing Northwestern Correspondence Polytechnic Institute.

Zhizhin, G. V. (1997b). Stationary waves reversible radical polymerization. *Chemical Physics*, *3*, 114–123.

Zhizhin, G. V. (2004a). *Self - regulating waves of chemical reactions and biological populations*. Science.

Zhizhin, G. V. (2004b). Model of a wave of ideal solid-flame combustion with a variable surface of chemical interaction. *Chemical Physics*, *40*(1), 95–102.

Zhizhin, G. V. (2005). *Dissipative structures in chemical, geological and ecological systems*. Science.

Zhizhin, G. V. (2008). *Combustion waves with distributed zones of chemical reactions (nonasymptotic combustion theory)*. Publishing Werner Regen.

Zhizhin, G. V. (2012). Mathematical model of a solitary wave of a population of beetles - bark beetles. *Proceedings of the St. Petersburg State Forestry Academy*, 196, 153 - 162.

Zhizhin, G. V. (2014). Mathematical model of flight of the gregarious locust. *Biosphere*, *6*(2), 112–117.

Zhizhin, G. V., & Bolshakova, N. N. (2000a). Single waves in populations of multicellular animal organisms. *Mathematical Modelling*, *12*(12), 55–65.

Zhizhin, G. V., & Larina, T. I. (1994). Standing waves gas chemical reactions in porous inert media. *Combust*, *4*, 11–20.

Zhizhin, G. V., & Poritskaya, I. J. (1994). Self-regulating chemical reactions of n - th order in condensed matter. *Combust*, *6*, 61–68.

Zhizhin, G. V., & Selikhovkin, A. V. (2012). *Mathematical modeling of the development and distribution of insect populations - stem pests in the forests of Russia.* S.-Petersburg: St. Petersburg State Forestry University.

KEY TERMS AND DEFINITIONS

Separatrix: a trajectory that separates qualitatively different types of trajectories from each other.

Solytary wave: wave motion, which at each time instant is localized in a finite region of space and rapidly decreases with distance from this region.

Special point: The point at which the first derivatives of all phase variables with respect to the independent variable are equal to zero.

The gregarious locust: the state of the locust in which the locust forms large flocks capable of long flights; the gregarious locust is characterized by bright coloration and aggressive behavior.

The semi - infinite reaction zone method: A method that assumes a low reaction rate to the reaction front and a significant reaction rate after the reaction front to infinity.

Zero isocline: The line at the points of which the first derivative of some phase variable with respect to the independent variable is zero.

Chapter 12
The Locust Wave

ABSTRACT

A mathematical model of locust swarm flight is developed. Its differential equations are studied qualitatively to find a solution relevant to swarm flight mode. An analytical approximation is found and numerically evaluated. The solution corresponds to a solitary wave where locust profile is consistent with that it represents a soliton, whereas nutritive base profile is found to be consistent with a damped nonlinear wave. Simple approximate solutions for calculating solitary wave velocity and characteristic length are obtained.

INTRODUCTION

The propagation of wave in populations of animal organisms complicated by many accompanying processes. Such processes include changes in the concentration of the nutritional resource for the organisms of the population, environmental pollution with waste products of organisms, the desire for the herd of animal organisms, the interaction of animal populations and populations of plant organisms. Mathematical models of waves in populations of multicellular organisms taking into account changes in the concentration of nutrient resources and environmental pollution by products of vital activity of organisms are constructed and investigated in the already mentioned monograph (Zhizhin, 2004) and in work (Zhizhin, Bolshakova, 2000). The influence of the bark beetle herd in the process of attacking the forests in the form of a solitary wave has been studied in works (Zhizhin, 2011; Zhizhin, & Selikhovkin, 2012). In this section, the flight pattern of the gregarious locust will be investigated, since the locust still causes enormous damage to vegetation in many parts of the earth.

Many works have been devoted to the description of mass locust breeding and migration of its flocks (Bei - Bienko, & Mishchenko, 1951; Kopaneva, & Stebaev, 1985; Chetverikov, 1905; Shcherbinovsky, 1952, 1958). Well - known entomologist

DOI: 10.4018/978-1-6684-7745-8.ch012

B.P. Uvarov in the 20s of the last centuries showed (Bey - Bienko, 1970) that the phenomenon of so - called phase variation is characteristic of the locust: depending on the density of the group and the concentration of food, either single forms or herd larvae form. A single form of larvae is formed when there is enough feed. This form of the locust characterized by inconspicuous coloring and calm behavior. With a lack of food and large crowding, a gregarious locust form is formed. This form is characterized by bright color, great mobility and aggressiveness. Processes, causing the dynamics of the number of any insects, associated with the interaction of the two kingdoms living organisms: plants and animals. External influences — abiotic, biotic, and man - made — are multifaceted, and the response of bio – geo – kenosis is difficult to foresee, which requires the use of mathematical methods to describe and analyze them.

Existing work on the mathematical description of the behavior of insects includes questions of mathematical modeling of the distribution of locust populations over the territory. Topaz C.M. et al. considered a model of the locust transition to the active phase (Topaz, C.M. et al., 2012). Taylor R.A.J. (1979) assumes that the locust migration is completely determined by the flow of air (Taylor, R.A.J.,1979). Topaz C.M. et al. raise of a model creeping locust migration (Topaz, C.M. et al., 2008).

12.1. MATHEMATICAL MODEL OF FLIGHT OF THE GREGARIOUS LOCUST

In this section, a mathematical model of the flight of the gregarious locust is built over long distances when the locust is removed from the earth's surface at distances significantly exceeding the size of insects, as this is probably the most dangerous option for the environment to spread locusts in space.

This regime is not consistent with the locust migration crawling model. A model that considers locusts at the flight stage as inert dust particles in an air stream contradicts the established notions of aggressiveness and high mobility of the herd locust. Following the concept of S.S. Chetverikov (Chetverikov, 1905) one will consider the process of locust propagation as the propagation of living matter in space in the form of waves of life. Then, to describe the flight of the locust flock, parabolic systems of equations can be used, as soon as they describe the propagation of nonlinear waves in space in various excitable media (Kolmogorov et al., 1937; Svirezhev, 1987; Zhizhin, 2004, 2005, 2008).

The system of non - stationary equations describing in a one - dimensional formulation the change in the concentration of the biological population $N(x,t)$ and the food supply $R(x,t)$, averaging the discrete flight process of the herd locust in space and time, can be written as

$$\frac{\partial N}{\partial t} = \frac{\partial}{\partial x}\mu\frac{\partial N}{\partial x} + F(N), \tag{12.1}$$

$$\frac{\partial R}{\partial t} = -\tau_f^{-1}N. \tag{12.2}$$

Here t is a time, x is a spatial coordinate, μ is locust mobility coefficient, τ_f is characteristic locust feeding time, $F(N)$ is a function describing a local change in locust population concentration.

A local decrease in the concentration of locust individuals in flight occurs for various reasons (diseases, their swallowing up by birds and even other individuals of the flock). Locusts do not breed during the flight, but locust has another way to combine the natural decline of the flock - this is attracting the locust to the flock from the environment through special chemical products released by insects.

In connection with this, equation (12.1) can be rewritten as (Zhizhin, 2014)

$$\frac{\partial N}{\partial t} = \frac{\partial}{\partial x}\mu\frac{\partial N}{\partial x} + (\tau_c^{-1} - \tau_e^{-1})N. \tag{12.3}$$

Here τ_e is the characteristic time of life of individuals of the population, τ_c is the characteristic time of replenishment of the population due to the desire of individuals of the locust to unite in a flock. We assume τ_e, τ_c, τ_f as constants. The mobility coefficient of locust individuals increases with a decrease in the concentration of food supply (Kopaneva, & Stebaev, 1985). This behavior of locusts can be described as a first approximation by an inversely proportional relationship, considering the degree of the concentration of the food supply to be equal to one, μ~A/R. The coefficient of proportionality A in this relationship must have the dimension of the product $[\mu][R]$. Therefore, A can be represented as a product of two constants, one of which has the dimension of the concentration of food supply, and the other has the dimension of mobility of locust individuals, that is,

$$\mu = \frac{\mu_0 K_R}{R}. \tag{12.4}$$

Let us call K_R the unit concentration of the food base and μ_0 the mobility coefficient of the individuals at a single concentration of the food base.

The constants in formulas (12.2) - (12.4) obviously depend on the locust species, as well as on temperature as a parameter.

One will consider the stationary mode of locust population spread by the wave. If to introduce the wave coordinate $z=ut-x$, where u is the unknown wave propagation velocity (determined during the solution of the problem), as well as dimensionless variables and parameters

$$r = \frac{R}{K_R}, n = \frac{N}{K_R}, \varsigma = z\tau_0\mu_0/2, w = u(\tau_0/\mu_0)^{1/2}, \varepsilon = \tau_0/\tau_f, \tau_0 = \frac{\tau_c\tau_e}{\tau_e - \tau_c},$$

then equations (12.3), (12.2) take the following form

$$w\frac{dn}{d\varsigma} = \frac{d}{d\varsigma}\left(\frac{1}{r}\frac{dn}{d\varsigma}\right) + n, \tag{12.5}$$

$$w\frac{dr}{d\varsigma} = -\varepsilon n. \tag{12.6}$$

Equation (12.5) with regard to (12.6) and the boundary condition $n=0$, $\frac{dn}{d\varsigma} = 0$ for $r=r\_$ (the initial concentration of the food supply) has the first integral

$$\frac{dn}{d\varsigma} = rw\left(n - \frac{r\_ - r}{\varepsilon}\right). \tag{12.7}$$

12.2. QUALITATIVE STUDY OF THE SYSTEM OF EQUATIONS

The system (12.7), (12.6) is autonomous and has in the phase plane (n, r) zero isoclines $r = 0, n = \frac{r\_ - r}{\varepsilon}$, on which $\frac{dn}{d\varsigma} = 0$, and $n=0$, on which $\frac{dr}{d\varsigma} = 0$, as well as two special points $\alpha(0,0)$, $\beta(0, r\_)$. From the discriminant of the system in the neighborhood of a point β one find the eigenvalues of this special point

$$\lambda_{1,2} = 0.5r\_w \pm \sqrt{(0.5r\_w)^2 - r\_}. \tag{12.8}$$

From (12.8) it follows that with $(0.5r\_w)^2 - r\_ > 0$ the point β has the type of an unstable node, and with $(0.5r\_w)^2 - r\_ < 0$ an unstable focus. Since we are looking for a solution that describes the change in the concentration of locust glands along the wave coordinate from zero to zero, we assume that the radical expression in (12.8) is positive. Then the point β will have the type of an unstable node, and from it the trajectories emanate from its own directions $\left(\frac{dn}{dr}\right)_{1,2} = -\lambda_{1,2}w/\varepsilon$, one of which corresponds to the desired wave solution.

From the discriminant of system (12.6), (12.7) in the vicinity of a point α, it follows that a point α has a saddle type with eigenvalues $\lambda_{3,4} = -r\_^{1/2}$, and the proper direction $\lambda_4 = -r\_^{1/2}$ in which the trajectory enters the point α is located in the positive quadrant of the plane. The sought solution leaves the point β along one of the separatrices of the node and enters the point α along the separatrix of the saddle (Figure 1), crossing the zero isocline $n = \frac{r\_ - r}{\varepsilon}$.

Figure 1. The phase plane of solitary wave of the locust population

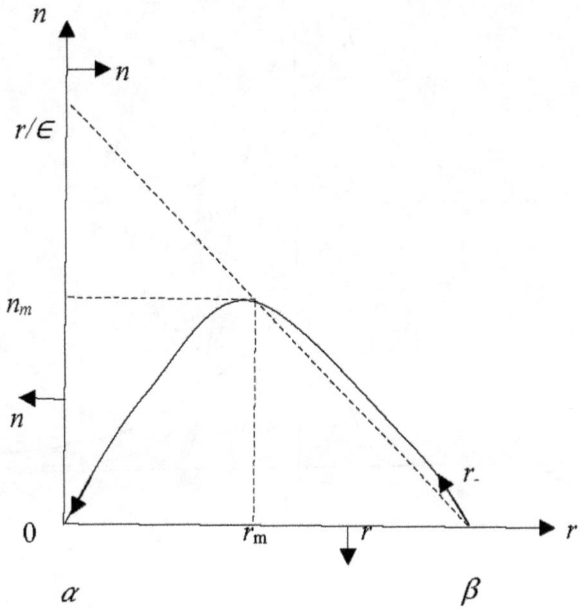

The arrows at the zero isoclines indicate the regions of positive values of the corresponding derivatives (Figure 1). The integral curves corresponding to this wave solution are shown in Figure 2.

Figure 2. Qualitative view of the integral curves of the system (12.6), (12.7)

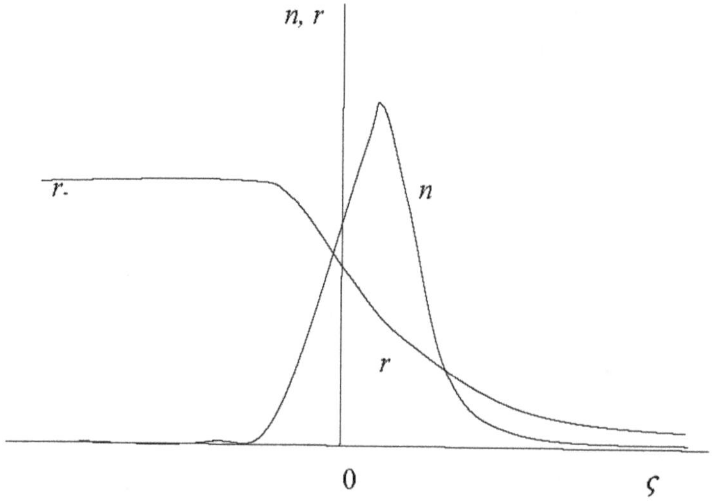

It can be seen that the concentration profile of the locust population has the form of a solitary wave (soliton), and the concentration profile of the food supply has the form of a nonlinear wave. Both proper directions in a point β are located above the zero isocline $n = \frac{r_- - r}{\varepsilon}$ and it is not possible to decide from qualitative analysis in which proper direction the wave solution leaves the special point β.

12.3. APPROXIMATE ANALYTICAL AND NUMERICAL SOLUTION OF THE SYSTEM OF EQUATIONS

To solve this question, we apply the semi - infinite reaction zone method. Imagine that a wave consists of five zones. In the first zone($-\infty < \varsigma < 0$), we consider the solution of the system (12.6), (12.7) in the neighborhood of the point β to be fair. In the second zone($\Delta < \varsigma < +\infty$), we consider the solution of the system in the neighborhood of the point α to be fair. In the third zone ($0 < \varsigma < \Delta$), due to its limited nature, one will take a linear dependence of the concentration of the food base on the spatial coordinate. Then the equations for the concentrations in the first zone have the form

$$r = r_- - (r_- - r_f)\exp(\lambda_{1,2}\varsigma), \tag{12.9}$$

$$n = n_f\exp(\lambda_{1,2}\varsigma), \tag{12.10}$$

and in the second zone this form

$$r = r_\Delta \exp(-r_-^{1/2}(\varsigma - \Delta)), \tag{12.11}$$

$$n = n_\Delta \exp(-r_-^{1/2}(\varsigma - \Delta)). \tag{12.12}$$

Here r_f, n_f are concentrations of the foot base and locust population on the front wave for $\varsigma = 0$; r_f, n_f are concentrations of the foot base and locust population on the bound second and third zones for $\varsigma = \Delta$. In the third zone equation for concentration of the foot base have form

$$r = r_f - \frac{(r_f - r_\Delta)\varsigma}{\Delta}. \tag{12.13}$$

In order to obtain an equation for locust population concentration in the third zone, expand the right - hand side of equation (12.7) into a Taylor series in the vicinity of the soliton maximum ($n=n_m$, $r=r_m$) and confine our-selves to the first term of the expansion

$$\frac{dn}{d\varsigma} = r_m n_m (\varsigma - \varsigma_m). \tag{12.14}$$

With (12.7) we have

$$n_m = \frac{r_- - r_m}{\varepsilon}. \tag{12.15}$$

Integrating (12.13), can get

$$n = n_m - 0.5\, r_m n_m (\varsigma - \varsigma_m)^2. \tag{12.16}$$

The set of boundary conditions includes:

1) the equalities of concentrations of the locust population on both sides of the boundaries of the first and third zones at $\varsigma=0$, as well as the boundaries of the second and second and third zones at $\varsigma=\Delta$
2) $n_f = n_m - 0.5\, r_m n_m \varsigma_m^2$, (12.17)

$$n_\Delta = n_m - 0.5\, r_m n_m (\Delta - \varsigma_m)^2 \ ; \tag{12.18}$$

3) the equalities of the first derivatives of the concentrations of the components on ς with both sides of the boundaries of the first and third zones, as well as the boundaries of the second and third zones

$$(r_- - r_f)\lambda_{1,2} = \frac{r_f - r_\Delta}{\Delta}, \tag{12.19}$$

$$n_f\lambda_{1,2} = r_m n_m \varsigma_m, \tag{12.20}$$

$$r_\Delta r_-^{1/2} = \frac{r_f - r_\Delta}{\Delta}, \tag{12.21}$$

$$n_\Delta r_-^{1/2} = r_m n_m(\Delta - \varsigma_m); \tag{12.22}$$

4) the equality of the second derivative to zero of the n on ς in equation (12.5) in point ς=0, which leads to equation

$$w = \frac{r_f - r_\Delta}{\Delta\, r_f^2} + \frac{n_f}{r_m n_m \varsigma_m}. \tag{12.23}$$

In addition, from the equation (12.13) follows

$$r_m = r_f - \frac{(r_f - r_\Delta)\varsigma_m}{\Delta}. \tag{12.24}$$

The system of ten algebraic equations (12.8), (12.15), (12.77) – (12.24) is closed with respect to ten unknown parameters $\lambda_{1,2}, w, \Delta, r_m, r_\Delta, n_m, n_\Delta, r_f, n_f, \varsigma_m$. .

Solving this system (Zhizhin, 2014), can obtain an equation for calculating the wave propagation velocity

$$[r_f - (r_- - r_f)\lambda_{1,2}\varsigma_m](\varsigma_m/\lambda_{1,2} + 0.5\varsigma_m^2) = 1. \tag{12.25}$$

Here

$$\varsigma_m = b\frac{r_-^{-1/2} + 0.5b}{b + r_-^{-1/2} + \lambda_{1,2}^{-1}}, \tag{12.26}$$

$$b = \frac{r_f - (r_- - r_f)\lambda_{1,2} r_-^{-1/2}}{(r_- - r_f)\lambda_{1,2}}, \tag{12.27}$$

$$r_f = 0.5\lambda_{1,2}\frac{-\lambda_{1,2} + \sqrt{\lambda_{1,2}^2 + 4r_-(w\lambda_{1,2} - 1)}}{w\lambda_{1,2} - 1}. \tag{12.28}$$

To calculate the system, it is necessary to set the values of two parameters characterizing the initial concentration of the feed base r_- and the ratio of the characteristic process times ϵ. By equation (12.8), setting the speed values w, we determine the eigenvalues $\lambda_{1,2}$, then by (12.28) we calculate r_f, then by (12.27) we find b and by (12.26) we find ς_m. The found values of the parameters are substituted into the right side of (12.25). We vary the values of the parameters until both parts of equality (12.25) are aligned. A well-studied study showed that equation (12.25) has a solution if one of two eigenvalues $\lambda_{1,2}$ is chosen λ_2. Thus, the desired wave solution has its own number λ_2 and goes in its own direction $n = \frac{r_- - r}{\varepsilon}$, closest to the zero isocline. The remaining parameters are determined by the equations

$$r_\Delta = (r_- - r_f)\lambda_2 r_-^{-1/2}, \Delta = (r_f - r_\Delta) r_-^{-1/2} r_\Delta^{-1},$$
$$n_\Delta = n_m r_m(\Delta - \varsigma_m) r_-^{-1/2}, n_f = \varsigma_m n_m r_m \lambda_2^{-1}.$$

Numerical studies have shown that the results of calculations based on the obtained approximate analytical solution are approximated by simple power-law dependences of the parameters on r_- and ϵ. For example, the propagation velocity of a locust wave is approximated by the function

$$w = 2.6 r_-^{-1/2}. \tag{12.29}$$

The eigenvalue, with which the wave solution leaves the initial equilibrium position β, obeys the equality $\lambda_2 = 0.5 r_-^{1/2}$. (12.30)

Width of the intermediate zone in the soliton profile

$$\Delta = 2.3 r_-^{-1/2}. \tag{12.31}$$

Population concentration at the soliton maximum

$$n_m = 0.55 r_- / \varepsilon. \tag{12.32}$$

Figure 1 shows as an example the graphs of the functions calculated by the solution obtained.

Figure 3. Dependencies of the characteristics of the solitary locust wave on the initial concentration of the food supply (ε = 0.1): a) speed, b) own number of the initial equilibrium position, c) locust concentration at the soliton maximum

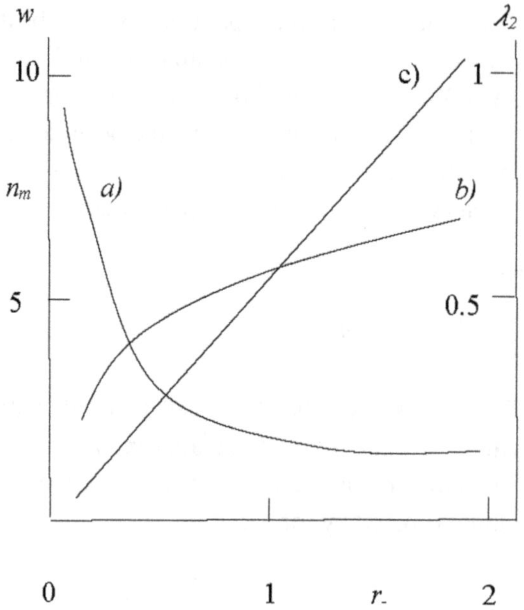

Figure 4 presents the results of the calculation of locust concentration profiles and the food resource in the wave.

Figure 4. Concentration profiles in the secluded locust population wave

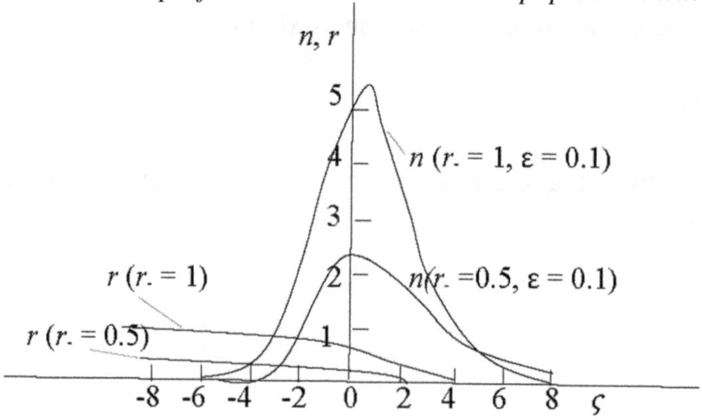

It can be seen that the locust concentration profile has the form of a soliton, and the food resource concentration profile has the form of a nonlinear switching wave from the initial final value to zero. Thus, the resulting solution is a solitary wave of the biological population.

Numerical studies have shown that the rate of a solitary population wave of locusts increases dramatically with a decrease in the initial concentration of the food supply. While the wave profile is stretched. This effect was also observed in other solitary population waves (Zhizhin, 2005).

CONCLUSION

Approximate analytical solution was obtained describing the propagation of a wave of locust population. This solution has self-similarity, since the dimensionless velocity of propagation of a solitary locust population wave is a function of only the dimensionless initial food base, while the dimension velocity depends on many parameters of the problem. Simple power-law dependences of the main characteristics of a solitary locust wave on the control parameters r_- and ϵ are obtained. Numerical studies have shown that the speed of a solitary locust population wave increases sharply (function (12.29)) with a decrease in the initial concentration of the food base r_-. In this case, the wave profile is stretched. Such an effect was also observed in other solitary waves (Zhizhin, 2004). An increase in the speed of a solitary locust wave with a decrease in the initial food supply is consistent with the inversely proportional dependence of locust mobility on the initial concentration of the food supply (12.4). The concentration profiles of the locust population in the wave significantly depend on the parameter ϵ characterizing the rate of kinetic processes in the wave.

Using the expression for the concentration at the maximum of the soliton (12.32), we will conduct a qualitative analysis of the influence of kinetic parameters on the shape of the soliton. We represent (12.32), taking into account the dependence of ϵ on the characteristic times of the process (12.5), in the form

$$n_m \sim \frac{r_-(V_c - \frac{1}{t_e})}{V_f}, \tag{12.33}$$

where $V_c \sim 1/\tau_c$ is the rate of attracting locusts to the swarm, $V_f \sim 1/\tau_f$ is the rate of consumption of food resources by locusts.

It follows from (12.33) that with the same initial concentration of the food base, an increase in the rate of consumption of the food base leads to a decrease in the concentration of locusts at the soliton maximum (in order to eat it, a smaller amount of locusts is required). Increasing the speed of attracting locusts to the swarm increases the number of locusts at the maximum of the soliton, that is, the solitary wave of locusts becomes more pronounced. Reducing the characteristic lifetime of the locust reduces the numerator in (12.33), that is, reduces the number of locusts at the soliton maximum and leads to the degeneration of the solitary wave.

Thus, the solution obtained, provided that the values of the characteristic times included in the dimensionless control parameters are known, makes it possible to quickly quantify the characteristics of the population solitary locust wave. This can be used to predict locust infestations and catastrophic destruction of vegetation. To determine the values of the characteristic times, special biological and field studies are required.

It should be noted that the developed model cannot describe the flights of locust swarms over very long distances at high altitudes from the Earth, where there is neither food nor feeding of the locust swarm from the environment. For example, in 1954, these insects from Northwest Africa reached the British Isles, covering a distance of 24 000 km. At the same time, the flight of locusts took place at an altitude of 2000 km. It can be assumed that the developed flight model of gregarious locusts describes the stage preceding these gigantic flights, when the swarm has already formed and flies far away already in a non-stationary mode due to its own accumulated potential.

REFERENCES

Berezovskaya, E. S., & Karev, G. P. (1999). Bifurcations of traveling waves in population taxis models. *Успехи физических наук*, *169*(9), 1011–1024. DOI: 10.3367/UFNr.0169.199909d.1011

Bey - Bienko, G.Ya. (1970). B.P. Uvarov (1889 - 1970) and his contribution to science and practice. *Entomological Review*, *49*(4), 915–922.

Chetverikov, S. S. (1905). Waves of life. From lepidopetrological observations of the summer of 1903. *Diary of the Zoological Department of the Imperial Society of Lovers of Natural Science. Anthropology and Ethnography.*, *3*(6), 106–111.

Isaev, A. S. (2001). *Population dynamics of forest insects*. Science.

Isaev, A. S., Nedorezov, L. V., & Khlebopros, R. G. (1979). Mathematical model of escape in the interaction of a predator and a prey. In *Mathematical analysis of the components of forest biocenoses* (pp. 74–82). Science.

Kolmogorov, A. N., Petrovsky, I. G., & Piskunov, N. S. (1937). The study of the diffusion equation, combined with an increase in the amount of substance, and its application to a single biological problem. *MSU Bulletin, series A. Mathematics and Mechanics*, *1*(1), 1–26.

Kopaneva, L. M., & Stebaev, I. V. (1985). *Locust life*. Agropromizdat.

Polozhiy, G. N. (1964). *Equations of mathematical physics*. High School.

Scherbinovsky, N.S. (1952). *Desert Locust Schistocerka*. Moscow: State ed. rural - host lit.

Scherbinovsky, N. S. (1958). The main laws of mass breeding of the desert locust and the migration of its flocks. Moscow: *9 th International Conference on Plant Quarantine and Plant Protection from Pests and Diseases.*

Svirezhev, Yu. M. (1987). *Nonlinear waves, dissipative structures and disasters in ecology*. Science.

Taylor, R. A. J. (1979). A simulation model of locust migratory behavior. *Journal of Animal Ecology*, *48*(2), 577–602. DOI: 10.2307/4181

Topaz, C. M., Bernoff, A. J., Logan, S., & Toolson, W. (2008). A model for rolling swarms of locusts. *The European Physical Journal. Special Topics*, *157*(1), 92–109. DOI: 10.1140/epjst/e2008-00633-y

Topaz, C.M., & D`Orsogua, M.R., Edelstein – Keshet, L., Bernoff, J. (2012). Locust dynamics: Behavioral phase change and swarming. *PLOS Comput. Biol.* 8: e 1002642.

Zhizhin, G. V. (1982). The structure of the wave front of the polymerization. *Reports of the USSR Academy of Sciences*, *263*(6), 1399–1402.

Zhizhin, G. V. (1988). Autowave processes of distribution of chemical reactions in a dispersion medium. *Journal of Applied Mechanics and Technical Physics*, *6*, 35–43.

Zhizhin, G. V. (1992). *Macrokinetics of frontal polymerization reactors*. Polytechnica.

Zhizhin, G. V. (1997a). *Mathematical models of waves radical polymerization*. Publishing Northwestern Correspondence Polytechnic Institute.

Zhizhin, G. V. (1997b). Stationary waves reversible radical polymerization. *Chemical Physics*, *3*, 114–123.

Zhizhin, G. V. (2004a). *Self - regulating waves of chemical reactions and biological populations*. Science.

Zhizhin, G. V. (2004b). Model of a wave of ideal solid-flame combustion with a variable surface of chemical interaction. *Chemical Physics*, *40*(1), 95–102.

Zhizhin, G. V. (2005). *Dissipative structures in chemical, geological and ecological systems*. Science.

Zhizhin, G. V. (2008). *Combustion waves with distributed zones of chemical reactions (nonasymptotic combustion theory)*. Publishing Werner Regen.

Zhizhin, G. V. (2012). Mathematical model of a solitary wave of a population of beetles - bark beetles. *Proceedings of the St. Petersburg State Forestry Academy*, 196, 153 - 162.

Zhizhin, G. V. (2014). Mathematical model of flight of the gregarious locust. *Biosphere*, *6*(2), 112–117.

Zhizhin, G. V., & Bolshakova, N. N. (2000a). Single waves in populations of multicellular animal organisms. *Mathematical Modelling*, *12*(12), 55–65.

Zhizhin, G. V., & Larina, T. I. (1994). Standing waves gas chemical reactions in porous inert media. *Combust*, *4*, 11–20.

Zhizhin, G. V., & Poritskaya, I. J. (1994). Self-regulating chemical reactions of n - th order in condensed matter. *Combust*, *6*, 61–68.

Zhizhin, G. V., & Selikhovkin, A. V. (2012). *Mathematical modeling of the development and distribution of insect populations - stem pests in the forests of Russia*. S.-Petersburg: St. Petersburg State Forestry University.

Chapter 13
Waves of Beetles:
Bark Beetles

ABSTRACT

Using the methods of the non-asymptotic theory of waves, waves of bark beetles that spread through the forests of planet Earth and cause enormous damage to forests, undermining and destroying trees, are considered. Mathematical models of bark beetle waves are constructed, analytical solutions are obtained, and numerical studies are carried out. Research includes various stages of development of bark beetles: from departure from wintering places, flight in search of a forest stand suitable for settlement, establishment of a settlement, laying eggs, growth of larvae, formation of pupae, birth of a new generation of bark beetles, to wintering. A reliable way to combat bark beetles is to improve the state of the forest stand, able to withstand the attacks of bark beetles, achieved by forestry and forest protection measures. The created mathematical models are an important element for forecasting the development and distribution of bark beetle foci using the wave theory.

INTRODUCTION

The mathematical study of bark beetle populations was preceded by extensive field studies in the forests of Russia. The main field studies of bark beetles were concentrated in the Lesinsky educational and experimental forestry enterprise, in the pine forests of the western regions of the Leningrad region, in the Tikhvin district of the Leningrad region, and also in Siberia (Irkutsk region, Bratsk district and Novosibirsk region, Iskitim district) (Zhizhin, Selikhovkin, 2012).

To develop mathematical models of bark beetle population development, knowledge of average population characteristics is not enough. It is necessary to determine the values of many parameters characterizing bark beetle individuals at different

DOI: 10.4018/978-1-6684-7745-8.ch013

stages of their development. In this regard, special field studies were conducted in the Leningrad Region to solve the following problems:

1) determine the intensity of beetle mortality during the wintering period;
2) determine the bark beetle stock;
3) determine the summer period;
4) take into account the mortality factors during the beetle flight and their settlement of trees (resin accumulation, predators, diseases, microclimatic conditions, unspecified causes);
5) determine the mass of larvae before pupation;
6) determine the mass of larvae after hatching;
7) determine the mass of pupae;
8) determine the mass of young beetles before flying out;
9) determine the mass of beetles before leaving for wintering.

The main factor weakening forest stands, contributing to the reproduction of bark beetles in the regions of Siberia, is industrial air pollution. In the Bratsk region there is a powerful aluminum production plant - a source of fluoride pollution, a forestry complex and thermal power plants (TPP), emitting a number of compounds, among which sulfur dioxide predominates. In the Iskitim region, sulfur and dust pollution predominates, the source of which are enterprises producing cement, lime, chemical industry enterprises and (TPP). In both regions, five pollution zones were identified, characterized by a certain level and gradient of pollution. The concentration of pollution increases from zone 5 to zone 1. In zone 1, the level of pollution is the maximum possible, at which some species of woody plants are still preserved, but in a strongly depressed state. In the second zone, the impact of pollution is The main factor weakening forest stands, contributing to the reproduction of bark beetles in the regions of Siberia, is industrial air pollution. In the Bratsk region there is a powerful aluminum production plant - a source of fluoride pollution, a forestry complex and thermal power plants (TPP) emitting a number of compounds, among which sulfur dioxide predominates. In the Iskitim region, sulfur and dust pollution predominates, the source of which are enterprises producing cement, lime, chemical industry enterprises and (TPP). In both regions, five pollution zones were identified, characterized by a certain level and gradient of pollution. The concentration of pollution increases from zone 5 to zone 1. In zone 1, the level of pollution is the maximum possible, at which some species of woody plants are still preserved, but in a strongly depressed state. In the second zone, the impact of pollution is a factor determining the death of forest stands. In the third and fourth zones, forest stands are weakened due to the impact of pollutants. In zone 5, external signs of weakening of woody plants are not expressed, and the level of pollution is close to the background.

A significant part of scientific publications is devoted to assessing the role of forest pests in a complex study of the state of plantations (Morreti, Barbalat, 2004; Parker, Clancy, Mathiansen, 2006; Isaev, Nedorezov, Khlebopros, 1982; Selikhovkin, 1992; Kataev, Osetrov, Popovichev, Selikhovkin, 2001; Selikhovkin, Bondarenko, Popovichev, 2010). Migration of biological populations across the territory in search of better living conditions is carried out in nature, as a rule, in a wave mode. It was shown that waves of biological populations can be of two types (Zhizhin, 2004, 2005). The first type of wave of a biological population is characterized by the filling of the territory with individuals of the population with a certain density, starting from the moving wave front and, in principle, to infinity along the spatial coordinate in the direction opposite to the direction of wave movement. This type of wave was first theoretically studied in the work of Kolmogorov, A.N., Petrovsky, I.G., & Piskunov, N.S. (1937). Many works on waves of biological populations rely on this study without sufficient grounds (Svirezhev, 1987, Kovalev, Vechernin, 1986, and others), although the model adopted in this study is very simple and does not reflect the features of many wave processes in biological populations. Taking into account the features of this type of waves, with the help of which individuals of the population capture the entire territory over which the wave ran, this type of waves is called conquest waves (Zhizhin, 2004). Another type of waves of biological populations has a profile of change in the density of distribution of individuals with a maximum, to the right and left of which the density of distribution of individuals in the population decreases to zero. This kind of waves is called solitary waves of the biological population (Keller, Segel, 1971; Kennedy, Aris, 1980; Ivanov, 1984). In physics, this type of wave is called solitons. It is known that populations of bark beetles also migrate across the territory in search of better conditions for their existence. It is of interest to determine the conditions for the possibility of spreading populations of beetles - bark beetles, both in waves of conquest and solitary waves. To do this, it is necessary to construct the corresponding mathematical models of the migration of the population of bark beetles, the solutions of which would be waves of the first or second type.

13.1. MATHEMATICAL MODEL OF THE DISTRIBUTION OF THE POPULATION OF BEETLES - BARK BEETLES BY A WAVE OF CONQUEST

The migration of beetles - bark beetles at the stage of flight after they leave their wintering grounds will be described by a one-dimensional parabolic diffusion equation:

$$\frac{\partial N}{\partial t} = \frac{\partial}{\partial x}\mu\frac{\partial N}{\partial x} + \left(\frac{1}{\tau_{c1}} - \frac{1}{\tau_{e1}}\right)N$$

, (13.1)

where N is the concentration of bark beetles in a flock; t is time; x is spatial coordinate; τ_{c1} is characteristic time of the increase in the flock of bark beetles due to the approach of beetles to the flock from wintering places; τ_{e1} is characteristic life time of bark beetles in a flock serving as food for other biological populations (other insects, birds); μ is mobility of bark beetles, which characterizes their ability to chaotic movement.

Let us assume that bark beetles in their movement react to a pheromone gradient emanating from trees, according to which bark beetles determine the direction of convective movement towards trees suitable for settlement. These trees are weakened to a certain extent, so that they cannot actively resist the colonization of bark beetles in them, but still contain enough food for the breeding of bark beetle offspring. Let us direct the spatial variable x towards the weakened forest stand and take the mobility of bark beetles μ as a constant value.

The characteristic time of the increase in the flock of bark beetles due to their desire to τ_{c1} will also be considered a constant value. The reciprocal of τ_{c1} is the rate of growth of the flock. It can be assumed that this rate is proportional to the concentration of living bark beetles in wintering areas by the time of the beginning of flight, i.e.

$$\tau_{c1}{}^{-1} = k_1 N_0,$$

where N_0 is the concentration of live bark beetles in the wintering areas by the time of the beginning of the flight; k_1 is coefficient of proportionality.

The rate of death of bark beetles from their consumption by other biological populations will be considered proportional to the concentration of bark beetles in the flock, i.e.

$$\tau_{c1}{}^{-1} = k_2 N,,$$

where k_2 is a proportionality coefficient that takes into account the concentration of individuals in biological populations that eat bark beetles.

Under the assumptions made, equation (12.1) takes the form

$$\frac{\partial N}{\partial t} = \mu\frac{\partial^2 N}{\partial x^2} + k_2(N_+ - N)N$$

, (13.2)

$N_+ = (k_1/k_2)N_0.$

We assume that the wintering area of this population of bark beetles is large enough for the formation of a stationary wave of flying bark beetles over the territory, i.e. a traveling wave of a constant shape is formed over the territory, moving at a constant speed. Let us show that equation (13.2) leads to a stationary wave of the conquest type (see Figure 1).

Figure 1. The wave of conquest of the population of bark beetles: N is concentration of bark beetles; u is wave propagation speed; x is spatial coordinate

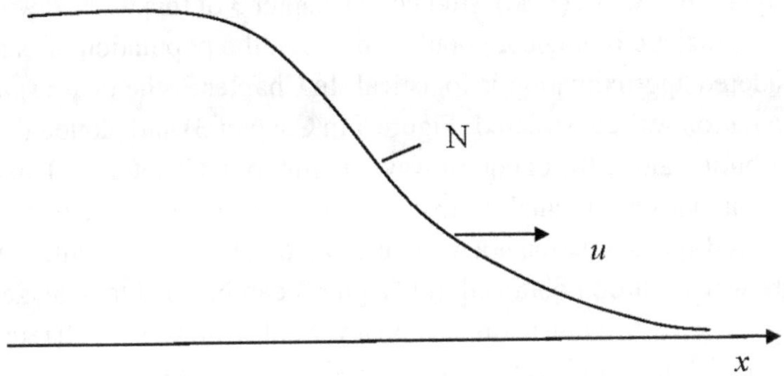

Let's introduce the wave coordinate $z=ut-x$. Then, passing in equation (13.2) to wave coordinate z we have

$$\left(\frac{\partial}{\partial t} = \frac{d}{dz}\frac{\partial z}{\partial t} = u\frac{d}{dz}, \frac{\partial}{\partial x} = \frac{d}{dz}\frac{\partial z}{\partial x} = -\frac{d}{dz}\right)$$

$$u\frac{dN}{dz} = \mu\frac{d^2N}{dz^2} + k_2(N_+ - N)N$$

. (13.3)

Let us introduce dimensionless variables

$$n = N/N_+, \zeta = z(k_2N_+/\mu)^{1/2}, w = u(k_2N_+\mu)^{-1/2}.$$

Then equation (13.3) takes the form

$$w\frac{dn}{d\zeta} = \frac{d^2n}{d\zeta^2} + (1-n)n$$

. (13.4)

This equation reduces to the system of first order equations

$$\frac{dp}{d\zeta} = wp - (1 - n)n, \tag{13.5}$$

$$\frac{dn}{d\zeta} = p$$

. (13.6)

The system of equations (13.5), (13.6) is autonomous, it coincides with the system of equations (3.55), (3.56), studied in Chapter 3 of this work, devoted to the analysis of the logistic biological population. Thus, the population of bark beetles in the considered approximation is logistical. In Chapter 3, the phase space of the logistic population was constructed (Figure 5 in Chapter 3) and plotted the concentration distribution along the conquest wave (Figure 6 in Chapter 3). This schedule completely coincides with schedule 13.1. This proves that the wave of bark beetles in the considered approximation is a wave of conquest. The dimensionless analytical approximate wave solution obtained in Chapter 3 can be used to analyze the bark beetle conquest wave. Moving to dimensional variables, however, will require using the definition of dimensionless variables introduced in this section.

13.2. MATHEMATICAL MODEL OF THE DISTRIBUTION OF A POPULATION OF BARK BEETLES BY A SOLITARY WAVE

However, over a long period of time, the process of propagation of a wave of a population of bark beetles cannot be stationary (i.e., with a constant structure and propagation velocity). First, if no tree stand suitable for their settlement is found in the wintering area of the bark beetles, then the bark beetles fly outside the wintering area. This makes it impossible for the insect population to be fed in flight by bark beetles from the wintering area. In this case, only the term associated with the death of bark beetles will remain in the source terms of equation (13.3). The concentration of bark beetles in the flying population will begin to decrease. This will lead to a limitation of the effective length of the wave. The wave will take the form of a solitary wave (soliton) (see Figure 2). The process of formation of a solitary wave of a population of bark beetles is, in principle, non-stationary. Secondly, at some point in time, a suitable stand for the settlement of bark beetles will be found.

Figure 2. Concentration profile in a solitary wave of a population of bark beetles

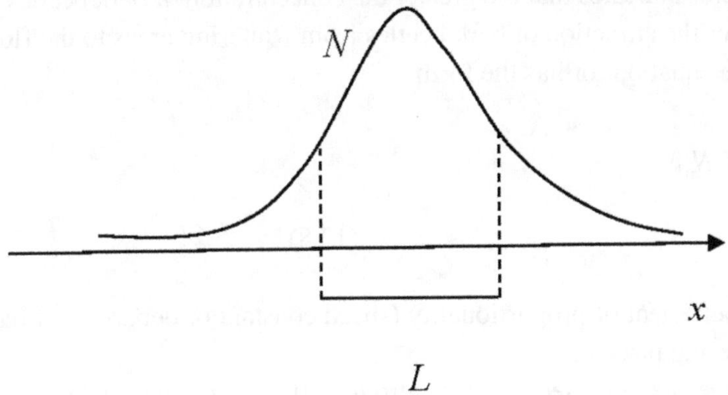

From this moment on, a flying flock of bark beetles will begin to settle on selected trees and intrude into them.

It is of interest to determine the characteristics of a stationary solitary wave of a population of bark beetles, which will lead to a non-stationary process of its formation.

We will assume that the rate of increase in the concentration of bark beetles in a flock τ_{c1}^{-1} is a variable. Since in the previous section this speed was defined as a value proportional to the concentration of live bark beetles in a given wintering place at the time of departure k_1N_0, the assumption of speed variability τ_{c1}^{-1}, provided that the proportionality coefficient k_1 is constant, is equivalent to the assumption of variability N_0. The latter can be explained by the following reasoning. Let us assume that before the start of the flight of bark beetles throughout the wintering area, live bark beetles are evenly distributed over the wintering area with a concentration of N_{00}. It should be considered that bark beetles cannot instantly fly out all at once. Therefore, by the time the wave of the population of bark beetles passed over the given place, part of the bark beetles had already flown from this place. Therefore, the concentration of live bark beetles in wintering areas during the flight is variable. Under the accepted condition, the diffusion equation (13.1) has the form

$$\frac{\partial N}{\partial t} = \mu \frac{\partial^2 N}{\partial x^2} + (k_1 N_0 - k_2 N) N$$

. (13.7)

Equation (13.7) must be supplemented with an equation for the variable N_0. We assume that the rate of change N_0 is proportional to the product NN_0. The first factor in this product indicates that the speed of emergence of bark beetles from wintering areas is proportional to the concentration of bark beetles in wintering areas. The second factor indicates that the greater the concentration of bark beetles in a flock, the stronger the attraction of bark beetles from wintering areas to the flock. In this regard, the equation for has the form

$$\frac{\partial N_0}{\partial t} = -k_0 N_0 N$$

, (13.8)

k_0 is coefficient of proportionality (speed constant of departure of bark beetles from wintering places).

As in the previous section, we introduce the wave coordinate $z = ut - x$ and dimensionless quantities

$$n = N/N_+, \zeta = z(k_2 N_+/\mu)^{1/2}, w = u(k_2 N_+ \mu)^{-1/2}, N_+ = (k_1/k_2) N_{00}$$

. (5.3)

In addition, we introduce $n_0 = N_0/N_{00}$ - a dimensionless concentration of bark beetles in wintering areas. Then the system of equations (13.7), 13.8) can be written as

$$\frac{dn}{d\zeta} = p$$

, (13.10)

$$\frac{dp}{d\zeta} = wp - n(n_0 - n)$$

, (13.11)

$$\frac{dn_0}{d\zeta} = -\frac{k_0 n_0 n}{k_2 w}$$

. (13.12)

13.3. QUALITATIVE STUDY OF THE SOLITARY WAVE DISTRIBUTION MODEL OF THE BARK BEETLE POPULATION

The system of equations (13.10) - (13.12) is autonomous and its solutions can be represented by trajectories in the three-dimensional phase space (p,n,n_0) (Zhizhin, 2008; Zhizhin, Selikhovkin, 2012). The axis n_0 in this space is a set of equilibrium positions of the system. The location of the trajectories in space (p,n,n_0) is determined by zero surfaces: the plane $p = 0$, on which the derivative $dn/d\zeta=0$; planes $n=0$, $n_0=0$, on which the derivative $dn_0/d\zeta=0$; surface $p=w^{-1}(n_0-n)n$, on which the derivative $dp/d\zeta=0$. These zero surfaces divide the phase space into regions with constant signs of the derivatives of the phase coordinates. In Figure 3, the areas of positive values of the derivatives of the corresponding phase coordinates are indicated by arrows near the zero surfaces.

Figure 3. Phase space of the system of equations (13.10) - (13.12)

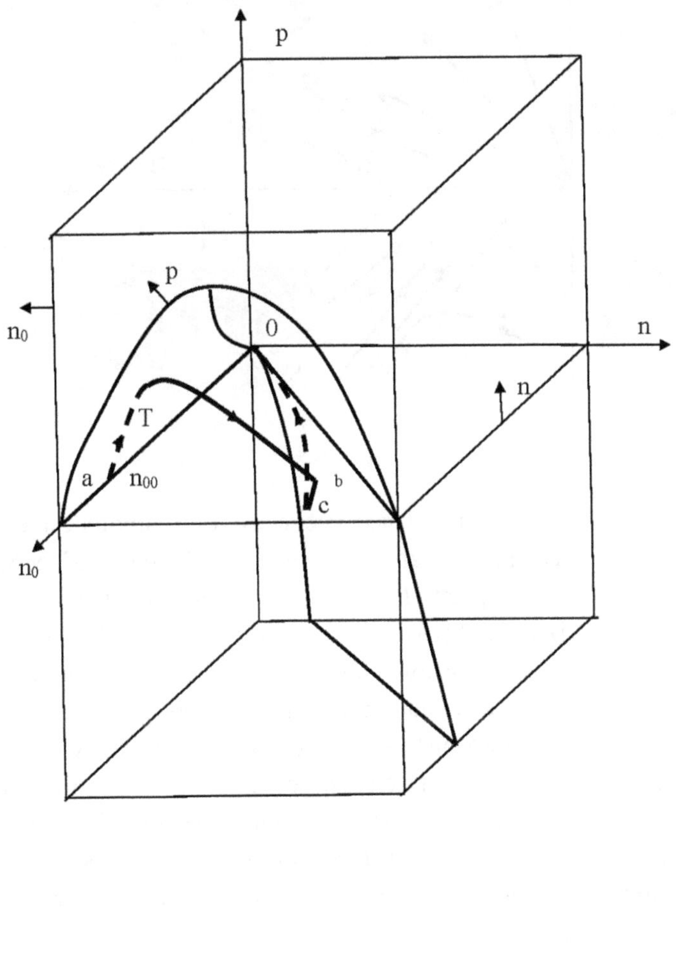

The coordinate plane n_0=0 is an integral plane, since on it the system of equations (13.10) - (13.12) degenerates into a system of two equations

$$\frac{dn}{d\zeta} = p, \frac{dp}{d\zeta} = wp + n^2$$

. (13.13)

Solutions to system (13.13) can be represented by trajectories in the phase plane (p,n) . It has a complex singular point at the origin (p=0, n=0). The zero isocline $p= -n^2/w$, located in the negative quadrant, and the zero isocline p=0, pass through the origin. The location of the trajectories in the phase plane is shown in Figure 4.

Figure 4. The phase plain of the system (13.13)

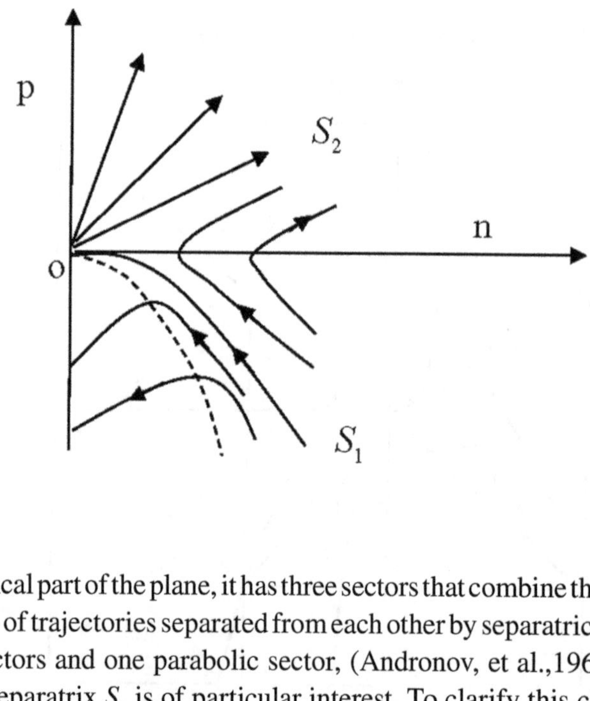

In the physical part of the plane, it has three sectors that combine three qualitatively different types of trajectories separated from each other by separatrices S_1 and S_2 (two hyperbolic sectors and one parabolic sector, (Andronov, et al.,1966)). For further analysis, the separatrix S_1 is of particular interest. To clarify this circumstance, let us again turn to the three-dimensional phase space (p,n,n_0). In accordance with the signs of the derivatives indicated in Figure 3, trajectories into the positive octant of the phase space over the zero surface $p=w^{-1}(n_0-n)n$ can go out of the equilibrium positions of the axis n_0(n_0>0). Let's choose some position of equilibrium a on this axis with the initial value of concentration n_{00}. The trajectory T leaving this equilibrium position can intersect the zero surface $p=w^{-1}(n_0-n)n$ and end up under the zero surface, but above the plane p = 0. After crossing the zero surface $p=w^{-1}(n_0-n)n$, the trajectory begins to approach the plane $p = 0$ and intersects this plane at some point b (Figure 3). After that, the trajectory goes into the negative octant of the phase space and turns around n (n begins to fall). It's accompanied approaching the trajectory to the zero surface $p=w^{-1}(n_0-n)n$ and the trajectory intersects this surface at some point c.

In the future, the trajectory begins to rise to the plane $p = 0$ (p increases). In this case, since n_0 it decreases all the time, the trajectory approaches the plane $n_0 = 0$. Of all the possible such trajectories, only the trajectory that approaches the separatrix S_1 in Figure 3 and enters a complex equilibrium position, the origin of coordinates of the three-dimensional phase space, makes sense.

The change in the values of the phase coordinates along the trajectory T with an increase ς in the wave coordinate is shown in Figure 5.

Figure 5. Solitary wave of the bark beetle population

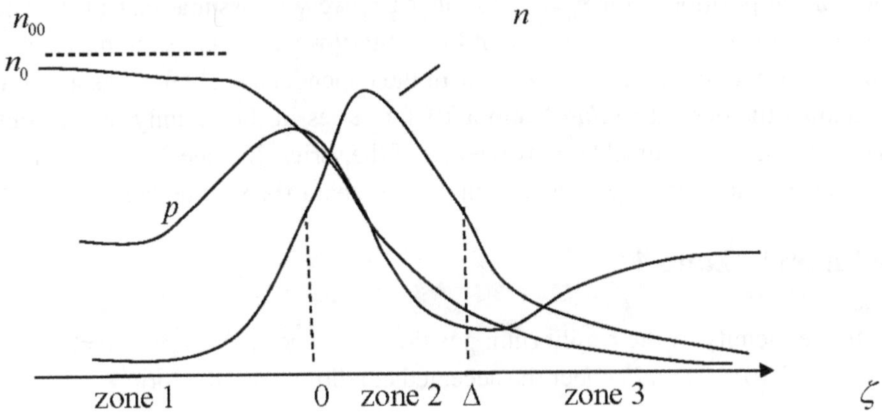

Figure 5 shows that the solution T is a solitary wave of the bark beetle population (soliton). In the laboratory (fixed) coordinate system, this wave moves in the direction of the axis x (towards the forest stand chosen for the settlement) with a certain speed u. At the same time, the initial concentration of bark beetles in wintering places vanishes.

The task is to find the conditions for the existence of such a wave solution and to obtain expressions for calculating the propagation velocity of a solitary wave and its structure (concentration change profiles along the wave) depending on the values of regime parameters.

It should be noted that in the work, devoted to modeling the distribution of a population of ragweed leaf beetles (Kovalev, Vechernin, 1986) the authors, assuming the existence of a solitary wave of this population, did not pay attention to the fact that the system of equations considered by them actually does not have solutions of the solitary wave type.

13.4. APPROXIMATE ANALYTICAL SOLUTION OF THE EQUATIONS OF PROPAGATION OF A POPULATION OF BARK BEETLES BY A SOLITARY WAVE

To obtain an approximate analytical solution, we apply of the infinite zone reaction method. Let us divide the solitary wave of the bark beetle population into three zones (see Figure 5). In the first zone ($-\infty<\zeta<0$), we consider the solution of the system of equations (13.10) – (13.12) linearized in the vicinity of the initial equilibrium position *a* for $n_0=n_{00}$. In this case, we will assume that in the entire first zone $n_0=n_{00}$. The second zone of the solitary wave contains the maximum of the soliton. We interpolate this zone with the dependences obtained as a result of expanding the derivative $dn/d\zeta$ into a Taylor series in the vicinity of the soliton maximum, limiting ourselves to two terms of the series. In zone 3, for the variable, we consider the solution corresponding to the separatrix S_1 to be fair (Figure 4).

Solution In Zone 1

In the vicinity of the equilibrium position with coordinates $p=0$, $n=0$, $n_0=n_{00}$, system (13.10) – (13.12) under the accepted conditions has the form

$$\frac{dn}{d\zeta} = p$$

, (13.14)

$$\frac{dp}{d\zeta} = wp - n n_{00}$$

. (13.15)

From the discriminant of system (13.14), (13.159) we find the eigenvalues of the initial equilibrium position

$$\lambda_{1,2} = \frac{w}{2} \pm \sqrt{\frac{w^2}{4} - n_{00}}$$

. (13.16)

It follows from (13.16) that the eigenvalues of the initial equilibrium position are either real and positive at $n_{00}<w^2/4$ (this corresponds to a singular point of the unstable knot type (Pontryagin, 1961)) or complex conjugate $n_{00}>w^2/4$ at (this corresponds to a singular point of the unstable focus type (Pontryagin, 1961)). Since trajectories leaving the equilibrium position are of interest for the analysis of the wave solution, the condition should be chosen $n_{00}<w^2/4$, i.e. a singular point of the unstable knot type.

We divide equation (13.15) by equation (13.14)

$$\frac{dp}{dn} = \frac{wp - nn_{00}}{p}$$

. (13.17)

We will reveal the uncertainty at a point $p=0$, $n=0$ according to the L'Hopital rule. We obtain a quadratic equation for eigendirections at a singular point

$$\left(\frac{dp}{dn}\right)^2 = w\frac{dp}{dn} - n_{00}$$

. (13.18)

From (13.18) we have

$$\left(\frac{dp}{dn}\right)_{1,2} = \frac{w}{2} \pm \sqrt{\frac{w^2}{4} - n_{00}} = \lambda_{1,2}$$

. (13.19)

In order for the trajectory to cross the zero surface $p=w^{-1}(n_0-n)n$, and this is necessary for the trajectory to be a wave solution (see Figure 3), given that n_0 it decreases all the time and at the same time the zero surface $p=w^{-1}(n_0-n)n$ is pressed against the plane p = 0, you should choose your own direction with the smallest slope . Therefore, from (13.19) we choose the second eigendirection (the minus sign in front of the root) and, accordingly, the eigenvalue λ_2.

Thus, the solution in the vicinity of the initial equilibrium position has the form

$$n = n_f\exp(\lambda_2\zeta), n_0 = n_{00}$$

, (13.20)

where n_f is the value n at the boundary of zone 1 at ζ=0.

Solution In Zone 2

Considering that zone 2 contains the maximum of the soliton, let us consider the system of equations (13.10) – (13.12) in the vicinity of the maximum. Let us expand the right-hand side of equation (13.10) in a Taylor series in the vicinity of the soliton maximum

$$P(\zeta)= P(\zeta_m)+\left.\frac{dP}{d\zeta}\right|_{\zeta=\zeta_m}(\zeta-\zeta_m)+.....$$

Since $P(\zeta_m)=0$, and according to (13.11) $dP/d\zeta=wP-\alpha$, where $\alpha=(n_0-n)n$, then

$$\left.\frac{dP}{d\zeta}\right|_{\zeta=\zeta_m} = -\alpha_m = -(n_{0m}-n_m)n_m:$$

the m indicates the values of the variables at the maximum of the soliton. Restricting ourselves to two terms in the Taylor series, we have an approximate equality

$$P(\zeta)= -(n_{0m}-n_m)n_m(\zeta-\zeta_m)\ .$$

Substituting this equality into equation (13.11) and integrating it, we obtain the law of change in the concentration of insects in the second zone of the solitary wave

$$n = n_m - 0,5(n_{0m}-n_m)n_m(\zeta-\zeta_m)^2$$

. (13.21)

Substituting (13.21) into equation (13.12) and integrating it taking into account the condition $n_0(\zeta_m)=n_{0m}$, we obtain the law of change in the concentration of insects in wintering areas along the second zone of a solitary wave

$$n_0 = n_{0m}\exp\left\{ -\frac{k_0 n_m}{k_2 w}(\zeta-\zeta_m)[1-\frac{n_{0m}-n_m}{6}(\zeta-\zeta_m)^2]\right\}$$

. (13.22)

Solution In Zone 3

In zone 3 n_0=0, and the profile $n(\zeta)$should be determined by the separatrix S_1 (see Figure 13.4). Let us introduce a variable y=n^2 in the system of equations (13.13). Then this system takes the form

$$\frac{dy}{d\zeta} = 2py^{1/2}$$

, (13.23)

$$\frac{dp}{d\zeta} = wp + y$$

. (13.24)

The phase diagram of the system (13.23), (13.24) is shown in Figure 6.

Figure 6. Phase diagram of the system (13.23), (13.24)

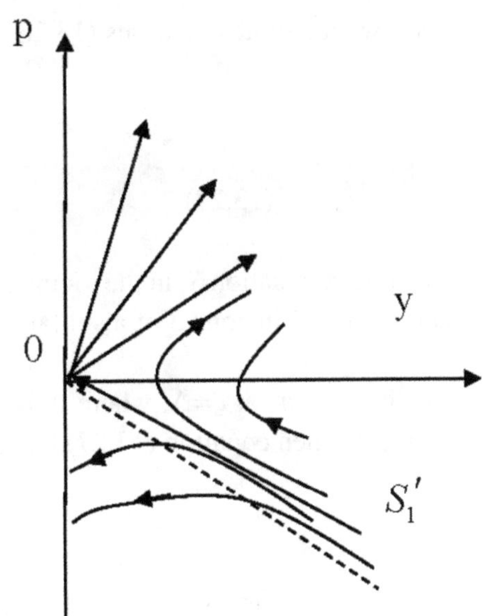

The difference between the phase diagram in Figure 6 and the phase diagram in Figure 4 is that the zero isocline in Figure 6 (dotted line) is a straight line $p= -y/w$. In its vicinity there is a separatrix S_1', which is the image of the separatrix S_1 in Figure 4. Divide equation (13.24) by equation (13.23)

$$\frac{dp}{dy} = \frac{wp + y}{2py^{1/2}}$$

. (13.25)

Expanding the uncertainty of the right side of (13.25) at the point p=0, y=0 according to L'Hospital's rule two times, we obtain an expression for the eigendirection at the singular point $dp/dy= -1/w$.

Integrating the last equation, taking into account that the proper direction passes through the origin, we obtain

$$p= -y/w, \qquad (13.26)$$

those, the separatrix equation S_1'approximately coincides with the zero isocline equation.

Returning to the variable n, we integrate equations (13.26)

$$n = \left(\frac{\zeta}{w} + K\right)^{-1}$$

. (13.27)

Equation (13.27) is the separatrix equation S_1 in the vicinity of the final equilibrium position, i.e. profile equation $n(\zeta)$ in zone 3 of a solitary wave (K is constant of integration).

We use the boundary condition $n=n_\Delta$ at $\zeta=\Delta$, where Δ is the width of zone 2 (i.e., the right boundary of zone 2), then equation (13.27) takes the form

$$n = \frac{n_\Delta}{1 + n_\Delta(\zeta - \Delta)/w}$$

. (13.28)

Substituting equality (13.28) into equation (13.12) and integrating, taking into account that $n_0=n_{0\Delta}$ at $\zeta=\Delta$, we obtain the profile equation $n_0(\zeta)$ in zone 3 of a solitary wave of insects

$$n_0 = \frac{n_{0\Delta}}{[1 + n_\Delta(\zeta - \Delta)/w]^{k_0/k_2}}$$

. (13.29)

The equations of the laws of change of variable concentrations along the wave coordinate of a solitary insect wave (13.25) - (13.29) include many parameters:

$n_f, n_\Delta, n_{0\Delta}, n_m, n_{0m}, n_{00}, \Delta, \lambda, w, \zeta_m, k_0/k_2$.

Two parameters from the above list n_{00} and k_0/k_2 should be considered known, since, in principle, they should be set before solving the problem. The values of the remaining nine parameters must be found in order to be able to use equations (13.25) - (13.29) when constructing the structure of a solitary insect wave. Therefore, one must have a system of nine algebraic equations with respect to these nine parameters. Moreover, this system must have a solution, i.e. must be resolvable, and must include the parameters n_{00}, k_0/k_2, as given parameters. So far, out of these nine equations, there is only one equation (13.19), which relates the eigenvalue λ_2 and the wave velocity w. The remaining eight equations can be obtained by setting certain conditions at the boundaries of the zones. The choice of conditions should ensure the solvability of the problem and be based on its meaning.

On the boundary between zones 1 and 2 (ζ=0) we set the condition for the continuity of the values of the variables *n*, n_0 and the continuity of the first derivative $dn/d\zeta$. The condition of continuity of variables by equations (13.21), (13.22) leads to the equalities

$$n_f = n_m - 0,5(n_{0m} - n_m) n_m \zeta_m^{\ 2}$$

, (13.30)

$$n_{00} = n_{0m} \exp\left[\frac{k_0 n_m}{k_2 w}\zeta_m\left(1 - \frac{n_{0m} - n_m}{6}\zeta_m^2\right)\right]$$

. (13.31)

Differentiating equations (13.20), (13.21) and equating the derivatives $dn/d\zeta$ at the boundary of the first and second zones at ζ=0, we have the equality

$$\lambda_2 n_f = (n_{0m} - n_m) n_m \zeta_m$$

. (13.32)

In addition, at the boundary of zones 1 and 2, we equate to zero the sum of the second derivatives $d^2n/d\zeta^2$ on both sides of the boundary. This simulates the existence of an inflection point in a real soliton profile. Differentiating equations (13.20), (13.21) two times, we obtain

$$\lambda_2^2 n_f = (n_{om} - n_m) n_m \quad . \qquad (13.33)$$

On the border between zones 3 and 2 ($\zeta=\Delta$), we set the condition for the continuity of the values of the variables n, n_0 and the continuity of the first derivatives $dn/d\zeta$, $dn_0/d\zeta$. The condition of continuity of variables on this boundary by equations (13.21), (13.22) leads to the equalities

$$n_\Delta = n_m - 0,5(n_{0m} - n_m) n_m (\Delta - \zeta_m)^2 \quad , \qquad (13.34)$$

$$n_{0\Delta} = n_{0m} \exp\left[-\frac{k_0 n_m}{k_2 w}(\Delta - \zeta_m)(1 - \frac{n_{0m} - n_m}{6}(\Delta - \zeta_m)^2\right] \quad . \qquad (13.35)$$

Differentiating equations (13.28), (13.21) and equating the derivatives $dn/d\zeta$ at the boundary of the third and second zones at $\zeta=\Delta$, we have the equality

$$n_\Delta{}^2/w = (n_{0m} - n_m) n_m (\Delta - \zeta_m) \quad . \qquad (13.36)$$

Differentiating equations (13.29), (13.22) and equating the derivatives $dn_0/d\zeta$ at the boundary of the third and second zones at $\zeta=\Delta$, we have the equality

$$n_{0\Delta}^2 = n_{0m} n_m [1 - 0,5(\Delta - \zeta_m)^2] \exp\left\{ -\frac{k_0 n_m}{k_2 w}(\Delta - \zeta_m)[1 - \frac{n_{0m} - n_m}{6}(\Delta - \zeta_m)]\right\} \quad . \qquad (13.37)$$

So we have nine non-linear algebraic equations (13.19), (13.30) - (13.37) with respect to nine unknowns $n_f, n_\Delta, n_{0\Delta}, n_m, n_{0m}, \Delta, \lambda_2, w, \zeta_m$. Despite the complexity of this system of equations, by successive elimination of unknowns, it can be reduced to a system of two equations for the wave velocity w and the width of the second zone Δ

$$n_{00} = 2\lambda_2 \varphi \exp\frac{16k_0 y}{3k_2(3-y^2)^2}$$

, (13.38)

$$\varphi\frac{3-y^2}{yw} = \exp\frac{k_0 y^2(9-y^2)}{k_2(3-y^2)^2}$$

, (13.39)

where $y = \Delta\lambda_2 - 1, \varphi = \frac{\lambda_2}{3} + \frac{3wy}{(3-y^2)^2}, \lambda_2$- is determined by equation (13.19).

The model of the solitary wave of the population of bark beetles, in comparison with the conquest wave, more adequately corresponds to the observations of the process of migration of bark beetles and indicates not only the speed of the wave movement, but also its characteristic length depending on the dimensionless concentration of bark beetles in the wintering areas by the time of the beginning emergence of bark beetles from wintering areas and kinetic parameters characterizing the rate of development and death of the population of bark beetles.

Numerical Study of the Analytical Solution of Wave Equations Describing the Distribution of a Population of Bark Beetles by a Solitary Wave

For given values n_{00}, k_0/k_2 the system of equations (13.38), (13.39) is solved numerically. As a result, the speed of the solitary wave w and the width Δ of the second zone are found. After that, λ_2 it is determined by (13.19), n_m- by the equation

$$n_m = \frac{6wy\lambda_2}{(3-y^2)^2}$$

, (13.40)

n_{0m} - according to the equation

$n_{0m} = n_m + \lambda_2 2/3$, (13.41),

$\zeta_m = 1/\lambda_2$, (13.42)

n_f calculated according to equation (13.30), n_Δ- according to equation (13.34), $n_{0\Delta}=n_\Delta$.

Based on the obtained values of the parameters, the concentration profiles along the solitary wave are constructed according to equations (13.20) - (13.22), (13.28), (13.29).

As an example, in Figure 7 shows the results of the calculation of dependencies $y_1(w)$ according to the equation (13.38) and $y_2(w)$ according to the equation (13.39) at n_{00}=0,8; k_0/k_2=1. In Figure 8 shows according to structure solitary wave of bark beetles.

Figure 7. Solving equations (13.38), (13.39)

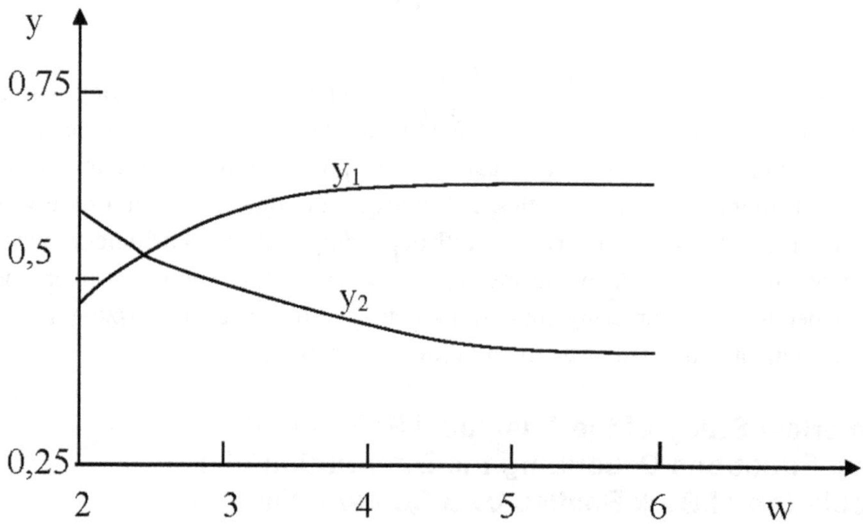

Figure 8. The structure of a solitary wave of a population of bark beetles at n_{00}=0,8; k_0/k_2=1

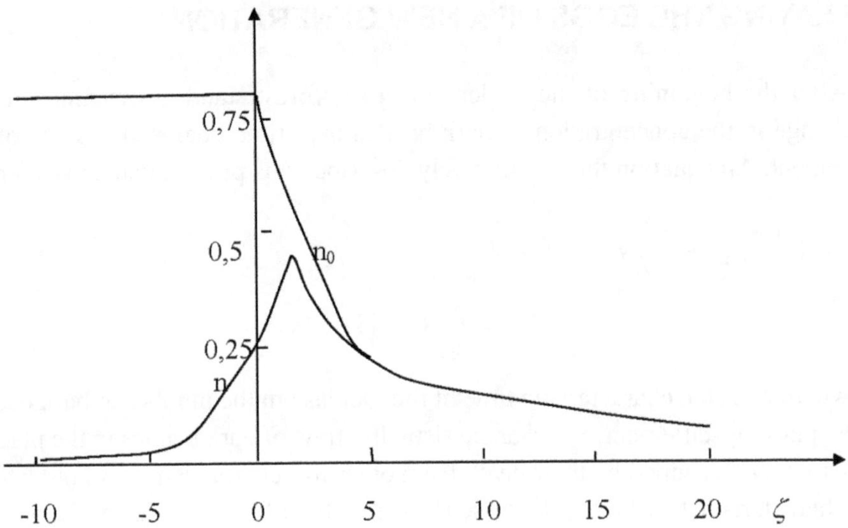

The parameter values corresponding to the intersection point of the graphs in Figure 7 are equal to

$$w = 2,45; \lambda_2 = 0,387966; \Delta = 3,93075; n_m = 0,4034011 ;$$
$$n_{0m} = 0,503746; n_f = 0,26818; n_\Delta = n_{0\Delta} = 0,3663386; \zeta_m = 2,5775418 \cdot$$

From the analysis of the constructed graphs, it follows that the speed of the solitary wave is greater than 2 and the decrease in insect concentrations in flight and in wintering places occurs in the third zone of the solitary wave at k_0/k_2=1 according to one law. A detailed numerical study of the influence of the initial concentration of beetles - bark beetles in wintering areas and the ratio of velocity constants k_0/k_2=1 on the characteristics of a solitary wave can be carried out after a reasonable determination of the values of these parameters.

It should be noted that the model of a solitary wave of bark beetles constructed in this chapter differs significantly from the model proposed in work (Zhizhin, 2011), since it takes into account the nonlinear nature of changes in the concentration of bark beetles in wintering areas in the second zone of a solitary wave and the possibility of incomplete disappearance of bark beetles. wintering grounds in this area.

13.5. MATHEMATICAL MODEL OF THE FORMATION OF THE CENTER OF THE SETTLEMENT OF THE POPULATION OF BARK BEETLES UNTIL THE MOMENT OF LAYING THE EGGS OF A NEW GENERATION

After the beginning of the settlement in the forest stand, a dynamic process of change in the concentration of bark beetles over time begins at the site of the settlement. An equation that qualitatively describes this process can be written as

$$\frac{dN}{dt} = \left(\frac{1}{\tau_{c2}} - \frac{1}{\tau_{e2}} - \frac{1}{\tau_a}\right)N$$

, (13.43)

where τ_{c2} is the characteristic time of the increase in the number of bark beetles in the place of settlement; τ_{e2}- characteristic life time of bark beetles in the place of settlement, determined by their death from other insects (parasites and predators); τ_a - characteristic time of the decrease in the number of bark beetles in the place of settlement, determined by the resistance of the tree, the effects of competition with a large accumulation of individuals, i.e. lack of space, lack of food, a change in the chemical situation in the settlement, a lot of noise, leading to the flight of bark beetles from the place of the settlement in search of more free places.

The rate of death of bark beetles from insects at the stage of settlement, as well as at the stage of flight, will be considered proportional to the concentration of bark beetles, but with its own specific coefficient of proportionality k_3

$$1/\tau_{e2} = k_3 N$$

. (13.44)

The rate of decrease in the number of bark beetles due to the resistance of the tree and the flight of bark beetles from the place of settlement, associated with the effects of competition, assuming that the primary cause of the effects of competition is the lack of space, we set the expression

$$\frac{1}{\tau_a} = \frac{1}{\tau_{a0}}\exp\frac{1-\varphi}{\varphi}$$

, (13.45)

where $1/\tau_{a0}$ is the rate of decrease in the number of bark beetles from the place of settlement with an excess of space (tree resistance), φ is the proportion of free space unoccupied by passages carved by settled bark beetles ($0 \le \varphi \le 1$).

Since in equation (13.43) to change the concentration of bark beetles, a variable proportion of free space entered φ, then equation (13.43) must be supplemented with an equation describing the change φ. It can be written in the form

$$\frac{d\varphi}{dt} = -\eta N/\tau_a, \qquad (13.46)$$

where η is a coefficient that characterizes the influence of the rate of change of free space on the process of development of the population of bark beetles.

Taking into account the accepted conditions, the system of equations for the process of formation of a settlement has the form

$$\frac{dN}{dt} = \left(\frac{1}{\tau_{c2}} - k_3 N - \frac{1}{\tau_{a0}} \exp\frac{1-\varphi}{\varphi}\right) N, \qquad (13.47)$$

$$\frac{d\varphi}{dt} = -N\frac{\eta}{\tau_{a0}} \exp\frac{1-\varphi}{\varphi}. \qquad (13.48)$$

Let us introduce dimensionless variables and parameters

$$\tau = t/\tau_{c2}, n = k_3 \tau_{c2} N, \beta = \tau_{c2}/\tau_{a0}, \gamma = \eta/\left(\tau_{a0} k_3\right). \qquad (13.49)$$

Then equations (13.47), (13.49) take the form

$$\frac{dn}{d\tau} = \left(1 - n - \beta \exp\frac{1-\varphi}{\varphi}\right) n, \qquad (13.50)$$

$$\frac{d\varphi}{d\tau} = -\gamma n \exp\frac{1-\varphi}{\varphi}. \qquad (13.51)$$

The system of equations (13.50), (13.51) is autonomous and its solutions can be represented by trajectories in the phase plane (n,φ).

The straight-line $n = 0$ on this plane is a set of equilibrium positions of the system. Moreover, it makes sense to consider only the case when $\beta < 1$, since then there are stable and unstable equilibrium positions on the axis φ (see Figure 9).

Figure 9. Phase diagram of the system (13.50), (13.51)

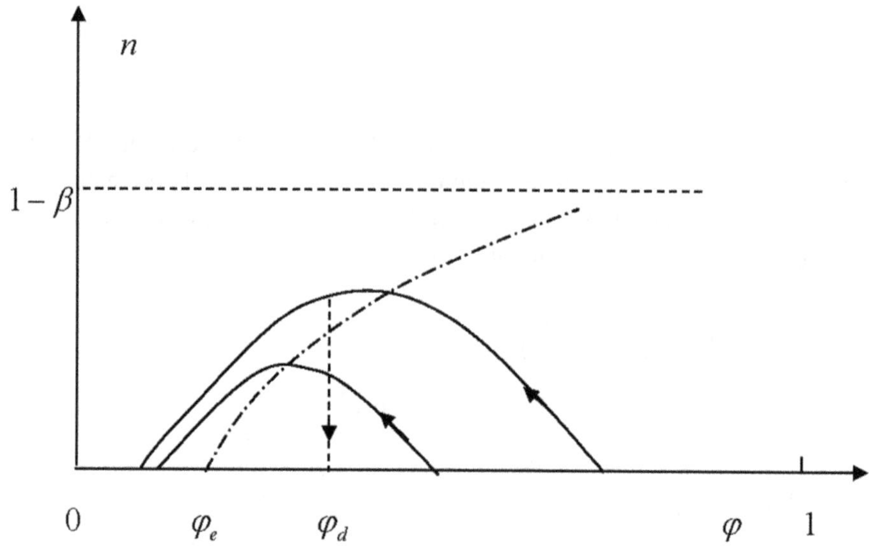

Stable equilibrium positions are located on the axis φ at $\varphi<\varphi_e$. Unstable equilibrium positions are located on the axis φ at $\varphi>\varphi_e$. The trajectories go from unstable equilibrium positions to stable equilibrium positions through the maximum concentration of bark beetles at the moment the trajectories cross the zero isocline $n = 1 - \beta \exp(\frac{1-\varphi}{\varphi})$ (dashed line in Figure 9), on which $dn/d\tau=0$. Thus, immediately after the beginning of the settlement of bark beetles in the forest stand, the concentration of bark beetles in the settlement first grows, and then begins to decrease due to death and as a result of the fact that some of the bark beetles leave the settlement due to its overpopulation. The tendency to zero content of bark beetles in the settlement (the extinction of the settlement), according to the phase diagram in Figure 9, is asymptotic. However, this asymptotic process of extinction is superimposed by an abrupt process of dying of bark beetles after laying eggs of a new generation of bark beetles at some point in time τ_d. For the beginning of the process of formation of the settlement, you can choose a certain moment τ_0 at which the front of the wave (either the wave of conquest or the solitary wave) reached the place of the settlement with a concentration of bark beetles on the front n_f. Graphs of time changes in the concentration of bark beetles and the free share of space in the place of settlement, according to the accepted model, have the form shown in Figure 10.

Figure 10. Graphs of changes in the concentration of bark beetles and the free share of space at the first stage of the settlement process

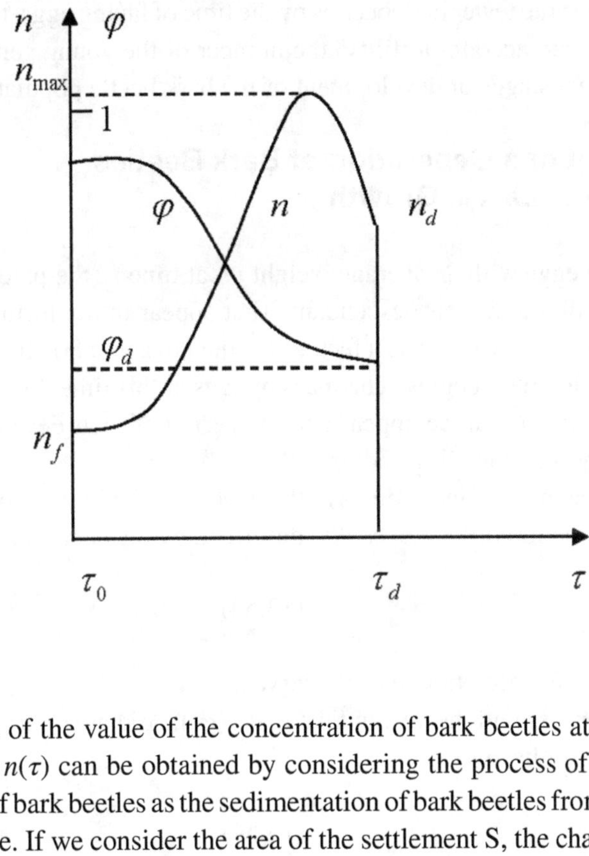

An estimate of the value of the concentration of bark beetles at the maximum of the function $n(\tau)$ can be obtained by considering the process of increasing the concentration of bark beetles as the sedimentation of bark beetles from the incoming population wave. If we consider the area of the settlement S, the characteristic size of the incoming wave of the population Δ with the concentration of bark beetles N with the speed u, then we get the ratio free share of space in the first stage of the settlement process $SN_{max}=hu\Delta N_+$, where h is the characteristic size of the population wave in height.

Then from the obtained relation we have $N_{max}=hu\Delta N_+/S$ or

$$n_{max} = k_3 \tau_{c2} u \Delta h N_+ / S \quad (13.52)$$

Relation (13.52) connects the characteristics of the population wave, the wintering area and the center of the settlement with each other. From relation (13.52) it follows that the smaller the area of the settlement, the greater the concentration of bark beetles at the maximum dependence $n(\tau)$ in the place of the settlement, i.e. the greater the concentration of bark beetles in the center of the settlement. This will lead to faster extinction of the population due to the effects of competition (lack of space, deterioration of living conditions, lack of food). The latter is of fundamental importance, since the fewer bark beetles by the time of laying eggs, the significantly smaller (taking into account fertility) the number of the young generation of bark beetles at the next stages of development of the bark beetle population.

Development of a Generation of Bark Beetles at the Stage of Larval Growth

After laying eggs with an average weight m_0 at time t_d, the parental generation of bark beetles dies and their descendants first appear in the form of eggs with a concentration $N_p = \nu N_d$, where ν is a fecundity equal to about 80÷100. The period of embryonic development begins. The mass of eggs at this time does not change. At the moment of time t_ε, larvae appear and the period of their development begins, accompanied by active feeding and growth.

Change in the mass of individuals at the post-embryonic stage of development, i.e. after the formation of larvae, can be described by the equation

$$dm/dt = a(\varphi), \qquad (13.53)$$

where $a(\varphi)$ is the growth rate of the larvae.

As a first approximation, we can take that $a(\varphi_d)$. And at $t = t_\varepsilon m = m_0$. Then, integrating (13.53), we have

$$m = a(t - t_\varepsilon) + m_0. \qquad (13.54)$$

At the end of the period of development of larvae at the moment of formation of pupae t_l, from (13.54) we have

$$m_l = a(t_l - t_\varepsilon) + m_0. \qquad (13.55)$$

The change in the number of larvae is associated with their death from insects and lack of space and food. This change can be described by the equation

$$\frac{dN}{dt} = -\frac{1}{\tau_{d1}} N$$

, (13.56)

Where $\frac{1}{\tau_{d1}}$ is the rate of death of bark beetles due to diseases, attacks by other insects, lack of space and food.

At this stage, there is no new arrival of larvae, except for their formation from eggs at the time moment t_{ε}, the larvae cannot leave the settlement on their own, and only their death, mainly due to lack of space, and therefore food during their active growth, affects the change in their number. Therefore, the rate of their death can be represented by the expression

$$\frac{1}{\tau_{d1}} = \frac{1}{\tau_{d10}}\exp\frac{1-\varphi}{\varphi}$$

, (6.15)

where $\frac{1}{\tau_{d10}}$ is the rate of death of larvae with an excess of space (φ=1).

The equation for changing the proportion of free space is

$$\frac{d\varphi}{dt} = -\eta_1 N/\tau_{d1}$$

, (13.58)

Where η_1 is a proportionality coefficient that takes into account the effect of a change in the share of free space on a change in the number of larvae at a given stage of population development.

The system of equations (13.56) - (13.58) has the first integral

$$N= \eta_1\varphi + A\ . \qquad (13.59)$$

We determine the integration constant A in equation (13.69) fr om the boundary condition

$$N= N_p = vN_d \text{ at } \varphi=\varphi_d\ . \qquad (13.60)$$

Then the function $N(\varphi)$ has the form

$$N= N_p - \eta_1(\varphi_d - \varphi)\ . \qquad (13.61)$$

It follows from equations (13.57), (13.58) that over time, the proportion of free space at the stage of larval development rapidly decreases with increasing speed, since, despite the decrease in the number of N, accompanied by a decrease in φ, the exponential dependence of the right side of equation (13.58) on φ is predominant, and it leads to an increase in the rate of fall φ.

Stage of Pupal Development

At the moment of transformation of the larva into a pupa t_l, the shell of the larva is shed, the mass of the individual changes abruptly and remains constant throughout the entire period of development of the pupa

$$m= m_p= m_l - \Delta m_l, \qquad (13.62)$$

where m_p is the mass of the pupa.

The change in the number of pupae occurs due to their death from insects, infections and invasions. This change can be described by the equation

$$\frac{dN}{dt} = -N/\tau_{e4}, \qquad (13.63)$$

$$\frac{dN}{dt} = -N/\tau_{e4}, \qquad (13.63)$$

where τ_{e4}^{-1} is the rate of death of pupae.

Integrating (13.63), we have

$$N = N_l \exp\left(-\frac{t - t_l}{\tau_{e4}}\right). \qquad (13.64)$$

From (13.64) we find the concentration of pupae at the moment of their transformation into young bark beetles

$$N_g = N_l \exp\left(-\frac{t_c - t_l}{\tau_{e4}}\right). \qquad (13.65)$$

Departure of Young Bark Beetles and Their Wintering

After the formation of young bark beetles, they feed either in the population settlement itself, if the resources for food still remain in the tree, or in the vicinity of the tree on young shoots, destroying them. Then the bark beetles leave for the winter. If young bark beetles fed in a tree, then they overwinter in a tree. If young bark beetles fed on young shoots, then they overwinter in fallen leaves in the vicinity of the settlement of the parental generation. During wintering, some of the bark beetles die so that by the next spring their concentration

$$N_{01} = N_g - \Delta N, \qquad (13.66)$$

where ΔN is the part of the population of bark beetles that died during wintering.

CONCLUSION

The chapter is devoted to the development of mathematical models for the development of populations of bark beetles at all stages of development: from departure from wintering places in spring, flight in search of a forest stand suitable for settlement, creation of a settlement, laying eggs, growth of larvae, formation of pupae, birth of a new young generation of bark beetles before leaving for the winter. Such end-to-end mathematical modeling of the development of bark beetle populations has been undertaken for the first time. Therefore, a certain schematization of mathematical models at each stage of development is natural. However, the sequence of transition of model solutions at one stage to model solutions at another stage makes it possible to assess the nature of the patterns of insect population development during the year and even for several years in a row. To carry out such assessments at a quantitative level, it is necessary to know the values of many parameters that characterize insects at each stage, i.e. insufficient knowledge of average population characteristics.

Mathematical modeling of the dynamics of forest insects is a complex area associated with the processes occurring in two kingdoms of living organisms, plants and animals. In forest biogeocenoses, there are many connections that constantly change direction in time and space and interact with different intensity. External influences (abiotic, biotic and anthropogenic) are multifaceted, and the responses of forest biogeocenoses are difficult to predict, because it is necessary to analyze the quantitative and qualitative changes in the relationships between the components of biocenoses in space and time. Mathematical modeling, in essence, allows you to create a simulation model of the development of an insect population, depending on the conditions and assumptions embedded in the model. In this case, the variety of possible options for the development of processes is represented by a set of different trajectories of the corresponding phase spaces of systems of differential equations. Analysis using phase spaces can be especially important in forecasting outbreaks of mass reproduction of insect pests in the forest. However, for this it is necessary to understand the mechanisms of outbreaks of insects and describe them by equations. Particularly frequent outbreaks of reproduction in forests under the influence of stress factors - recreational loads, ground fires, hurricanes, industrial pollution. Fires and gale-force winds, which have led, for example, to the formation of windblows and windbreaks on hundreds of thousands of hectares of coniferous plantations, create excellent conditions for outbreaks of mass reproduction of pests, since recently felled trees are an excellent food base for bark beetles, incapable of resisting entry into her insects. Bark beetles will begin to flock from the surrounding forests to windblows and windbreaks and breed on the available food base. As a result, their numbers in these places will become huge. In subsequent years, after the use of fallen and damaged trees, the surrounding forest areas will become the object of distribution of bark beetles. When a healthy standing tree is attacked by a large number of bark beetles, it also becomes incapable of adequate defense and dies. This process has been repeatedly observed in the forest biogeocenoses of the North-West of Russia. To be able to mathematically describe this process, in order to predict catastrophic consequences, it is necessary to consider systems of equations for the transfer of a population of bark beetles towards the food base. However, there were no studies of this kind with regard to populations of bark beetles until this work, although migration processes in populations of bark beetles play a key role. It is known that biological populations spread over the territory in waves. It has been shown that biological population waves can be of two types: conquest waves and solitary waves. The first type of wave of a biological population is characterized by the filling of the territory with individuals of the population with a certain density. Another type of population waves has a density distribution profile of individuals with a maximum, to the right and left of which the distribution density of individuals in the population decreases to zero.

When there is a wave of conquest of the population of bark beetles, all suitable trees in the territory over which the wave ran are populated by bark beetles. As shown in this chapter, for the existence of such a wave, a constant feeding of the wave by insects flying out of the wintering grounds at the time of the passage of the wave over this part of the territory is necessary. If a population wave in search of a food base flies over a territory free from wintering bark beetles, then as a result of the natural death of some of the beetles, the wave takes the form of a solitary wave. It is important that the considered systems of equations actually have wave solutions. In the previous chapter, a system of equations describing the process of locust propagation was obtained, and it was proved that this system of equations has wave solutions of the solitary wave type.

REFERENCES

Andronov, A. A., Leontovich, E. A., Gordon, I. I., & Mayer, A. G. (1966). Qualitative theory of dynamical systems of the second order. Moskow: Nauka.

Isaev, A. S., Nedorezov, L. V., & Khlebopros, R. G. (1982). *Influence of food transformation on the dynamics of insect abundance. Consortium relations of a tree and dendrophilic insects*. Nauka.

Ivanov, N.N. (1984). Solitary waves in the theory of populations. *News Academy of Sciences of the USSR. Ser. biol.* 5, 656 - 663.

Kataev, O. A., Osetrov, A. V., Popovichev, B. G., & Selikhovkin, A. V. (2001). Population density dynamics of bark beetles (Coleoptera, Scolytidae) in forest stands weakened by natural and anthropogenic factors. *Readings in memory of Nikolai Alexandrovich Kholodkovsky. Ed. S. Yu. Sinev. St. Petersburg: Russian Entomological Society*, Issue 54, 82.

Keller, E. F., & Segel, L. A. (1971). Traveling bands of chemotaxis bacteria: A theoretical analysis. *Journal of Theoretical Biology, 30*(2), 238–248. DOI: 10.1016/0022-5193(71)90051-8

Kennedy, C. R., & Aris, R. (1980). Traveling waves in a simple population model involving growth and death. *Bulletin of Mathematical Biology, 42*(3), 397–429. DOI: 10.1016/S0092-8240(80)80057-7 PMID: 7378612

Kolmogorov, A. N., Petrovsky, I. G., & Piskunov, N. S. (1937). Investigation of the diffusion equation associated with an increase in the amount of a substance and its application to one biological problem. *Bul. Moscow State University. Ser. A. Mathematics and mechanics*. 1(6), 333 - 358.

Kovalev, O. N., & Vechernin, V. V. (1986). Description of a new wave process in populations on the example of the introduction and dispersal of the ragweed leaf beetle Zygogramma suturalis F. (Coleoptera, Chrysomelidae). *Entomological Review, 65*(1), 21–38.

Moretti, M., & Barbalat, S. (2004). The effects of wildfires on wood-eating beetles in deciduous forests on the southern slope of the Swiss Alps. *Forest Ecology and Management, 187*(1), 85–103. DOI: 10.1016/S0378-1127(03)00314-1

Parker, T. J., Clancy, K. M., & Mathiasen, R. L. (2006). Interactions among fire, insects and pathogens in coniferous forests of interior western United Stated and Canada. *Agricultural and Forest Entomology, 8*(3), 167–189. DOI: 10.1111/j.1461-9563.2006.00305.x

Pontryagin, L.S. (1961). *Ordinary differential equations*. Moskow: Gosizdat fiz. - mat. lit.

Selikhovkin, A. V. (1992). Quantitative assessment of the degree of impact of dendrophagous insects on the state of forest stands. *Forestry*, *1*, 16–19.

Selikhovkin, A. V., Bondarenko, E. A., & Popovichev, B. G. (2010). Modern forest entomology: Directions of research, problems and prospects. *Entomological Review*, *LXXXIX*(2), 1, 62–84. DOI: 10.1134/S0013873810020053

Svirezhev, Yu. M. (1987). *Nonlinear waves, dissipative structures and catastrophes in ecology*. Moskow. *Science*.

Zhizhin, G. V. (2004). *Self-regulating waves of chemical reactions and biological populations*. Science.

Zhizhin, G. V. (2005). *Dissipative structures in chemical, geological and ecological systems*. Science.

Zhizhin, G. V. (2008). *Combustion waves with distributed zones of chemical reactions (Non -asymptotical theory of combustion)*. Werner Regen Publishing House.

Zhizhin, G. V. (2011). Mathematical model of a solitary wave of a population of beetles - bark beetles. *Proceedings of the St. Petersburg State Forest Engineering Academy*, 196, 153 - 162.

Zhizhin, G. V., & Bolshakova, N. N. (2000). Solitary waves in populations of multicellular animal organisms. *Mathematical Modelling*, *12*(12), 55–65.

Zhizhin, G.V., & Selikhovkin, A.V. (2012). *Mathematical modeling of the development and distribution of insect populations - stem pests in the forests of Russia*. St. Petersburg: St. Petersburg State Forestry Technical University named after S.M. Kirov.

Chapter 14
Ground Fire Wave

ABSTRACT

Using the methods of non-asymptotic theory of waves in excitable media, a mathematical model of a ground fire has been constructed and studied. A qualitative study of the corresponding system of differential equations has been carried out, an approximate analytical solution has been obtained, and numerical studies have been carried out. It has been established that the length of the zone of chemical reactions in the ground fire wave is significant. It plays an important role in the structure of the ground fire wave; its length sensitively reacts to any changes in the fire wave and in the environment. It can be considered that the length of the chemical reaction zone is the internal degree of freedom of the ground fire wave, which maintains the stability of the wave within certain limits.

INTRODUCTION

The vegetation cover of the Earth is of great importance for maintaining the chemical composition of the atmosphere suitable for the existence of life on Earth. Enormous damage to the vegetation cover of the Earth is caused by fires, often covering large areas of the earth's surface. To determine the possibilities of preventing and fighting fires, it is necessary to build and study physical and mathematical models of fires. Combustion is the main process in a fire. It is accompanied by various other physical and chemical processes, which significantly complicate the possibility of studying the patterns of fire propagation. Obviously. that in the analysis of fires it is necessary to use modern concepts of burning. In particular, the non-asymptotic theory of waves in excitable media developed in this monograph reflects an important feature of natural fires - the coverage of large areas by fires during the propagation of fires. At the same time, many guiding documents regulating the method of combating natural fires are based on the classical theory of combustion, which uses the

DOI: 10.4018/978-1-6684-7745-8.ch014

concept of a narrow reaction zone in a fire flame. This is a fundamentally wrong approach, which should be disposed of.

14.1. VEGETABLE COMBUSTIBLE MATERIALS AND PRODUCTS OF THEIR COMBUSTION

The vegetation cover is a complex of combustible materials that differ in their pyrological functions: combustion conductors; materials that support combustion, and materials that delay the spread of combustion. A particularly important role for natural fires is played by combustion conductors, i.e. those components of the canopy that enable the continuous propagation of flame over it. It is known that plants are built from organic and mineral compounds. The organic part consists of fiber, lignin and hemicellulose. Often it includes extractive and protein substances. Fiber, or cellulose, has a general formula $(C_6H_{10}O_5)_n$. Cellulose is the main material from which the shells of plant cells are built. Polysaccharides, called hemicelluloses, are close to them in composition. Along with six-carbon links $(C_6H_{10}O_5)_n$, their chains also contain five-carbon links $(C_5H_8O_4)_n$. Hemicelluloses are involved in the construction of cell walls and fill the space between cells. With age, aromatic substances, often called "woody matter" or lignin, appear in plant cells. Lignin has a complex structure, characterized by a relatively high carbon content with a lower oxygen content. Cell contents often include proteins and essential oils (terpenes). In some plants (for example, in fir), individual cells are almost completely filled with these oils. Resins are often dissolved in oils, which are released at the sites of plant damage after the oil has evaporated. Resins, waxes, fats and other plant components that can be extracted from plants belong to the group of extractives. One of the main pyrological parameters of plant materials is their moisture content, i.e. the ratio of the mass of water contained in plants to the dry mass of the plant.

The conductors of combustion include dead grass, lichens, fallen needles, fallen leaves, mosses, peat and bedding (Konev, 1975, 1977, 1979). Dead grass after drying becomes the most flammable material. It can occupy large spaces in the steppe and in the forest and reach a considerable height. The speed of flame propagation through dead grass can reach 20 km per hour. Lichens are bushes 15 - 20 cm high, they are characterized by a high content of carbohydrates. Small parts of trees, needles, dying fall to the ground, forming a layer of combustible material - litter. Deciduous tree species periodically shed dead leaves and form a layer of fresh fallen leaves 7 cm high. Under this layer follows a layer of last year's leaf litter 2-3 cm thick, and finally, a highly mineralized layer of last year's litter - litter. All plant combustible materials have approximately the following composition: 48-50% carbon, 38-42% oxygen, 5.5-6.5% hydrogen, 0.5-2.3% nitrogen. If we do not take into account elements with a low content, the composition of plant combustible materials approximately corresponds to the formula $C_6H_9O_4$. Ash content is usually understood as the ratio of the weight of ash to the absolutely dry weight of combustible material. In vegetable combustible materials, the ash content, as a rule, does not exceed 10%.

Combustion products include gaseous, liquid and solid substances formed during the combustion process. During complete combustion, the main products of plant materials are carbon dioxide and water. In most cases, however, combustion is not complete and is accompanied by underburning. Vegetable combustible materials have two types of underburning. One is due to incomplete combustion of gasification products. A sign of this underburning is the formation of smoke. The second type of underburning in the process of thermal decomposition of the carbonaceous residue (coke). There are two stages in the process of plant materials (Amosov, 1958): flame and flameless. It can be said that at the first stage the process of release and combustion of gaseous products prevails, and at the second stage the process of coke afterburning prevails.

14.2. TYPES OF WILDFIRES AND WAYS TO DESCRIBE THEM

Natural fires are subdivided into forest (grassroots, general riding, top riding) steppe, underground and mass. In case of grassroots forest and steppe fires, litter (fallen twigs, needles and leaves), soil cover (grasses, mosses, lichens), as well as shrubs are burned. In the case of a crown fire, a forest fire burns the crowns of trees (forest canopy). In the event of a wholesale crown forest fire, the forest canopy, shrubs, litter and ground cover are simultaneously burned. Underground fires burn peat bogs. In a massive fire, vegetable combustible substances are simultaneously ignited over vast areas. Such fires can occur as a result of powerful radiation in the atmosphere. Top forest fires are initiated, as a rule, by ground forest fires. The oxidizing agent in natural fires is atmospheric oxygen. The fire goes out with a lack of oxygen, even if the temperature of the combustible materials is high enough. Therefore, for ground fires it is necessary to blow vegetation with air. In underground fires, air enters the peat layer through an inlet in the soil (Artsybashev, 1974). Combustion occurs in the mode of oxygen starvation and the combustion front slowly spreads to tens and even hundreds of meters from the inlet. From the above brief enumeration of possible types of natural fires, it follows that the processes accompanying natural fires are very complex and diverse.

And despite the urgent need to determine the patterns of fire propagation, their mathematical description encounters serious obstacles. At present, a lot of empirical information has been accumulated about the individual elements of the overall process, which is a natural fire (Konev, 1975, 1977, 1979; Amosov, 1958; Artsybashev,

1974). For the purposes of describing natural fires, a lot of disparate information is involved on the patterns of heat and mass transfer, thermal decomposition, radiation, convective motion, and chemical reaction kinetics (Driesdel, 1990) . However, this disparate information, drawn from various textbooks, cannot form a single mathematical model of at least one of the types of natural fires. It is clear that the rate of spread of a natural fire is an integral and very important characteristic of a fire. It cannot be obtained as a result of some simple evaluation. At the same time, it cannot be found by trying to take into account all the stages and stages of a complex phenomenon, which is a natural fire. It is necessary to single out the determining processes, and this is the task of mathematical modeling.

The most general mathematical models of natural fires were developed by A.M. Grishin (Grishin, 1981, 2002, 2004, 2005). However, these models are too general and represent an eclectic combination of equations of various processes containing too many unknown parameters and variables. When trying to study natural fires on the basis of these models, the authors themselves involuntarily have to accept such strong simplifications, in which the essence of fires is thrown out - chemical combustion reactions (Burasov, Grishin, 2005).

Of the listed types of natural fires, the ground fire (forest or steppe) seems to be the most suitable for mathematical modeling, especially since the general crown forest fire is initiated by a ground fire. Top crown fire is difficult to model because it is not yet clear how to describe the transfer of a fire front from one tree to another. In an underground fire, the geometry of gas flows is very complex.

Let us turn to the ground fire, trying to determine the speed of its stationary propagation as an eigenvalue of the corresponding system of the parabolic system of equations.

14.3. ONE-DIMENSIONAL MATHEMATICAL MODEL OF GROUND FIRE

We assume that the ground fire spreads in a strip of width h a above the soil level. Moreover, plant combustible materials are distributed in this strip evenly with a certain density ρ_1. The strip h is blown at a certain speed v directed against the direction of the spatial variable x, coinciding with the direction of movement of the ground fire wave (Figure 1).

Figure 1. Qualitative view of the temperature profile in the ground fire wave

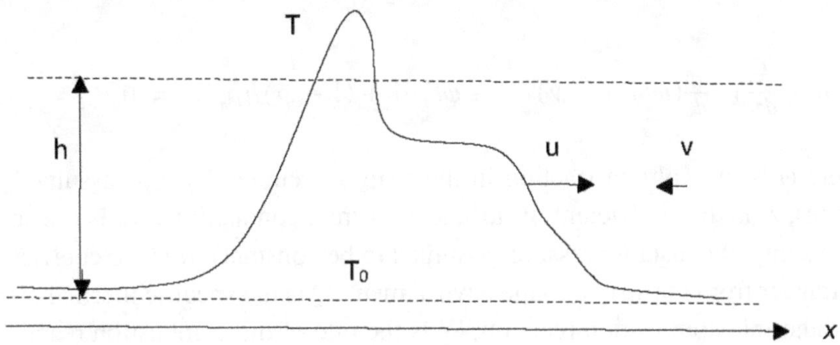

In a one-dimensional formulation, the details of the release of volatiles from plant combustible materials with the formation of local flames are not considered. We will assume that the cause of the spread of a fire is, first of all, the possibility of propagating with a certain wave speed u of a chemical combustion reaction of a gas component consisting of air and released combustible volatiles. After the passage of the combustion wave of the gas component in the stationary solid phase (coke, ash), slower processes of combustion of combustible materials in local stationary sources occur. The resulting solid smoke particles also slowly settle on hard surfaces. In the combustion wave, the processes of heating and drying of plant materials, their thermal decomposition with the release of volatiles, combustion of volatiles and then cooling of the combustion products to the initial ambient temperature take place T_0. Since we consider an open system of substances in the strip h, the processes of release and combustion of volatiles in this strip will be accompanied by a mass of gas recoil from the strip h to the area above the strip h.

In a stationary setting at constant velocities in coordinates related to the moving combustion wave, this corresponds to a variable gas flow through the strip h. The change in the gas flow rate under these conditions is associated with a change in the gas density ρ along the wave. The release of the gas component from the strip h leads to the formation of conjective columns (Grishin, 2005) above the strip of combustible plant materials, which can cause tree crown fires. It is natural to assume that the pressure P in the gas within the strip h is constant. Air is mixed with volatiles by turbulent diffusion. Turbulence is formed both due to the release of volatiles and due to the vibrations of plants when air flows around them. Heat is transferred along the spatial variable also by means of turbulent heat conduction.

The system of equations for the laws of conservation of energy and mass for the gas component of a stationary wave of a ground fire with a wave coordinate $z = (u - v)\, t - x$ under the accepted conditions can be written as

$$\varphi h\frac{d}{dz}(\lambda_t\frac{dT}{dz})-\frac{d}{dz}\left(\varphi h\rho(u-v)c_pT\right)-\alpha(T-T_0)+\varphi h\rho QW-(1-\varphi)\rho_1Q_1W_f = 0, \quad (14.1)$$

$$\varphi h\frac{d}{dz}(\rho D_t\frac{dC_f}{dz})-\frac{d}{dz}\left(\varphi h\rho(u-v)C_f\right)-\varphi h\rho W+(1-\varphi)h\rho_1W_f = 0. \quad (14.2)$$

Here φ is the volume fraction in the strip *h* occupied by gas (assumed to be constant); λ_t is the coefficient of turbulent thermal conductivity; c_p is the specific heat capacity at constant pressure (assumed to be constant); α is the coefficient of heat transfer from the strip h to the environment; Q is the specific thermal effect of the exothermic combustion reaction; W is the rate of the combustion reaction; Q_1 is the specific thermal effect of the endothermic process of volatile release; W_1 is the rate of release of volatiles; D_f there is a turbulent diffusion coefficient; C_f is the mass fraction of volatiles in the gas.

In equation (14.1), radiation is neglected, since the radiation in the heat balance during the combustion of plant materials is no more than 10% (Konev, 1977). The combustion reaction rate W in (14.1), (14.2) is the rate of chemical interaction of combustible volatiles and air oxygen. Since the loss of volatiles in the combustion process is proportional to the loss of oxygen, we represent the rate of combustion reaction as a function of the mass fraction of volatiles and temperature according to the Arrhenius law

$$W = k_0C_f^k\exp(-\frac{E}{R_0T})$$

, (14.3)

where k_0 is the pre-exponent; k is the order of gross reaction by mass fraction of volatiles; E is the activation energy of the combustion reaction; R_0 is the universal gas constant.

Thermodynamic parameters are related by the ideal gas equation of state

$$P=\rho RT \quad (14.4)$$

R is the specific gas constant.

For turbulent transport coefficients, the same ratios are accepted as for their laminar counterparts

$$\lambda_t = \lambda_{t0}(T/T_0)^{0.67}, D_t = D_{t0}(T/T_0)^{1.67} \quad (14.5)$$

The convective term in equation (14.1) due to the accepted conditions and equation (14.4) is equal to zero

$$\frac{d}{dz}\left(\varphi h\rho(u-v)c_pT\right) = 0, \tag{14.6}$$

i.e. the change in gas flow $\varphi h\rho(u–v)$ through the strip h is inversely proportional to the change in gas temperature. The convective term in equation (14.2) splits into two terms

$$\varphi h(u-v)\frac{d}{dz}(\rho C_f) = -(u-v)\frac{\varphi hP}{RT^2}\frac{dT}{dz} + (u-v)\varphi h\frac{P}{RT}\frac{dC_f}{dz}. \tag{14.7}$$

Equations (14.1), (14.2), taking into account equalities (14.3) - (14.7), after the introduction of dimensionless variables, take the form

$$\frac{d}{d\zeta}(\theta^{0.67}\frac{d\theta}{d\zeta}) - \overline{\alpha}(\theta-1) + C_f^k\theta^{-1}\exp(-\sigma/\theta) + \overline{W}_fQ_1/Q = 0, \tag{14.8}$$

$$Le\tau\frac{d}{d\zeta}(\theta^{0.67}\frac{dC_f}{d\zeta}) - V\frac{dC_f}{\theta d\zeta} + VC_f\frac{d\theta}{\theta^2 d\zeta} - \frac{C_f^k}{\theta}\exp(-\sigma/\theta) + \overline{W}_f = 0, \tag{14.9}$$

$$\zeta = (\frac{PQk_0}{\lambda_{t0}T_0^2 0R})^{1/2}z, Le = \frac{D_{t0}\rho_0 c_p}{\lambda_{t0}}, \tau = \frac{Q}{c_pT_0}, \theta = \frac{T}{T_0},$$

$$V = \frac{u-v}{k_0}(\frac{PQk_0}{\lambda_{t0}T_0^2R})^{1/2}, \overline{W}_f = \frac{1-\varphi}{\varphi}\frac{\rho_1 W_f RT_0}{Pk_0}, \overline{\alpha} = \frac{\alpha T_0^2 R}{\varphi hPQk_0}, \sigma = \frac{E}{R_0T_0}.$$

The system of equations (14.8), (14.9) can be reduced to an autonomous system of four first-order differential equations

$$\frac{d\eta}{d\zeta} = \overline{\alpha}\frac{\theta-1}{\theta^{0.67}} - 0.67\frac{\eta^2}{\theta} - \frac{C_f^k}{\theta^{1.67}}\exp(-\sigma/\theta) - \frac{Q_1\overline{W}_f}{Q\theta^{0.67}}, \tag{14.10}$$

$$\frac{dn}{d\zeta} = \frac{Vn}{Le\theta^{1.67}\tau} - \frac{0.67\eta n}{\theta} - \frac{VC_f\eta}{Le\theta^{2.67}\tau} + \frac{C_f^k}{\theta^{1.67}Le\tau}\exp(-\sigma/\theta) - \frac{\overline{W}_f}{Le\tau\theta^{0.67}}, \tag{14.11}$$

$$\frac{d\theta}{d\zeta} = \eta, \tag{14.12}$$

$$\frac{dC_f}{d\zeta} = n. \tag{14.13}$$

For further analysis, in accordance with the methods of the non-asymptotic theory of waves in excitable media (Zhizhin, 2004, 2005, 2008), we divide the combustion wave into four zones (Figure 2).

Figure 2. Splitting a ground fire wave into zones

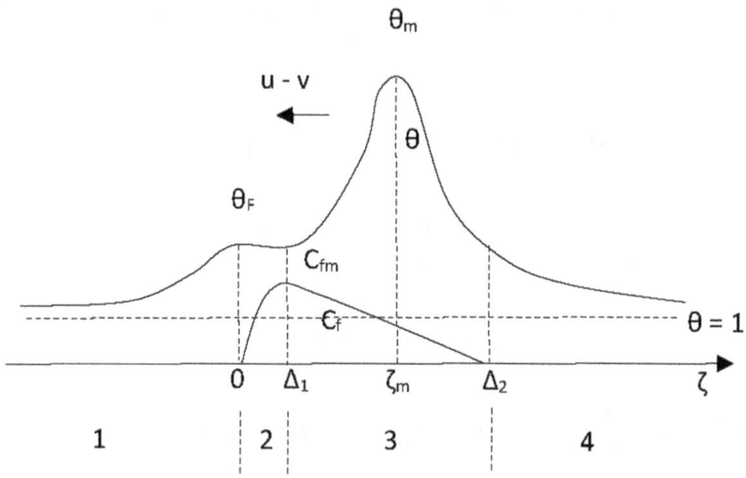

In the first zone (1) ($-\infty<\zeta\leq 0$) the substance is heated. In the second zone (2) ($0\leq\zeta\leq\Delta_1$), volatiles are released at a practically constant temperature. In the third zone (3) ($\Delta_1\leq\zeta\leq\Delta_2$) there are chemical reactions of combustion of volatiles. In the fourth zone (4) ($\Delta_2\leq\zeta<+\infty$), the reaction products are cooled to the initial temperature $\theta = 1$.

14.4. APPROXIMATE ANALYTICAL SOLUTION OF THE PROBLEM OF GROUND FIRE WAVE PROPAGATION

In zones 1, 4, the mass fraction of volatiles is equal to zero (in zone 1, volatiles have not yet formed, in zone 4, volatiles have partially burned out, and the rest has left the strip *h*, flying into the area above the strip *h*). Therefore, equation (14.10) takes the form

$$\frac{d\eta}{d\zeta} = -0.67\frac{\eta^2}{\theta} + \overline{\alpha}\frac{\theta - 1}{\theta^{0.67}}. \tag{14.14}$$

Equation (14.14) together with equation (14.12) form a closed system of equations. It has an equilibrium position a with coordinates $\eta = 0$, $\theta = 1$. Solutions to system (14.14), (14.12) can be represented by trajectories in the phase plane (η, θ). Their behavior is determined by the location of zero isoclines $\eta = 0$, on which $d\theta / d\zeta = 0$, and $\eta = \pm\sqrt{\bar{\alpha}(\theta - 1)\theta^{0.33}/0.67}$ (dotted line), on which $d\eta / d\zeta = 0$ (Figure 3).

Figure 3. Phase plane of system (14.14), (14.12)

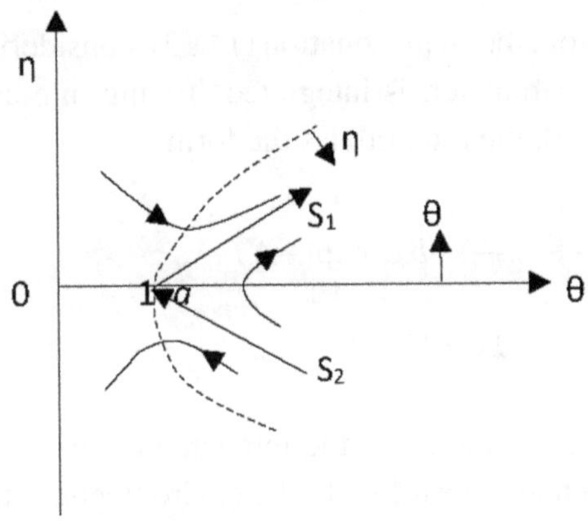

Arrows at zero isoclines indicate areas of increase in the corresponding phase coordinates. The eigenvalues of the equilibrium position a can be found from the discriminant of the system (14.14), (14.12) linearized in the vicinity of the equilibrium position a,

$$\lambda_{1,2} = \pm\bar{\alpha}^{1/2}. \quad (14.15)$$

It follows from (14.15) that in the linear approximation the equilibrium position a has the saddle type (Pontryagin, 1961), i.e. in the range of physical values of the variable $\theta \geq 1$, one separatrix S_1 leaves the equilibrium position, and the other separatrix S_2 enters the equilibrium position (Figure 3).

The separatrix S_1 corresponds to the solution of the system (14.10) - (14.13) in zone 1, and the separatrix S_2 corresponds to the solution of the system (14.10) - (14.13) in zone 4. Taking into account the boundary conditions on $\pm\infty$, these solutions have, respectively, the form

$$\theta = 1 + (\theta_F - 1)\exp(\lambda_1\zeta), -\infty < \zeta \leq 0, \quad (14.16)$$

$$\theta = 1 + (\theta_{\Delta_2} - 1)\exp(\lambda_2(\zeta - \Delta_2)), \Delta_2 \leq \zeta < +\infty, \quad (14.17)$$

where θ_F is value of temperature on rite the boundary of zone 1 at $\zeta = 0$, θ_{Δ_2} is value of temperature on left the boundary of zone 4 at $\zeta = \Delta_2$.

In zone 2, considering the release of volatiles as a phase transition, we assume that the temperature is constant and equal θ_F. Then $\eta = 0$ and equation (14.11), given that there are no chemical combustion reactions in zone 2, takes the form

$$\frac{dn}{d\zeta} = \frac{Vn\theta_F^{-1} - W_f}{Le\tau\theta_F^{0.67}}. \quad (14.18)$$

Equation (14.18), together with equation (14.13), considering the rate of volatile release as a constant parameter, is integrated. Taking into account the boundary condition $C_f = 0$ at $\zeta = 0$, the integral has the form

$$C_f = bg\zeta - bg^2\exp(-\frac{\Delta_1 - \zeta}{g}) + bg^2\exp(-\Delta_1/g), \quad (14.19)$$

$$g = \tau Le\theta_F^{1.67}/V, b = W_F(\tau Le\theta_F^{0.67})^{-1}.$$

In the chemical reaction zone 3, the temperature first rises, and then, as the volatiles burn out and due to heat loss to the environment, it falls.

Expanding in a Taylor series the right side of equation (14.12) in the vicinity of the temperature maximum, limiting ourselves to the first non-zero term of the series, using the expression of the right side of equation (14.10) at the maximum point at $\eta = 0$, we obtain

$$\frac{d\theta}{d\zeta} = f_m(\zeta - \zeta_m), f_m = \bar{\alpha}(\theta_m - 1)\theta_m^{-0.67} - C_{fe}^k\theta_m^{-1.67}\exp(-\sigma/\theta_m), \quad (14.20)$$

C_{fe} is the value of the mass fraction of volatiles at the maximum coordinate $\zeta = \zeta_m$. Integrating (14.20), we have

$$\theta = \theta_m + 0.5f_m(\zeta - \zeta_m)^2. \quad (14.21)$$

At the boundaries of zone 3, the equalities

$$\theta_{\Delta_1} = \theta_m + 0.5f_m(\Delta_1 - \zeta_m)^2 = \theta_F, \quad (14.22)$$

$$\theta_{\Delta_2} = \theta_m + 0.5f_m(\Delta_2 - \zeta_m)^2. \quad (14.23)$$

Accepting the condition of smoothness of the function θ (ζ), we equate the derivatives of temperature with respect to ζ on both sides of the boundary of zones 4 and 3 at $\zeta = \Delta_2$

$$\lambda_2(\theta_{\Delta_2} - 1) = f_m(\Delta_2 - \zeta_m). \tag{14.24}$$

Let us also equate the derivatives of temperature on both sides of zone 2

$$\lambda_1(\theta_F - 1) = f_m(\Delta_1 - \zeta_m). \tag{14.25}$$

It follows from equation (14.19) that the function $C_f(\zeta)$ in zone 2 has a maximum when it reaches the boundary $\zeta = \Delta_1$, i.e. zero derivative $dC_f/d\zeta$. Therefore, on the other side of the boundary $\zeta = \Delta_1$, we also require the zero value of the derivative $dC_f/d\zeta$ for the smoothness of the function $C_f(\zeta)$. Expanding the right side of equation (14.13) into a Taylor series in the vicinity of the maximum C_{fm}, i.e. at $\zeta = \Delta_1$, restricting ourselves to the first non-zero member of the series, using the value of the right side of (14.11) at $\zeta = \Delta_1$ and n = 0, given that there is no volatile emission in zone 3, we obtain

$$\frac{dC_f}{d\zeta} = \psi_m(\zeta - \Delta_1), \psi_m = \left[C_{fm}^k e^{-\sigma/\theta_F} + VC_{fm} f_m \frac{\zeta_m - \Delta_1}{\theta_F}\right] \frac{1}{Le\tau\theta_F^{1.67}}. \tag{14.26}$$

Integrating (14.26) we have

$$C_f = C_{fm} + 0.5\psi_m(\zeta - \Delta_1)^2. \tag{14.27}$$

On the left boundary of zone 3 at $\zeta = \Delta_2$ it is $C_f = 0$, then

$$C_{fm} = -0.5\psi_m(\Delta_2 - \Delta_1)^2. \tag{14.28}$$

From (14.27) we have

$$C_{fe} = C_{fm} + 0.5\psi_m(\zeta_m - \Delta_1)^2, \tag{14.29}$$

and from (14.19) we have

$$C_{fm} = bg\Delta_1 - bg^2 + bg^2\exp(-\Delta_1/g). \tag{14.30}$$

We assume that after the release of volatiles, they ignite in the ignition mode. Therefore, between the temperatures and mass fractions of volatiles at the beginning of zone 3 and at the point of maximum temperature, one can use the known relationships for the self-ignition regime (Zel'dovich, Bareblatt, Librovich, Makhviladze, 1980; Warnatz, Maas, Dibble, 2001). In the notation of this chapter, these relations have the form

$$\theta_m = (\theta_F + \tau)(1 - k\frac{\theta_F + \tau}{\sigma}), \tag{14.31}$$

$$C_{fe} = C_{fm}\theta_F{}^2 k\sigma^{-1}\tau^{-1}. \tag{14.32}$$

It can be seen that the resulting algebraic equations represent a closed system with respect to the unknown values of the parameters V, C_{fm}, C_{fe}, θ_m, θ_F, Δ_1, Δ_2, ζ_m, θ_{Δ_2}. Solving this system of equations, we obtain an equation for the propagation velocity of a ground fire wave

$$V = \frac{\theta_F^{2.67}(\theta_F - 1)\lambda_1 Le\tau}{8(\theta_m - \theta_F)^{\ 2}} + \frac{C_{fm}^{k-1}\theta_F}{\lambda_1(\theta_F - 1)} exp(-\frac{\sigma}{\theta_F}), \tag{14.33}$$

$$\theta_F = 0.5(3\sigma\tau/k)^{\ 1/2}, \tag{14.34}$$

$$C_{fm} = \frac{4}{3}\lambda_1^{2/k}\theta_m^{1/k}\exp(\frac{\sigma}{\theta_m k})\left[\frac{(\theta_F - 1)^2\theta_m^{0.67}}{2(\theta_m - \theta_F)} + \theta_m - 1\right]^{\ 1/k}, \tag{14.35}$$

θ_m located at (14.31).

For the obvious inequality $\theta_m > \theta_F$ to be satisfied, it is necessary that

$$(\frac{\sigma}{k\tau})^{\ 1/2} > \frac{2}{2 - \sqrt{3}}. \tag{14.36}$$

In addition, from the resolution of the algebraic system of equations it follows that

$$\theta_{\Delta_2} = \theta_F,\ C_{fe} = 3/4\ C_{fm}, \tag{14.37}$$

combustion reaction zone width

$$\Delta_2 - \Delta_1 = 4\frac{\theta_m - \theta_F}{\lambda_1(\theta_F - 1)}. \tag{14.38}$$

The coordinate of the border of zones 2 and 3 Δ_1 is found from (14.30), after that you can find ζ_m from the equation

$$\zeta_m = \Delta_1 + 2(\theta_m - \theta_F)(\theta_F - 1)\lambda_1. \tag{14.39}$$

After determining the unknown values of the parameters according to (14.31) - (14.39), it is possible to plot the temperature and mass fraction of volatiles in zones 1 - 4 using equations (14.16), (14.17), (14.19), ((14.21), (14.27).

14.5. NUMERICAL STUDY OF THE ANALYTICAL SOLUTION OF THE GROUND FIRE WAVE

For the numerical study of the obtained analytical solution, the second order of the chemical combustion reaction k = 2 was taken. Due to insufficient knowledge of the kinetic and thermophysical parameters, the numerical analysis was carried out in a dimensionless form. Some values of the parameter τ were taken, and for each value of τ, using inequality (14.36), the minimum value of the parameter σ was found. After that, several values of the parameter σ were set, greater than this minimum value, so that inequality (14.36) was certainly satisfied. Then the temperatures θ_F, θ_m were found, as well as the maximum values of the natural number λ_1 (i.e., the heat transfer coefficient $\overline{\alpha}$), at which the maximum value of the mass fraction of volatiles is equal to 1 (equation (14.35)). Naturally, for the calculation, the values of λ_1 were set λ_1 less than these maximum values of λ_1. Another degree of freedom in the calculation is the variation in the rate of release of volatiles. An analysis of equation (14.30) shows that after calculating the values V, C_{fm}, θ_F the length of the volatile emission zone is uniquely determined by (14.30). Moreover, the root of equation (14.30) always exists, and the length of the selection zone tends to zero when the wave speed tends to zero. Indeed, we write equation (14.30) in this form

$$Y_1 = Y_2, Y_1 = \exp(-\Delta_1/g), Y_2 = C_{fm}(bg^2)^{-1} + 1 - \Delta_1/g. \tag{14.40}$$

The function graphs Y_1, Y_2 are shown in Figure 4. These graphs always intersect at one point. Moreover, the slopes of the functions at Δ_1/g =0 are the same.

Figure 4. Graphical solution of equation (14.30)

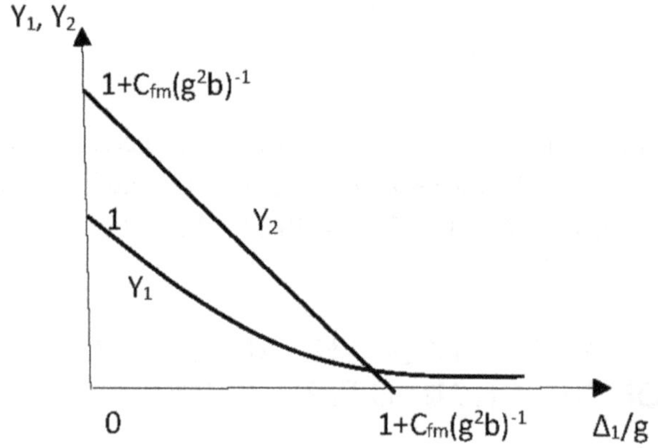

From Figure 3 it follows that the greater the rate of volatile release $\overline{W}_f$ (greater b), the lower Δ_1 under other conditions. The lower the wave speed V (the greater g), the lower the Δ_1. Calculation options are presented in (Zhizhin, 2008).

Summarizing the results of the calculation of ground fire waves, the following can be noted. For given parameters σ, τ, θ_m an increase in the natural number λ_1 (i.e., the heat transfer coefficient) leads to an increase in C_{fm}, i.e. to an increase in the mass fraction of volatiles necessary to compensate for heat losses to the environment. In this case, not only the maximum value of the mass fraction of volatiles in the combustion wave increases, but also the mass fraction of volatiles at the beginning of the reaction zone. In general, this causes an increase in the burning rate in the reaction zone and the length of the reaction zone is shortened $\Delta_2 - \Delta_1$.

An increase in the rate of chemical reactions simultaneously also leads to an increase in the speed of propagation of a ground fire wave V. Under these conditions, an increase in the rate of volatilization $\overline{W}_f$ naturally leads to a reduction in the length of the zone of volatilization. But since an increase in heat transfer requires an increase in the mass fraction of volatiles, the zone of volatile release at a constant release rate increases.

Increasing the σ parameter, i.e. activation energy of a chemical combustion reaction requires a significant increase in the pre-exponential k_0 of the reaction rate for the existence of a stationary fire wave, since an increase in the activation energy sharply reduces the value of the Arrhenius exponent. Otherwise, the stationary mode of combustion wave propagation is disrupted. An increase in the heat transfer coefficient under these conditions also requires an increase in the mass fraction of volatiles. However, the stationary regime of combustion wave propagation exists only up to a certain limit, after which the stationary regime is absent, and the stationary regime is disrupted (Semenov, 2005). The length of the chemical reaction zone with an increase in σ increases significantly, primarily due to an increase in k_0, since $\zeta \sim k_0^{1/2}$.

The dimensionless velocity of propagation of a ground fire wave decreases sharply with increasing σ, since it is inversely proportional to k_0. At the same time, the rate of release of volatiles also sharply decreases, since it is inversely proportional to k_0.

It is important that the length of the volatile emission zone is always significantly (several orders of magnitude) less than the length of the zone of chemical combustion reactions. Moreover, the ratio of the length of the volatile emission zone to the length of the zone of chemical reactions sharply decreases with increasing σ, since an increase in the activation energy exponentially reduces the rate of chemical reactions, which leads to an increase in the length of the chemical reaction zone. parameters are easier to illustrate in tabular form than graphically, due to the incomparability of the lengths of the zones of release of volatile and chemical reactions. The noted features of ground fire waves are consistent with the results of the study of nonadiabatic combustion waves in this monograph using the methods of the non - asymptotic theory of waves in excitable media, fundamentally different from the assumptions of the asymptotic theory of combustion, which assumes an infinitely small extension of the reaction zone in the combustion wave. It follows from the conducted studies that the length of the zone of chemical reactions in the ground fire wave is significant and plays an important role in the structure of the ground fire wave. The length of the chemical reaction zone is very sensitive to changes in thermophysical and kinetic parameters, thus being the internal degree of freedom of the surface fire wave structure.

The analysis carried out allows us to formulate the condition for the disruption of the stationary mode of ground fire wave propagation. It is obtained from equality (14.35) under condition $C_{fm} < 1$.

Disruption of the stationary regime of the spread of a ground fire occurs if

$$\overline{\alpha} > \overline{\alpha^*}, \overline{\alpha^*} = \frac{(3/4)^k \theta_m \exp(-\sigma/\theta_m)}{(\theta_F - 1)^2 (\theta_m - \theta_F)^{-1} \theta_m^{0.67} + \theta_m - 1},$$

and the values of θ_F and θ_m are, respectively, according to (14.34) and (14.31). It is clear that this estimate of the condition for the absence of a stationary mode of propagation of a ground fire is an estimate from above. Apparently, the disruption of the stationary mode of ground fire wave propagation will actually occur at lower values of the heat transfer coefficient.

CONCLUSION

A mathematical model of a ground fire wave is constructed, taking into account all the main features of a ground fire. The system of differential equations was studied qualitatively and an approximate analytical solution was obtained. Numerical studies of solutions depending on the values included in the system of parameters have been carried out. Analytical expressions are obtained for the wave velocity and characteristic lengths of the wave structure. It has been established that the length of the zone of chemical reactions in the ground fire wave is significant and plays an important role in its structure. Significant in the wave is the interdependence of the length of the volatile emission zone and the length of the zone of chemical reactions. It has been established that an attempt to cool the wave (removal of heat to the environment) leads to an increase in the mass fraction of volatiles and, as a result, to an increase in the rate of a chemical reaction and, accordingly, to an increase in the speed of propagation of a ground fire wave. The length of the zone of chemical reactions in a ground fire wave is very sensitive to changes in thermophysical and kinetic parameters. The ability of the length of the zone of chemical reactions to respond sharply to changes in the conditions of wave propagation allows us to assert that the length of the zone of chemical reactions is the internal degree of freedom of the ground fire wave, with the help of which the wave maintains its stability. The analysis carried out allows us to formulate the condition for the disruption of the stationary mode of ground fire wave propagation. Disruption of the stationary mode of ground fire wave propagation occurs in those cases when the heat transfer coefficient to the environment becomes more than a certain critical value.

The results obtained, along with their independent theoretical and applied significance, illustrate the essential importance of more complete correct formulations of the problem in the theory of combustion and, in particular, those associated with taking into account the variable length of the zone of chemical reactions in the combustion wave. This emphasizes the exceptional possibilities of the methods of the non-asymptotic theory of waves in excitable media.

REFERENCES

Amosov, G. A. (1958). *Some features of combustion during forest fires*. Leningrad Research Institute of Forestry.

Artsybashev, E.S. (1974). *Forest fires and fighting them*. Moscow: Timber industry.

Burasov, D. M., & Grishin, A. M. (2005). Determination of the propagation rate and ecological consequences of a steppe fire. *Abstracts of the 13th Symposium on combustion and explosion*. Chernogolovka. 80 - 81.

Drysdale, D. (1990). *Introduction to fire dynamics*. Stroyizdat.

Grishin, A. M. (1981). *Mathematical models of forest fires*. Tomsk State Publishing House. University.

Grishin, A.M. (2002). General mathematical models of forest and peat fires and their application. *Mechanical advances*. 12, 41 - 89.

Grishin, A.M. (2004). General mathematical model of steppe fires. *Ecological systems and devices.*12, 25 - 29.

Grishin, A. M. (2005). General mathematical models of natural fires. *Abstracts of the 13th Symposium on combustion and explosion*. Chernogolovka, 73 - 74.

Konev, E. V. (1975). On the ignition of forest combustible materials. *Problems of forest pyro - logy.* Rep. ed. N.P. Kurbatsky, E.V. Konev. Krasnoyarsk: Institute of Forest and Wood named after V.N. Sukachev, Siberian Branch of the USSR Academy of Sciences. 80 - 92.

Konev, E. V. (1977). *Physical bases of combustion of plant materials*. Science.

Konev, E. V. (1979). Results of the study of combustion processes in forest ground fires. Burning in the forest. *Materials of the First All-Union Scientific and Technical Conference*. Rep. editor. N.P. Kurbatsky. Krasnoyarsk: Institute of Forest and Wood named after V.N. Sukachev of the Siberian Branch of the Academy of Sciences of the USSR. Part 2. *New developments in the problem of forest fires*. 53 - 63.

Pontryagin, L.S. (1961). *Ordinary differential equations*. Moskow: Gosizdat fiz. - mat. lit.

Semenov, N. N. (2005). Selected Works.: Vol. 3. *About some problems of chemical kinetics and reactivity*. Nauka.

Warnatz, J., Maas, U., & Dibble, R. W. (2001). *COMBUSTION. Physical and Chemical Fundamentals, Modeling and Simulations, Experiments, Pollutant Formation*. Springer.

Zeldovich, Ya. B., Barenblatt, G. I., Librovich, V. B., & Makhviladze, G. M. (1980). *Mathematical theory of combustion*. Nauka.

Zhizhin, G. V. (2004). *Self-regulating waves of chemical reactions and biological populations*. Science.

Zhizhin, G. V. (2005). *Dissipative structures in chemical, geological and ecological systems*. Science.

Zhizhin, G. V. (2008). *Combustion waves with distributed zones of chemical reactions (non-asymptotic theory of combustion)*. Werner Regen Publishing House.

Chapter 15
Proteranthies on the Sun

ABSTRACT

The motion of gas in prominences on the Sun, emanating from the chromosphere and returning to the Sun`s chromosphere, is investigated. The study showed for the first time that there is no continuous transition of the gas velocity through the speed of sound in prominences emanating from the solar chromosphere. Moreover, one of the limiting final states in the flow of prominence gas emanating from the chromosphere is simultaneously the initial limiting state of the prominence gas flow returning to the solar chromosphere. A hypothesis is expressed about the possibility of creating a aircraft with an engine. Using the phenomenon of nuclear detonation.

INTRODUCTION

The processes on the Sun, as the star closest to the planet Earth, have a great influence on everything that happens on Earth, in particular on the Earth's biosphere. Visible manifestations of the state of processes on the Sun are prominences. The term prominence is used to refer to a large number of objects (jets) appearing in the chromosphere or corona of the Sun, which differ from the surrounding matter in higher density and lower temperature. the surrounding matter with higher density and lower temperature (Tandberg-Hagen, 1980). The prominences consist of filaments about 1000 km in diameter and reach lengths of up to 200,000 km. Moreover, one part of the prominences leaves the chromosphere in the corona of the Sun, and the other part of the prominences turn around and return to the chromosphere (Tandberg - Hagen, 1980; Monin, 1980). The movement of gas in prominences is accompanied by a transition through the speed of sound, the formation of shock waves and detonation waves. Very strong shock waves can cause nuclear reactions. During periods of solar activity during flares, powerful streams of charged particles from the solar surface are formed in the solar atmosphere, accompanied by strong magnetohydrodynamic shock waves (Pikelner, 1961). Moreover, due to the large loss

DOI: 10.4018/978-1-6684-7745-8.ch015

of heat by radiation, nuclear reactions behind the shock wave front are not initiated by the thermal motion of plasma particles but occur due to the powerful acceleration of particles in the magnetohydrodynamic shock wave (Syrovatsky, 1975). It should be noted that the idea that nuclear reactions can take place not only in the core of the Sun, but also in its atmosphere was first expressed by I.S. Shklovsky (Shklovsky, 1955). Later, the mechanism of particle acceleration in the vicinity of the shock wave was elucidated, leading to nuclear reactions in the solar atmosphere. It is associated with the existence of a zero line of the magnetic field and with the formation of a neutral layer of electric current in the vicinity of this line. When this layer is ruptured by a shock wave, there is a desire to heal the gap, and this happens with the help of particle acceleration. Moreover, the speed of movement of shock waves accompanied by nuclear reactions is supersonic. Taking into account that the temperature in the Sun's atmosphere reaches several million degrees, the speed of sound in the Sun's atmosphere reaches about 100 km/s, and the speed of propagation of shock waves is 500 km/s. Waves of nuclear detonation occurring in the atmosphere of the Sun have a strong global impact on ecological systems on Earth. Back in 1915, A.L. Chizhevsky drew attention to the cyclic dependence of some epidemics on the activity of the Sun (Chizhevsky, 1927). The influence of the Sun at the level of various systems in a living organism (cardiovascular, nervous, etc.) occurs due to various types of radiation and electromagnetic fields (Miroshnichenko, 1981). During the periods of activity of the Sun, the streams of particles coming from the Sun cover the Earth. These streams carry "frozen" magnetic fields with them. Thus, during periods of activity, the solar magnetic fields, as it were, capture the Earth. For example, it was found that the incidence of leukopenia (a decrease in the concentration of leukocytes in the blood) is proportional to the level of solar activity. The latter is responsible for spasm of blood vessels. Magnetic storms on the Sun lead to a significant increase in deaths from myocardial infarction. The number of biological populations changes under the influence of solar activity (Gnevyshev, 1971; Bychkov and Seleverstov, 1978; Korytin, 1972).

15.1. General Equations of the Law of Reversal of Influences

Since prominences are jets emanating from the solar chromosphere, it is of interest to describe the motion of gas in these jets. Approximately, the gas motions in these jets can be considered one-dimensional; then, to describe these motions, one can use the law of reversal of influences established by L.A. Vulis (Vulis, 1946 a, b; 1947).

Let us consider a stationary flow of a compressible medium along a channel (along of coordinate x), assuming this flow to be one-dimensional. We write the equation of motion of the medium and the equation of conservation of energy in the form

$$\frac{dP}{dx} + \rho u\frac{du}{dx} = \rho\frac{dL_{ex}}{dx} - \rho\frac{dL_f}{dx}, \qquad (15.1)$$

$$\frac{dh}{dx} + u\frac{du}{dx} = \frac{dQ}{dx} + \frac{dL_{ex}}{dx}, \qquad (15.2)$$

u is speed of motion of the medium; $h = c_p T$ is enthalpy of the medium; $\frac{dQ}{dx}$ is derivate of thermal energy transmitted through the channel wall; $\frac{dL_{ex}}{dx}$ is derivate of the work done by the gas when it moves against external forces; $\frac{dL_f}{dx}$ is derivate of the work of the medium against frictional forces against the channel walls; P is pressure in the medium; ρ is density of the medium; T is temperature; c_p is specific heat of at constant pressure.

The mass conservation equation has the form

$$\rho u F = G, \qquad (15.3)$$

F is channel cross-sectional area; G is mass flow through the cross-section.

Let us supplement equations (15.1) – (15.3) with the equation of state

$$P = \rho RT, \qquad (15.4)$$

R is the specific gas constant.

Equations (15.1) – (15.4) can be converted to equations of the law of reversal of actions

$$(1 - M^2)\frac{du}{udx} = -\frac{dF}{Fdx} + \frac{dQ}{c_p Tdx} + \frac{dL_f}{RTdx} - \frac{dL_{ex}}{\gamma RTdx} + \frac{dG}{Gdx}, \qquad (15.5)$$

$$(1 - M^2)\frac{d\rho}{\rho dx} = M^2\frac{dF}{Fdx} - \frac{dQ}{c_p Tdx} - \frac{dL_f}{RTdx} + \frac{dL_{ex}}{\gamma RTdx} - M^2\frac{dG}{Gdx}, \qquad (15.6)$$

$$(1 - M^2)\frac{dT}{Tdx} = (\gamma - 1)M^2\frac{dF}{Fdx} + (1 - \gamma M^2)\frac{dQ}{c_p Tdx} - (\gamma - 1)M^2\frac{dL_f}{RTdx} + (\gamma - 1)\frac{dL_{ex}}{\gamma RTdx} - (\gamma - 1)M^2\frac{dG}{Gdx}, \qquad (15.7)$$

$$(1 - M^2)\frac{dP}{Pdx} = \gamma M^2\frac{dF}{Fdx} - \gamma M^2\frac{dQ}{c_p Tdx} - [(\gamma - 1)M^2 + 1]\frac{dL_f}{RTdx} + \frac{\gamma dL_{ex}}{\gamma RTdx} - \gamma M^2\frac{dG}{Gdx}, \qquad (15.8)$$

γ is gas adiabat exponent, $M = u/a$ is the Mach number, and $a = \sqrt{\gamma RT}$ is the speed of sound in the gas.

It follows from the equations that with a single external action on the flow for a continuous transition of the gas flow velocity through the speed of sound, it is necessary to change the sign of this action. It seems possible to modify the law of reversal of influences, formulating it in relation to the sum of influences, since often several external influences act simultaneously on the gas flow (Zhizhin, 2004).

Let us designations $w_g = -\frac{dF}{Fdx}$ - geometric effect, $w_t = -\frac{dQ}{c_p Tdx}$ - thermal effect, $w_f = -\frac{dL_f}{RTdx}$ - friction effect, $w_{ex} = -\frac{dL_{ex}}{\gamma RTdx}$ - work effect, $w_r = \frac{dG}{Gdx}$ - expenditure effect. Let us denote η by the sum of external influences the entire right-hand side of the equation (15.5) for the gas flow velocity. Then the equations (15.5) – (15.8) are reduced to the form

$$(1 - M^2)\frac{du}{udx} = \eta, \tag{15.9}$$

$$(1 - M^2)\frac{d\rho}{\rho dx} = -\eta + (1 - M^2)(w_r + w_g), \tag{15.10}$$

$$(1 - M^2)\frac{dT}{Tdx} = -(\gamma - 1)M^2\eta + (1 - M^2)[w_t - (\gamma - 1)w_{ex}], \tag{15.11}$$

$$(1 - M^2)\frac{dP}{Pdx} = -\gamma M^2\eta - (1 - M^2)(w_f + \gamma w_{ex}). \tag{15.12}$$

From equations (15.9) – (15.12) it follows that for a continuous transition of the gas velocity through the value of the speed of sound under the action of several external influences, it is necessary that at the moment of reaching the speed of sound the sign of the sum of external influences changes. It should be noted that this condition is necessary, but not sufficient. Moreover, as follows from these equations, the coefficients of the influence of the actions do not depend on the type of action, but on the considered gas parameter. This follows from the fact that the coefficients of the sum of actions in the right-hand sides of equations (15.9) – (15.12) depend only on the considered gas parameter. Denoting the coefficients of influence by letter A with the index of the gas parameter we have

$$A_u = 1, A_\rho = -1, A_T = -(\gamma - 1)M^2, A_P = -\gamma M^2, A_{M^2} = 2 + (\gamma - 1)M^2.$$

The second terms in the right-hand sides of equations (15.9) – (15.12) vanish due to the vanishing of the difference; therefore, the factors of this difference do not have significant value when the right-hand sides of the equations pass through zero.

15.2. Gas Movement in Prominences on the Sun

Sun processes have a significant impact on the Earth, in particular, on human health. Prominence is the name for a large number of objects appearing in the chromosphere or corona of the Sun (rye), which differ from the surrounding matter in greater density and lower temperature (Tanberg-Hangen, 1977).Moreover, one part of the prominences leaves the chromosphere further from the center of the Sun into the corona, and the other part of the prominences unfolds and returns to the chromosphere (Tanberg-Hagen, 1977; Monin, 1980). At present, there is no mathematical description of the gas movement in the prominences, corresponding to the main observed features of the prominences. Only opinions are expressed that the motion of the gas in the prominences is controlled by the magnetic field of the Sun and the trajectories along which the gas moves, apparently, are close to the lines of force of the magnetic field. 63 years ago, an attempt was made to qualitatively describe the motion of gas in prominences on the basis of the law of reversal of influences (Vulis, Stanyukovich, 1960) (10 years later this work was completely reprinted in (Stanyukovich, 1971) without reference to previous work). The main conclusion, obtained in (Tanberg-Hangen, 1977), is that a continuous transition through the speed of sound is possible in the flow of gas outflowing from the chromosphere. However, this description did not use mathematical methods of analysis, did not specify the law of gas radiation in general, and did not consider the trajectories of system of equations under study. In this Chapter (see Zhizhin, 2021), the indicated drawbacks it is corrected, and it is shown that is no continuous transition through the speed of sound in the gas flow of a prominence outflowing from the chromosphere.

Assuming that with powerful gas emissions from the interior of the Sun, the motion occurs almost one-dimensional and for a long enough time (i.e. the process is quasi-stationary), we will use the general equations of the law of reversal of actions (15.5) – (15.8) in the presence of a gravity field and radiation. In these equations, in this case, it is necessary to leave two influences: mechanical and thermal. In addition, it is sufficient to leave only two equations for the gas velocity and temperature. It will separately consider two movements: from the of the Sun and to the center of the Sun. Let us denote the spatial coordinate x_1 when the gas moves from the center of the Sun, and the spatial coordinate when moving to the center of the Sun x_2. Radiation in both coordinate systems leads to a loss of thermal energy, therefore $\frac{dQ}{dx_1} < 0, \frac{dQ}{dx_2} < 0$ in both cases.

When gas moves from the center of the Sun, the moves against the force of gravity, i.e., wastes work, hence $\frac{dL_{ex}}{dx_1} < 0$. When gas moves to the center of the Sun, the gas moves under the influence of gravity, i.e., $\frac{dL_{ex}}{dx_2} > 0$. Radiation is a highly nonlinear process. Let`s take it into account using Kirchhoff`s law

$$\frac{dQ}{c_p T dx_{1,2}} = -\frac{aT^3}{c_p}, a$$

is the constant. (15.28)

The differential of gas work in the gravity field has the form, (Vulis, Stanyukovich, 1960; Stanyukovich, 1971)

$$\frac{dL_{ex}}{\gamma RTdx_1} = -\frac{fm}{\gamma RTdx_1}, \frac{dL_{ex}}{\gamma RTdx_2} = \frac{fm}{\gamma RTdx_2}, \tag{15.29}$$

m is the mass of the Sun, γ is the gravitational constant.

Let`s introduce dimensionless variables $\theta = \frac{a^{1/3}T}{c_p^{1/3}}, \chi_{1,2} = x_{1,2}\sqrt{\frac{R^3 c_p}{af^3 m^3}}, w_{1,2} = \theta\chi_{1,2}^2$.

Then, for the motion of gas from the center of the Sun from equations (15.4), (15.5), (15.7) one obtain an autonomous of differential equations

$$(1 - M^2)\frac{dM^2}{M^2 d\chi_1} = -(1 + \gamma M^2)\theta^3 + \frac{\gamma + 1}{\gamma w_1}, \tag{15.30}$$

$$(1 - M^2)\frac{d\theta}{\theta d\chi_1} = -(1 - \gamma M^2)\theta^3 - \frac{\gamma - 1}{\gamma w_1}, \tag{15.31}$$

$$(1 - M^2)\frac{dw_1}{d\chi_1} = -(1 - \gamma M^2)\theta^3 w_1 - \frac{\gamma - 1}{\gamma} + 2(1 - M^2)\theta^{1/2} w_1^{1/2}. \tag{15.32}$$

The right-hand sides of equations (15.30) – (15.32) can be transformed by separating the dimensionless sum of external influences $\eta = \theta^3 - \frac{1}{\gamma w_1}$

$$(1 - M^2)\frac{dM^2}{d\chi_1} = -M^2(1 - M^2)\gamma\theta^3 - M^2(1 + \gamma)\eta, \tag{15.33}$$

$$(1 - M^2)\frac{d\theta}{d\chi_1} = -\gamma(1 - M^2)\theta^4 + (\gamma - 1)\theta\eta, \tag{15.34}$$

$$(1 - M^2)\frac{dw_1}{d\chi_1} = (1 - M^2)(2\theta^{1/2} w_1^{\ 1/2} - \gamma\theta w_1) + (\gamma - 1) w_1 \eta. \tag{15.35}$$

From equation (15.33) (15.35) it follows that the right-hand sides of these equations vanish on a line $\eta = 0$ located on a plane $M^2 = 1$. This line is the locus of singular peculiarity points of the system, at which the derivatives have uncertainty 0 / 0. Since the singular points form a line in three-dimensional (M^2, θ, w_1) space, the trajectories in the vicinity of each such point are located in the two-dimensional plane (Poincare, 1947). Therefore, can consider the projection of this plane onto the coordinate plane (M^2, θ) in the vicinity of each point of the line η = 0 on the plane $M^2 = 1$. To do this, consider two equations (15.33), (15. 34) in the vicinity of some point on the line η = 0, denote this point (w_{1*}, θ_*). Taking into account that in this case $\frac{1}{w_{1*}\gamma} = \theta_*^3$, equation (15.33), (15.34) have the form

$$(1 - M^2)\frac{dM^2}{d\chi_1} = (1 - M^2)\gamma\theta_*{}^3 - (1 + \gamma)(\theta^3 - \theta_*^3), \tag{15.36}$$

$$(1 - M^2)\frac{d\theta}{d\chi_1} = -(1 - M^2)\gamma\theta_*{}^3 + (\gamma - 1)\theta_*(\theta^3 - \theta_*^3). \tag{15.37}$$

One divide equation (15.36) into equation (15.37)

$$\frac{dM^2}{d\theta} = \frac{(1 - M^2)\gamma\theta_*^3 - (1 + \gamma)(\theta^3 - \theta_*^3)}{-\gamma(1 - M^2)\theta_*^4 + (\gamma - 1)\theta_*(\theta^3 - \theta_*^3)}. \tag{15.38}$$

Equation (15.38) has uncertainty 0/0 at the point ($M^2 = 1, \theta = \theta_*$). We will reveal this uncertainty according to L`Hopital`s rule

$$\frac{dM^2}{d\theta} = \frac{-\frac{dM^2}{d\theta}\gamma\theta_* - (1 + \gamma)3}{\frac{dM^2}{d\theta}\gamma\theta_*^2 + (\gamma - 1)3\theta_*}. \tag{15.39}$$

From (15.39) can obtain a quadratic equation for the derivative $\frac{dM^2}{d\theta}$ (proper direction at a point $M^2 = 1$, η = 0), solving which one find

$$\left(\frac{dM^2}{d\theta}\right)_{1,2} = \frac{-(4\gamma - 3) \pm \sqrt{4\gamma^2 - 36\gamma + 9}}{2\gamma\theta_*}. \tag{15.40}$$

The radical expression $4\gamma^2 - 36\gamma + 9$ in (15.40) is negative for real values $\gamma > 1$. This means that there are no real values of the eigen direction $\frac{dM^2}{d\theta}$ at the points of the line η = 0 on the plane $M^2 = 1$. Consequently, there are no trajectories passing through the speed of sound in prominences emanating from the solar chromosphere.

It is important that the analysis carried out takes into account the nonlinearity of radiation (the nonlinearity index entered the radical expression in formula (15.40)). In addition, it is important that the value of θ_* is only in the denominator of formula (15.40). This means that the complex value of the derivative $\frac{dM^2}{d\theta}$ is preserved for all solution points η = 0 in the plane $M^2 = 1$. Thus, the result obtained in (Vulis, Stanykovich, 1960) on the possibility of a continuous any of the gas velocity through the value of the sound velocity under the combined actions of radiation and gravity in the gas stream from the stars should be considered erroneous. This is a consequence of neglecting the nonlinearity of radiation and the absence of a mathematical analysis of the behavior of the trajectories in the vicinity of the point, corresponding to the compensation of mechanical and thermal effects at $M^2 = 1$.

The system of trajectories of equations (15.36), (15.37) in the vicinity of a point ($M^2 = 1, \theta = \theta_*$)

On a line $\eta = 0$ in projection onto the coordinate plane (M^2, θ) is shown in Figure 1.

Figure 1. Phase diagram of trajectories of equation (15.36), (15.37) in a neighborhood of a singular point

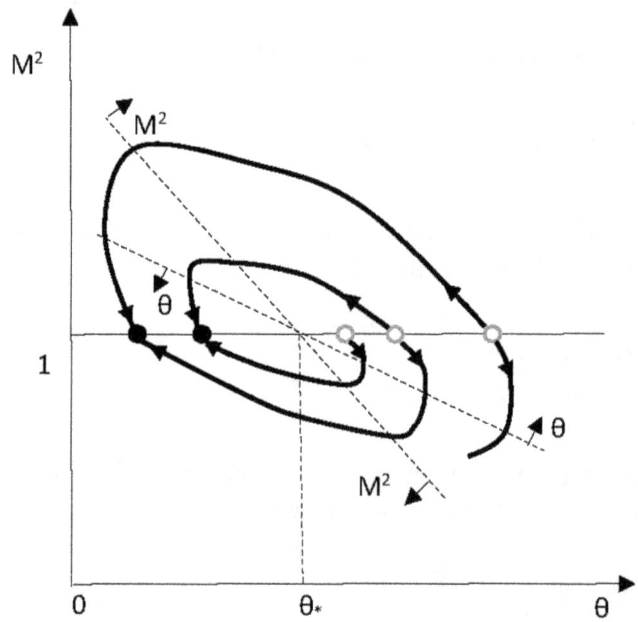

The dotted lines in Figure 15.1 denoted zero lines at the points of which the right-hand sides of equations (15.36), (15.37) equal zero. These lines, together with the line $M^2 = 1$, divide the phase plane into regions of constant sign of the derivatives $\frac{dM^2}{d\chi_1}, \frac{d\theta}{d\chi_1}$. The positive sign of the derivatives is indicated by the arrows at the corresponding zero lines. The singular point ($M^2 = 1, \theta = \theta_*$) divides the line into two parts. To the right of the singular point are the initial limit states (indicated by the empty circle). To the left of the singular point are the final limit states (indicated by a solid circle). According to the assumption, the gas in the prominence, proceeding from the solar chromosphere, has a supersonic speed. This corresponds to trajectories located on the phase plane above the straight line $M^2 = 1$. These trajectories start from the initial limit states and tend to the final limit states to the left of the singular point. With this movement at the beginning of the path, the speed increases, and the temperature drops. After crossing the zero lines, the speed begins to decrease, and then the temperature rises. There is no continuous transition through the speed of sound.

Let us now represent the entire set of trajectories of the phase space (M^2, θ, w_1) when the gas moves from the center of the Sun. For this, it is necessary to take into account three equations of the system (15.30) – (15.32), i.e. take into account the change in the variable w_1. Solutions in this case can be represented in form of spirals around a line $\eta = 0$ on a plane $M^2 = 1$ composed of in the supersonic and subsonic regions of space (M^2, θ, w_1) (Figure 2).

Figure 2. The phase space of system (15.30) – (15.32)

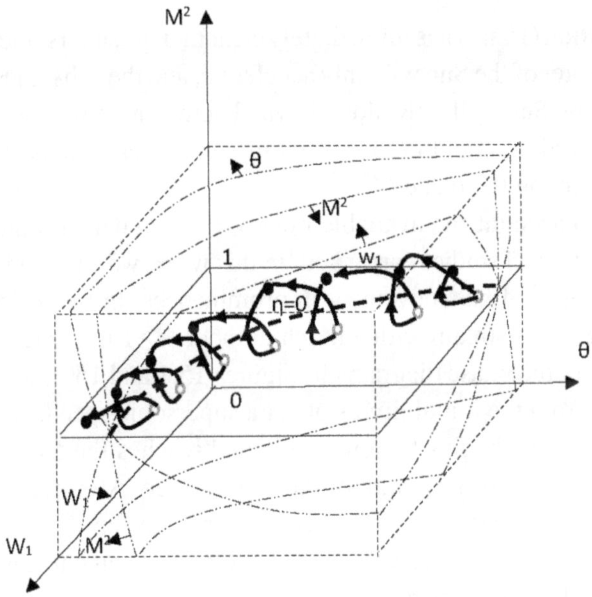

In this figure, zero surfaces, i.e., surfaces at the points of which the derivative of the corresponding phase coordinates M^2, θ, w_1 equal zero are shown by thin dashed lines. Zero surfaces together with the plane $M^2 = 1$ divide the phase space into regions with constant signs of derivatives of the phase coordinates (areas with positive values of the corresponding derivatives are indicated by arrows at the zero surfaces). Solutions go from the start limit point on the plain $M^2 = 1$ (to the right of the line $\eta = 0$) to the end limit point (to the left of the line $\eta = 0$). There is no continuous transition through the speed of sound. Pieces of trajectories in the supersonic part of the phase space correspond to the solutions corresponding to the two-way gas movement from the center of the Sun.

Let`s now consider the motion of the prominence to the center of the Sun after its rotation. Taking into account expressions (15.28), (15.29), the equations of the law of reversal of actions with coordinate χ_2 leads to an autonomous system of equations

$$(1 - M^2)\frac{dM^2}{M^2 d\chi_2} = -(1 + \gamma M^2)\theta^3 - (1 + \gamma)\gamma^{-1} w_2^{-1}, \tag{15.41}$$

$$(1 - M^2)\frac{d\theta}{\theta d\chi_2} = -(1 - \gamma M^2)\theta^3 + (\gamma - 1)\gamma^{-1} w_2^{-1}, \tag{15.42}$$

$$(1 - M^2)\frac{dw_2}{d\chi_2} = -(1 - \gamma M^2)\theta^3 w_2 + 2(1 - M^2)\theta^{1/2} w_2^{1/2} + (\gamma - 1)\gamma^{-1}. \tag{15.43}$$

From equation (15.41) it is immediately clear that the supersonic gas flow moving towards the center of the Sun will only accelerate, and the subsonic gas flow towards the center of the Sun will only slow down. There can be no continuous transition through the speed of sound in these streams. In this case, only the initial limiting states are present on the plane $M^2 = 1$.

Taking into account the available observations under prominence on the Sun and the mathematical studies carried out in this work, we can propose the following sequence of stages of gas movement in prominences. In an explosion in the solar chromosphere, a gas stream with a Mach number $M \geq 1$ is ejected from it. Then, in accordance with the phase diagrams in Figures 15.1 and 15.2, the gas in the prominence moves from the initial limit state at a supersonic speed, at first accelerating even more, and then begins to slow down and reaches the final limiting state on the plane $M^2 = 1$ at a finite length of the prominence. After this, the gas begins to return to the Sun in the subsonic flow region, gradually slowing down until it reaches the solar chromosphere. The rotation of the gas movement can be accompanied by a change in the gas temperature due to the energy consumption for the rotation. Therefore, the transition to return motion in phase diagrams should be regarded as a discontinuous motion. There is also no continuous transition through the speed of sound.

From supersonic gas flow in prominences, jump-like transitions to subsonic regions of motion are naturally possible, which are not considered in this work. Apparently, the influence of the solar magnetic field affects the specific form of the location of the prominence in space, orienting its filaments.

CONCLUSION

This chapter considers the motion of gas in prominences on the Sun. Previous studies of this issue were carried out in the works of L.A. Vulis and K.P. Stanyukovich (Vulis and Snanyukovich, 1960; Stanyukovich, 1971). These works argued that the gas in the prominence, leaving the chromosphere of the Sun, moving from the center of the Sun under the influence of two influences: thermal (gas heat loss due to radiation) and mechanical (doing work against gravity), accelerating can become supersonic, continuously passing through the speed of sound. However, in these works, the emission of gas in a prominence was not specified by any law, i.e. the dependence of the radiation intensity on the parameters of the gas was not taken into account. In addition, the question of the possibility of a continuous transition through the speed of sound in the flow has not been studied mathematically, remaining at the level of possible values of the impact differentials. In this paper, the gas radiation was specified by the well-known Kirchhoff law, and the corresponding system of differential equations was studied qualitatively in a three-dimensional phase space. It is found that there are no points in the phase space with a continuous transition of the velocity of gas in the prominence through the value of the speed of sound. Moreover, there are no such points either when gas moves from the center of the Sun, or when gas moves towards the center of the Sun. This essential fact testifies to the inaccuracy of the known formulations of the law of reversal of influences. The vanishing of the sum of the effects on the gas flow is a necessary but not sufficient condition for the possibility of a continuous transition through the speed of sound. A sufficient condition for a continuous transition through the speed of sound is the existence of a trajectory continuously passing through a singular point in the phase space, in which the Mach number is equal to one. In this case, there is no such trajectory in the entire set of points at which the Mach number is equal to one. It is essential that the absence of such treacories is determined by the coefficients of influence of thermal and mechanical effects in the law of reversal of effects (Figure 1). It is they that lead to the fact that in the vicinity of singular points the trajectories look like a focus in the projection onto the plane, and, consequently, there are no trajectories with a continuous transition through the speed of sound. Moreover, this situation does not depend on the impact values themselves. Let us assume that in a prominence, i.e. in the flow of gas from the Sun's chromosphere, spontaneous nuclear reactions continue to occur (Sudek, 1956), leading to the release of heat from nuclear reactions. The analysis shows that in this case, too, due to the conservation of the coefficients of influence of thermal and mechanical actions in the law of reversal of actions, there are no trajectories with a continuous transition of the gas velocity through the speed of sound. Apparently, such transitions can appear if nuclear collision reactions in a magnetic field take place in the gas flow

(Syrovatsky, 1975). The question of the implementation of such reactions under terrestrial conditions is of interest in connection with the possibility of creating an engine that uses the phenomenon of nuclear detonation. The speed of an aircraft with such an engine, taking into account the known estimates of the speed of sound in the solar chromosphere, can exceed 500 km per second.

REFERENCES

Bychkov, Yu. A., & Selevestrov, A. I. (1978). Solar activity and population dynamics of herring. *Fish Farmer, 1*, 17–20.

Chizhevsky, A.L. (1927). On the relationship between the periodic activity of the Sun and epidemics of cholera and influenza. *Russian-German medical journal.* 9, 511-539.

Gnevyshev, M. N. (Ed.). (1971). *Influence of solar activity on the atmosphere and bisphere of the Earth.* Nauka.

Korytin, S.A. (1972). On the correlation of changes in the numbers of dark and light sables with solar activity. *Collection of scientific and technical information All-Russian Research Institute of Hunting and Fur Breeding.* 36, 14 - 22.

Miroshnichenko, L. I. (1981). *Solar activity and the Earth.* Nauka.

Monin, A. S. (1980). *Solar Cycle.* Gidromethioizdat.

Pikelner, S. B. (1961). *Fundamentals of space electrodynamics.* Moscow: State. Ed. Phys. - Mat. Literature.

Poincaré, A. (1947). *On Curves Defined by Differential Equations.* State Publishing House of Techno-Theoretical Literature.

Shipa, G. (1956). *Getting energy in nuclear reactions. In aerodynamics of large speeds and reactive equipment. Combustion processes. Redists B. Lyuis, R.N. Piz, H.S. Taylor. PRINCETON.* Princeton University Press.

Shklovsky, I. S. (1955). On the question of the content of lithium, beryllium and deuterium in the solar atmosphere. *Reports of the Academy of Sciences of the USSR, 105*(5), 931–934.

Stanyukovich, K. P. (1971). *Unsteady Motions of a Continuous Medium.* Science.

Syrovatsky, S. I. (1975). Acceleration of infected particles in processes like solar flares. *Proceedings of the Academy of Sciences of the USSR. Physical series.* 39(2), 359-374.

Tanberg-Hangen, E. (1977). Prominences. In A. Broucek, C. Durand (Editors). *Solar and solar-terrestrial physics.* Illustrated Dictionary of Terms. Preiburg: D. Rider Publishing Company, 100 – 107.

Vulis, L.A. (1946a). On the transition through the speed in a gas flow. *Reports of the USSR Academy of Sciences*, LIV, 8, 669 – 672.

Vulis, L.A. (1946b). On the effect of friction on the transition through the speed of sound. *Reports of the USSR Academy of Sciences*, LIV, 9, 773 – 775.

Vulis, L.A. (1947). On the Law of Reversal of Effects in Real Gas Flow. *Reports of the USSR Academy of Sciences*, LVI, 8, 799 – 801.

Vulis, L. A., & Stanyukovich, K. P. (1960). On the motion of gas in a gravity field. *Proceedings of the Kazakh State University*, 2, 25–31.

Zhizhin, G.V. (2004). Synergetics of flows. *Life and safety. Ecology*. 2–3 a, 366–385.

Zhizhin, G. V. (2021). Synergetics of Gas Streams. *Journal of Nature Science and Sustainable Technology*, *15*(3), 127–144.

Afterword

Based on the analysis of experimental data on the propagation of waves in excitable media (chemical and biological), it is shown that the asymptotic methods for describing these waves do not agree with the experimental data and in many cases contradict them. In this regard, non-asymptotic methods for analyzing waves in excitable media are considered in detail, and then these methods are applied to describe and analyze waves in combustion processes (cold and hot flames, solid-flame combustion, gunpowder combustion, etc.) and biological populations (bacteria, locusts, bark beetles, etc.). In each particular case, approximate analytical solutions are obtained, numerical studies are carried out, and it is shown that non-asymptotic methods for analyzing waves make it possible to discover fundamentally new modes of propagation of these waves. For example, in the study of combustion waves, in which both diffusion processes and heat conduction processes are significant at Lewis numbers greater than one, the dependence of the velocity on the Lewis number in the non-asymptotic theory is directly opposite to the nature of this dependence in the asymptotic theory. In particular, according to the asymptotic theory, the combustion wave velocity increases without limit as the Lewis number increases, but non-asymptotic methods prove that the wave velocity decreases monotonically with increasing Lewis number. Non-asymptotic methods have discovered a new fundamental property - the flame has an internal degree of freedom, which ensures the stability of the flame when heat is lost to the environment. To compensate for heat loss, the flame increases the length of the chemical reaction zone. This is especially important to keep in mind when fighting fires.

When analyzing detonation waves, it was found that the limitation of the gas velocity at the end of the detonation wave by the local value of the speed of sound, adopted in the classical theory of detonation, limits the depth of the chemical reaction in the wave (no more than 30 percent). The non-asymptotic theory, which does not impose this limitation a priori, proves that significant degrees of burnout can be achieved in a detonation wave, and very extended zones of chemical reactions are formed. It is proved that the estimates of the parameters of adiabatic detonation waves according to the Chapman-Jouguet theory are very far from the solutions according to the non-asymptotic theory of detonation and numerical solutions according to differential equations. Back in 1940 Ya.B. Zel'dovich raised the question of studying the possibility of passing through the speed of sound in a detonation wave after a shock wave from the region of subsonic flows. However, in his works, based on asymptotic methods, it was impossible to consider this issue. This is done in this paper using non-asymptotic methods. It is shown that in the detonation wave after the shock wave, it is possible to go over the speed of sound and achieve complete burnout of the combustible reagent if the gas contains chemically inert particles. The structure of such detonation waves consists of six qualitatively different zones.

Non-asymptotic methods find natural application in the analysis of hard-flame combustion waves with condensed particles having a variable chemical interaction surface. Here, an analytical solution is obtained, which indicates that stationary solutions of such waves exist for small values of the parameter m, which characterizes the particles. Although earlier, without using non-asymptotic methods, it was stated that such solutions exist for large values of the parameter m. It was established that there is a limiting critical powder combustion rate at which an increase in the temperature of the condensed phase, caused by an increase in the temperature of the hot end of the condensed phase, is compensated by a decrease in the temperature of the condensed phase, caused by an increase in the speed burning. The conditions for the transition through the critical speed of combustion of gunpowder with the achievement of supercritical growths are determined.

Solitary waves of unicellular organisms (for example, bacteria) have been studied by non-asymptotic methods. The greatest variety of wave regimes is found in populations of multicellular organisms, in which the fertility function is determined not by the physiological parameters of individual individuals, but by the probability of meeting partners. These systems have a large number of equilibrium positions. The bifurcation diagram of these systems contains hyperbolic-umbilic catastrophe points. An analytical solution describing the propagation of locust waves was obtained by non-asymptotic methods. As a result of processing this solution, simple power dependences of the solitary locust wave on the control parameters are obtained. It is shown that the propagation velocity of a solitary locust wave increases sharply with a decrease in the initial concentration of the food supply. In this case, the wave profile is strongly stretched.

Mathematical models have been developed for the development of populations of bark beetles at all stages of development from departure from wintering places in spring, search for a suitable forest stand, establishment of a settlement, oviposition, growth of larvae, formation of pupae, birth of a new young generation before leaving for wintering. This makes it possible to assess the nature of population development over the course of a year and even several years.

Previously, it was argued that the gas in protuberances on the Sun, moving from the center of the Sun against the action of gravity, radiating heat, can go through the speed of sound. However, in this paper, by examining the corresponding system of differential equations, it is shown that this is not the case. There is no continuous transition through the speed of sound either in the movements of gas from the Sun, or in the movements of gas towards the Sun. Such transitions can appear, apparently, only by the motion of charged plasma particles in the magnetic field of the Sun. Considering that the speed of sound in the solar chromosphere reaches 500 km per second, the speed of gas in such streams can have the same values or even higher. It is of interest to carry out such a process under terrestrial conditions. This would lead to the possibility of creating aircraft with similar speeds.

Thus, when studying waves in excitable media of animate and inanimate nature, it was found that these waves have the ability to self-regulate, their structures are complex and consist of several qualitatively different zones. This is consistent with the statements of the author's previous monograph "Non-Euclidean geometry of living and non-living matter in a higher-dimensional space" (Zhizhin, 2022) that living and non-living nature exist in the process of interaction. Living nature is a product of the development of inanimate nature. The designation of inanimate nature by the term "inert" should be considered incorrect.

Compilation of References

Adler, J. (1966). Chemotaxis in bacteria. *Science, 153*(3737), 708–716. DOI: 10.1126/science.153.3737.708 PMID: 4957395

Adler, J., & Dahl, M. M. (1967). Method for Measuring the Motility of Bacteria and for Comparing Random and Non-random Motility. *Journal of General Microbiology, 46*(2), 161–173. DOI: 10.1099/00221287-46-2-161 PMID: 5339791

Adler, J., & Templeton, B. (1967). The Effect of Environmental Conditions on the Motility of Escherichia coli. *Journal of General Microbiology, 46*(2), 175–184. DOI: 10.1099/00221287-46-2-175 PMID: 4961758

Albeverio, S., & Fensteid, J., Heeg-Kran, R., Lindström, T. (1990). *Non-standard methods in stochastic analysis and mathematical physics*. Moscow: Mir.

Aldushin, A.P., & Martemyanova, T.M., Merzhanov, A.G., Khaikin, B.I., Shkadinsky, K.G. (1972). Propagation of the front of an exothermic reaction in condensed mixtures during the interaction of components through a layer of a refractory product. *Physics of combustion and explosion,* 2, 202 - 212.

Aldushin, A. P., Lugovoy, V. D., & Merzhanov, A. G. (1978). Conditions for the degeneracy of a stationary combustion wave. *Reports AN USSR, 243*(6), 1434–1442.

Allee, W. C. (1951). *Cooperation among animals with human implications*. N.Y. Shuman.

Amosov, G. A. (1958). *Some features of combustion during forest fires*. Leningrad Research Institute of Forestry.

Andreev, K. K. (1966). *Thermal decomposition and combustion of explosives*. Nauka.

Andronov, A. A., Leontovich, E. A., Gordon, I. I., & Mayer, A. G. (1966). Qualitative theory of dynamical systems of the second order. Moskow: Nauka.

Artsybashev, E.S. (1974). *Forest fires and fighting them*. Moscow: Timber industry.

Ayvazyan, R. G., & Sinelnikova, T. A. (1994). The mechanism of linear and quadratic branching of reaction mixtures and the role of HO2 radicals in chain combustion. *Kinetics and Catalysis*, *35*(1), 22–29.

Azatyan, V. V., ... (1976). Isothermal propagation of a rarefied nitrogen trifluoride flame. *Proceedings of the Academy of Sciences of the USSR. Chemical series.* 7, 1459 - 1461.

Azatyan, V. V. (2002). Fast reactions of substitution and displacement of atoms from polyatomic molecules and solid salts by thermally equilibrium atomic reagents. *Kinetics and Catalysis*, *42*(2), 165–181.

Azatyan, V. V., Novozhilov, B. V., Posvyansky, V. S., & Semyonov, N. N. (1976). Speed and limits of propagation of the cold flame of carbon disulphide oxidation. *Kinetics and Catalysis*, *17*(6), 1386–1395.

Bachelis, R. D., & Melamed, V. G. (1965). On the non-uniqueness of stationary solutions for the system of equations of combustion theory with piecewise constant rate constant, thermal conductivity and diffusion coefficients. *Reports AN USSR*, *163*(6), 1338–1341.

Bachelis, R. D., & Melamed, V. G. (1966). On the non-uniqueness of the stationary solution of the system of equations of the theory of combustion. *Applied Mathematics and Mechanics*, *30*(2), 368–374.

Bagdasaryan, Kh. S. (1966). *Theory of radical polymerization*. Nauka.

Barenblatt, G. I. (1978). *Similarity, self-similarity, intermediate asymptotics*. Gidrometioizdat.

Bei-Bienko, G. Ya. (1970). B.P. Uvarov (1889-1970) and his contribution to science and practice. *Entomological Review*, *49*(4), 915–922.

Belyaev, A. F., Bobolev, V. K., Korotkov, A. I., Sulimov, A. A., & Chuiko, S. V. (1973). *Transition of combustion of condensed systems into explosion*. Nauka Publishing House.

Berezovskaya, F. S., & Karev, G. P. (1999). Traveling wave bifurcations in population models with taxis. *Успехи физических наук*, *169*(9), 1011–1024. DOI: 10.3367/UFNr.0169.199909d.1011

Berman, V. S., & Ryazantsev, Yu. S. (1972). Application of the theory of matched asymptotic expansions to the calculation of stationary thermal propagation of an exothermic reaction front in a condensed medium. *Journal of Applied Mechanics and Technical Physics*, *5*, 106–112.

Berman, V. S., & Ryazantsev, Yu. S. (1972). On the analysis of the problem of the thermal propagation of a flame by the method of matched asymptotic expansions. *Applied Mathematics and Mechanics*, *36*(4), 659–666.

Bey - Bienko, G.Ya. (1970). B.P. Uvarov (1889 - 1970) and his contribution to science and practice. *Entomological Review*, *49*(4), 915–922.

Brown, M., Dollimore, D., & Galvey, A. (1980). *Reaction in the Solid State*. Elsevier.

Buckmaster, J. (1976). The quenching of deflagration waves. *Combustion and Flame*, *26*(2), 151–162. DOI: 10.1016/0010-2180(76)90068-7

Burasov, D. M., & Grishin, A. M. (2005). Determination of the propagation rate and ecological consequences of a steppe fire. *Abstracts of the 13th Symposium on combustion and explosion*. Chernogolovka. 80 - 81.

Bush, W. B., & Fendell, F. E. (1970). Asymptotic analysis of laminar flame propagation for general Lewis number. *Combustion Science and Technology*, *1*(6), 421–428. DOI: 10.1080/00102206908952222

Butakov, A.A., & Zanin, A.M. (1978). Experimental study of the regularities of the flow of an exothermic reaction with a changing viscosity of the reagent in a tubular reactor. *Physics of combustion and explosion,* 14(5), 91-95.

Bychkov, Yu. A., & Selevestrov, A. I. (1978). Solar activity and population dynamics of herring. *Fish Farmer*, *1*, 17–20.

Chapman, D. L. (1899).. . *Philosophical Magazine*, *47*, 90.

Chauvin, R. (1963). *Les Socie'tes Animales. De L'Abeille av Gorille*. PLON.

Chebyshev, P. L. (1947). *Mathematical analysis* (Vol. 2). The theory of mechanisms known as parallelograms. Complete works. Publishing House of the Academy of Sciences of the USSR.

Chechilo, N. M., & Enikolopyan, N. S. (1974). On the structure of the polymerization wave front and the mechanism of propagation of the polymerization reaction. *Reports AN USSR*, *214*(5), 1131–1133.

Chechilo, N. M., & Enikolopyan, N. S. (1975). Influence of the concentration and nature of initiators on the propagation of the polymerization reaction. *Reports AN USSR*, *221*(5), 1140–1142.

Chechilo, N. M., & Enikolopyan, N. S. (1976). On the role of pressure and initial temperature of the reaction mixture during the propagation of the polymerization reaction. *Reports AN USSR*, *230*(1), 160–163.

Chechilo, N. M., Khvilivitsky, R. Ya., & Enikolopyan, N. S. (1972). On the phenomenon of propagation of the polymerization reaction. *Reports AN USSR, 204*(5), 1180–1181.

Chetverikov, S. S. (1905). Waves of life. From lepidopetrological observations for the summer of 1903. Diary of the Zoological Department of the Imperial Society of Lovers of Natural Science. *Anthropology and Ethnography, 3*(6), 106–111.

Chetverikov, S. S. (1905). Waves of life. From lepidopetrological observations of the summer of 1903. *Diary of the Zoological Department of the Imperial Society of Lovers of Natural Science. Anthropology and Ethnography., 3*(6), 106–111.

Chizhevsky, A.L. (1927). On the relationship between the periodic activity of the Sun and epidemics of cholera and influenza. *Russian-German medical journal.* 9, 511-539.

Davis, R., & Evans, J. C. (1974). Report on the Brookhaven Solar Neutrino Experiment. *Proceedings of the 6th Leningrad International Seminar "Particle Acceleration and Nuclear Reactions in Space".*

Drysdale, D. (1990). *Introduction to fire dynamics.* Stroyizdat.

Faraday, M. (1861). *The chemical history of the candle.* Royal Institution of Great Britain. DOI: 10.1038/scientificamerican03301861-194

Fedotov, V. G., & Chaikina, A. M. (1972). Investigation of the cold flame of the reaction of fluorine with hydrogen in a flow. *Reports of the Academy of Sciences of the USSR, 203*(2), 406–408.

Frank-Kamenetsky, D. A. (1967). *Diffusion and heat transfer in chemical chemistry.* Nauka.

Geidon, A. G., & Wolfgard, H. G. (1959). *Flame and its structure, radiation and temperature.* Metallurgizdat.

Gnevyshev, M. N. (Ed.). (1971). *Influence of solar activity on the atmosphere and bisphere of the Earth.* Nauka.

Grishin, A.M. (2002). General mathematical models of forest and peat fires and their application. *Mechanical advances.* 12, 41 - 89.

Grishin, A.M. (2004). General mathematical model of steppe fires. *Ecological systems and devices.*12, 25 - 29.

Grishin, A. M. (1981). *Mathematical models of forest fires.* Tomsk State Publishing House. University.

Grishin, A. M. (2005). General mathematical models of natural fires. *Abstracts of the 13th Symposium on combustion and explosion*. Chernogolovka, 73 - 74.

Ham, D. (Ed.). (1972). *Polymerization of vinyl monomers*. Chemistry.

Isaev, A. S. (2001). *Population dynamics of forest insects*. Science.

Isaev, A. S., Nedorezov, L. V., & Khlebopros, R. G. (1979). Mathematical model of escape in the interaction of a predator and a prey. In *Mathematical analysis of the components of forest biocenoses* (pp. 74–82). Science.

Isaev, A. S., Nedorezov, L. V., & Khlebopros, R. G. (1982). *Influence of food transformation on the dynamics of insect abundance. Consortium relations of a tree and dendrophilic insects*. Nauka.

Ivanitsky, G. R., Medvedinsky, A. B., & Tsyganov, M. A. (1991). From disorder to ordering – on the example of the movement of microorganisms. *Успехи физических наук*, *161*(3), 13–71. DOI: 10.3367/UFNr.0161.199104b.0013

Ivanov, N. N. (1984). Solitary waves in population theory. *Proceedings of the Academy of Sciences of the USSR. Biological series*. 5, 656-663.

Ivanov, N.N. (1984). Solitary waves in the theory of populations. *News Academy of Sciences of the USSR. Ser. biol.* 5, 656 - 663.

Ivanov, N.N. (1984). Solitary waves in the theory of populations. *News of the USSR Academy of Sciences. Biological series*, 5, 656 – 663.

Jouguet, E. (1905). *Jorn. De Mathematique*. 347.

Joulin, G., & Clavin, P. (1976). Asymptotic analysis of conditions of extinctions for luminary flames. *Acta Astronautica*, *3*(3/4), 223–240. DOI: 10.1016/0094-5765(76)90049-7

Kataev, O. A., Osetrov, A. V., Popovichev, B. G., & Selikhovkin, A. V. (2001). Population density dynamics of bark beetles (Coleoptera, Scolytidae) in forest stands weakened by natural and anthropogenic factors. *Readings in memory of Nikolai Alexandrovich Kholodkovsky. Ed. S. Yu. Sinev. St. Petersburg: Russian Entomological Society*, Issue 54, 82.

Keller, E. F., & Segel, L. A. (1971). Traveling bands of chemotaxis bacteria: A theoretical analysis. *Journal of Theoretical Biology*, *42*(3), 397–429.

Kennedy, C. R., & Aris, R. (1980). Traveling waves in a simple population model involving growth and death. *Bulletin of Mathematical Biology*, *42*(3), 397–429. DOI: 10.1016/S0092-8240(80)80057-7 PMID: 7378612

Kolmogorov, A. N., Petrovsky, I. G., & Piskunov, N. S. (1937). Investigation of the diffusion equation associated with an increase in the amount of a substance and its application to one biological problem. *Bul. Moscow State University. Ser. A. Mathematics and mechanics.* 1(6), 333 - 358.

Kolmogorov, A. N., Petrovsky, I. G., & Piskunov, N. S. (1937). The study of the diffusion equation, combined with an increase in the amount of substance, and its application to a single biological problem. *MSU Bulletin, series A. Mathematics and Mechanics*, *1*(1), 1–26.

Kolmogorov, A. N., Petrovsky, I. G., & Piskunov, N. S. (1937). The study of the diffusion equation, combined with the increase in the amount of matter and its application to one biological problem. *Bulletin of Moscow State University. Series A.Mathematics and Mechanics*, *1*(6), 1–26.

Kompaneets, A. S. (1976). *Laws of physical statistics. Shock waves. Super dense substance.* Science.

Konev, E. V. (1975). On the ignition of forest combustible materials. *Problems of forest pyro - logy.* Rep. ed. N.P. Kurbatsky, E.V. Konev. Krasnoyarsk: Institute of Forest and Wood named after V.N. Sukachev, Siberian Branch of the USSR Academy of Sciences. 80 - 92.

Konev, E. V. (1979). Results of the study of combustion processes in forest ground fires. Burning in the forest. *Materials of the First All-Union Scientific and Technical Conference.* Rep. editor. N.P. Kurbatsky. Krasnoyarsk: Institute of Forest and Wood named after V.N. Sukachev of the Siberian Branch of the Academy of Sciences of the USSR. Part 2. *New developments in the problem of forest fires.* 53 - 63.

Konev, E. V. (1977). *Physical bases of combustion of plant materials.* Science.

Kopaneva, L. M., & Stebaev, I. V. (1985). *Locust life.* Agropromizdat.

Korytin, S.A. (1972). On the correlation of changes in the numbers of dark and light sables with solar activity. *Collection of scientific and technical information All-Russian Research Institute of Hunting and Fur Breeding.* 36, 14 - 22.

Kovalev, O. N., & Vechernin, V. V. (1986). Description of a new wave process in populations on the example of the introduction and dispersal of the ragweed leaf beetle Zygogramma suturalis F. (Coleoptera, Chrysomelidae). *Entomological Review*, *65*(1), 21–38.

Kovalev, O. N., & Vechernin, V. V. (1986). Description of a new wave process in populations on the example of the introduction and dispersal of the ragweed leaf beetle. *Entomological Review*, *65*(1), 21–38.

Krinsky, V. I. (1968). Fibrillation in excitable media. *Problems of Cybernetics, 20*, 59–80.

Kulikovsky, A. G., & Slobodkina, F. A. (1967). On the stability of arbitrary stationary flows in the vicinity of transition points through the speed of sound. *Applied Mathematics and Mechanics, 31*(4), 5–21.

Lauffenburger, D., Kennedy, R., & Aris, R. (1984). Traveling bands of Chemotaxis Bacteria in the Context of Population Growth. *Bulletin of Mathematical Biology, 46*(1), 19–40. DOI: 10.1016/S0092-8240(84)80033-6

Lewis, B., & Elbe, G. (1938). *Combustion*. Flames and Explosions of Gases.

Lewis, B., Pease, R. N., & Taylor, H. S. (1956). *Combustion Processes*. Princeton University Press. DOI: 10.1515/9781400877027

Loskutov, A. Yu., & Mikhailov, A. S. (1990). *Introduction to synergetics*. Nauka.

Lovachev, L. A., Basevich, V. Ya., Kogan, A. M., & Posvyansky, V. S. (1970). To the calculation of the stationary reaction zone on the example of the H2 + O2 flame. *Proceedings of the Academy of Sciences of the USSR. Chemical series*. 6, 1256 - 1263.

Luzin, N. N. (1966). *Differential Calculus*. Higher School.

Malkin, A. Ya., & Zhirkov, P. V. (1984). Experimental investigation of flow in a tubular polymerization reactor. *Polym. Prok. Eng*, *2*(213), 207–218.

Manod, J. (1950). La technique de culture continue, theorie et applicationes. *Ann. Inst. Paster.*, 79.

Merzhanov, A. G., Borovinskaya, I. P., & Shkiro, V. M. (1972). *France Patent N° 2088668.*

Merzhanov, A. G., Borovinskaya, I. P., & Shkiro, V. M. (1973). *US Patent N°3726643.*

Merzhanov, A. G., Borovinskaya, I. P., & Shkiro, V. M. (1982). *Japan Patent N°1098839.*

Merzhanov, A.G., & Borovinskaya, I.P., Shkiro, V.M. (1967). *The phenomenon of wave localization of self-braking solid-phase reactions. Diploma No. 287.* USSR. Priority from 07/05/1967. Bulletin of Inventions, 1984, No. 32, 3. Bulletin of the Academy of Sciences of the USSR, 1984, 10, 141.

Merzhanov, A.G., & Borovinskaya, I.P., Shkiro, V.M. (1974). *England Patent N°1321084ю*

Merzhanov, A. G. (2000). *Solid – Flame Cjmbustion* [in Russian]. Izd. ISMAN.

Merzhanov, A. G. (2005). *Better to be needed than free....* Chernogolovka: Territory.
Zel'dovich, Ya. B., & Kamenetskii, D.A. Theory of thermal propagation of flames. (1938). *Zh. Fiz. Khim, 12*(1), 100–105.

Michelson, V. A. (1930). *On the normal rate of ignition of explosive gas mixtures. Collected works. 1.* New Agronom.

Miroshnichenko, L. I. (1981). *Solar activity and the Earth.* Nauka.

Monin, A. S. (1980). *Solar Cycle.* Gidromethioizdat.

Moretti, M., & Barbalat, S. (2004). The effects of wildfires on wood-eating beetles in deciduous forests on the southern slope of the Swiss Alps. *Forest Ecology and Management, 187*(1), 85–103. DOI: 10.1016/S0378-1127(03)00314-1

Naife, A.Kh. (1984). *An Introduction to Perturbation Methods.* Moscow: Mir.

Neumann, M.B. (1938). The role of peroxides in the formation of cold and hot hydrocarbon flames. *Good luck with chemistry.* 7(7), 341-384.

Novikov, S. S., & Ryazantsev, Yu. S. (1965). On the existence and uniqueness of the solution of the system of equations of the thermal theory of combustion. *Journal of Applied Mechanics and Technical Physics, 4*(4), 86–88. DOI: 10.1007/BF01565821

Novikov, S. S., & Ryazantsev, Yu. S. (1966). On the existence and uniqueness of the solution of the system of the thermal theory of combustion. *Applied Mathematics and Mechanics, 2*, 368–374.

Novozhilov, B.V., & Posvyansky, V.S. (1973). On the speed of propagation of a cold flame. *Physics of combustion and explosion.* 9(2), 225-230.

Novozhilov, B. V. (1961). The rate of propagation of the front of an exothermic reaction in the condensed phase. *Reports of the Academy of Sciences of the USSR, 441*(1), 151–153.

Novozhilov, B. V. (1965). Criterion of stability of the stationary mode of combustion of gunpowder. *Journal of Applied Mechanics and Technical Physics, 4*, 157–160.

Novozhilov, B. V. (1966). Nonlinear fluctuations in the burning rate of gunpowder. *Journal of Applied Mechanics and Technical Physics, 5*, 31–41.

Novozhilov, B. V. (1973). *Unsteady combustion of solid rocket propellants.* Nauka.

Novozhilov, B. V. (2004). Chaotization of the non-stationary burning rate of gunpowder. *Chemical Physics, 5*, 68–74.

Oudian, J. (1974). *Fundamentals of polymer chemistry.* Moscow: Mir.

Ovsyannikov, L. V. (1978). *Group analysis of differential equations*. Nauka.

Parker, T. J., Clancy, K. M., & Mathiasen, R. L. (2006). Interactions among fire, insects and pathogens in coniferous forests of interior western United Stated and Canada. *Agricultural and Forest Entomology*, *8*(3), 167–189. DOI: 10.1111/j.1461-9563.2006.00305.x

Pikelner, S. B. (1961). *Fundamentals of space electrodynamics*. Moscow: State. Ed. Phys. - Mat. Literature.

Pikelner, S.B. (1961). *Fundamentals of space electrodynamics*. Moscow: State Publishing House of Phys. - mat. literature.

Poincaré, A. (1947). *On Curves Defined by Differential Equations*. State Publishing House of Techno-Theoretical Literature.

Polozhiy, G. N. (1964). *Equations of mathematical physics*. High Schol.

Pontryagin, L.S. (1961). *Ordinary differential equations*. Moskow: Gosizdat fiz. - mat. lit.

Samara, A. A., Galaktionov, V. A., Kurdyumov, S. P., & Mikhailov, L. P. (1987). *Modes with aggravation in problems for quasilinear parabolic equations*. Science.

Samarsky, A. A., Galaktionov, V. A., Kurdyumov, S. P., & Mikhailov, A. P. (1987). *Blow-up Regimes in Problems for Quasilinear Parabolic Equations*. Nauka.

Scherbinovsky, N. S. (1958). The main laws of mass breeding of the desert locust and the migration of its flocks. Moscow: *9 th International Conference on Plant Quarantine and Plant Protection from Pests and Diseases.*

Scherbinovsky, N.S. (1952). *Desert Locust Schistocerka*. Moscow: State ed. rural - host lit.

Selikhovkin, A. V. (1992). Quantitative assessment of the degree of impact of dendrophagous insects on the state of forest stands. *Forestry*, *1*, 16–19.

Selikhovkin, A. V., Bondarenko, E. A., & Popovichev, B. G. (2010). Modern forest entomology: Directions of research, problems and prospects. *Entomological Review*, *LXXXIX*(2), 1, 62–84. DOI: 10.1134/S0013873810020053

Semenov, N. N. (2005). Selected Works.: Vol. 3. *About some problems of chemical kinetics and reactivity*. Nauka.

Shchelkin, K. I., & Troshin, Ya. G. (1963). *Gas dynamics of combustion*. Publishing House of the Academy of Sciences of the USSR.

Shipa, G. (1956). *Getting energy in nuclear reactions. In aerodynamics of large speeds and reactive equipment. Combustion processes. Redists B. Lyuis, R.N. Piz, H.S. Taylor. PRINCETON.* Princeton University Press.

Shkadinsky, G.V., & Khaikin, B.I., Merzhanov, A.G. (1971). Propagation of a pulsating front of an exothermic reaction in a condensed phase. *Physics of combustion and explosion.* 7(1), 19 - 28.

Shklovsky, I. S. (1955). On the question of the content of lithium, beryllium and deuterium in the solar atmosphere. *Reports of the Academy of Sciences of the USSR, 105*(5), 931–934.

Spalding, D.B. (1957). A theory of inflammability limits and flame – quenching. *Combust. Proc. Roy. Soc. London.* V.A 240, 1220, 83 – 100.

Stanyukovich, K. P. (1971). *Unsteady Motions of a Continuous Medium.* Science.

Svirezhev, Yu. M. (1987). *Nonlinear waves, dissipative structures and catastrophes in ecology.* Moskow. *Science.*

Svirezhev, Yu. M. (1987). *Nonlinear waves, dissipative structures and catastrophes in ecology.* Nauka.

Svirezhev, Yu. M. (1987). *Nonlinear waves, dissipative structures and disasters in ecology.* Science.

Syrovatsky, S. I. (1975). Acceleration of infected particles in processes like solar flares. *Proceedings of the Academy of Sciences of the USSR. Physical series.* 39(2), 359-374.

Syrovatsky, S. I. (1975b). Acceleration of charged particles in processes like solar flares. *Proceedings of the Academy of Sciences of the USSR. Physical series.* 39(2), 359-374.

Syrovatsky, S. I. (1975a). Characteristics of the current sheet and the thermal trigger of solar flares. *Letters to the Astronomical Journal, 2*(1), 1–4.

Tanberg-Hangen, E. (1977). Prominences. In A. Broucek, C. Durand (Editors). *Solar and solar-terrestrial physics.* Illustrated Dictionary of Terms. Preiburg: D. Rider Publishing Company, 100 – 107.

Taylor, R. A. J. (1979). A simulation model of locust migratory behavior. *Journal of Animal Ecology, 48*(2), 577–602. DOI: 10.2307/4181

Topaz, C.M., & D`Orsogua, M.R., Edelstein – Keshet, L., Bernoff, J. (2012). Locust dynamics: Behavioral phase change and swarming. *PLOS Comput. Biol.* 8: e 1002642.

Topaz, C.M., D`Orsogua, M.R., Edelstein – Keshet, L., & Bernoff, J. (2012). Locust dynamics: Behavioral phase change and swarming. *PLOS Comput. Biol.* 8: e 1002642.

Topaz, C. M., Bernoff, A. J., Logan, S., & Toolson, W. (2008). A model for rolling swarms of locusts. *The European Physical Journal. Special Topics, 157*(1), 92–109. DOI: 10.1140/epjst/e2008-00633-y

Ullyet, G. C. (1945). Oviposition by Ephestia kukniella. *Zell. J. Ent. Soc. S. Africa, 8*, 53–59.

Uspensky, V. A. (1987). *What is non-standard analysis?* Nauka.

Volpert, A. I. (1987). On the propagation of waves described by nonlinear parabolic equations. In Petrovsky, I. G. (Ed.), *Selected works. Differential equations. Probability theory*. Nauka.

Volterra, V. (1931). *Lecons sur la Theorie Mathematique de La Lutte Pour la Vie*. Gautheer – Villars.

Voronkov, V. G., & Semenov, N. N. (1939). Cold flame propagation in combustible mixtures containing 0.03% carbon disulfide. *Journal of Physical Chemistry, 13*(12), 1695–1727.

Vulis, L.A. (1946). On the transition through the speed of sound in a gas flow. *Reports of the Academy of Sciences of the USSR*, LIV, 8, 669 - 672.

Vulis, L.A. (1946a). On the transition through the speed in a gas flow. *Reports of the USSR Academy of Sciences*, LIV, 8, 669 – 672.

Vulis, L.A. (1946b). On the effect of friction on the transition through the speed of sound. *Reports of the USSR Academy of Sciences*, LIV, 9, 773 – 775.

Vulis, L.A. (1947). On the Law of Reversal of Effects in Real Gas Flow. *Reports of the USSR Academy of Sciences*, LVI, 8, 799 – 801.

Vulis, L.A. (1947). On the law of reversal of influences in a real gas flow. *Reports of the Academy of Sciences of the USSR*, LVI, 8, 799 - 801.

Vulis, L. A. (1946a). On the transition through the speed of sound in a gas flow. *Reports of the Academy of Sciences of the USSR*, (8), 669–672.

Vulis, L. A. (1946b). On the influence of friction on the transition through the speed of sound. *Reports of the Academy of Sciences of the USSR*, (9), 773–775.

Vulis, L. A. (1947). On the law of reversal of influences in a real gas flow. *Reports of the Academy of Sciences of the USSR*, (8), 799–801.

Vulis, L. A. (1950). *Thermodynamics of gas flows*. Gosenergoizdat.

Vulis, L. A., Gusika, P. L., & Zhizhin, G. V. (1972). Two-phase flows in a channel of constant cross section (qualitative study). *Journal of Applied Mechanics and Technical Physics*, *5*, 143–156.

Vulis, L. A., & Stanyukovich, K. P. (1960). On the motion of gas in a gravity field. *Proceedings of the Kazakh State University*, *2*, 25–31.

Warnatz, J. (1983). The mechanism of high temperature combustion of propane and butane. *Combustion Science and Technology*, *34*(1-6), 177–200. DOI: 10.1080/00102208308923692

Warnatz, J., Maas, U., & Dibble, R. W. (2001). *COMBUSTION. Physical and Chemical Fundamentals, Modeling and Simulations, Experiments, Pollutant Formation*. Springer.

Warnatz, J., Maas, U., & Dibble, R. W. (2001). *Combustion. Physical and Chemical Fundamentals, Modeling, and Simulations, Experiments, Pollutant formation*. Springer.

Warnatz, J., Mass, U., & Dibble, R. W. (2001). *Combustion. Physical and chemical fundamentals, modeling and simulations, experiments, pollutant formation*. Springer.

Westmoreland, F. R. et al. (2006). Study of the flame structure by molecular beam spectrometry using electron impact ionization and photon ionization by synchrotron radiation. *Physics of combustion and explosion,* 42(6), 58-63.

Williams, F. A. (1964). *Combustion Theory*. University of California.

Williams, F. A. (1971). *Theory of combustion*. Nauka.

Zaikin, A. I., & Zhabotinsky, A. M. (1977). Autowave processes in active concentration media. In Frank, G. M. (Ed.), *Bophysics of complex systems and radiation disturbances* (pp. 107–112). Science.

Zel'dovich, Ya. B., Barenblatt, G. I., Librovich, V. B., & Makhviladze, G. M. (1980). *Mathematical theory of combustion and explosion*. Nauka.

Zeldovich, Ya.B. (1944). *Theory of combustion and detonation of gases*. Moscow - Leningrad: Publishing houses of the Academy of Sciences of the USSR.

Zeldovich, Ya.B., & Raiser, Yu.P. *Physics of shock waves and high-temperature hydrodynamic phenomena*. Moscow: Nauka.

Zeldovich, Ya. B. (1940). On the theory of detonation propagation in gaseous systems. *Journal of Experimental and Theoretical Physics*, *10*(5), 542–568.

Zeldovich, Ya. B. (1941). The theory of the propagation limit of a silent flame. *Journal of Experimental and Theoretical Physics*, *11*(1), 159–168.

Zeldovich, Ya. B. (1942). On the theory of combustion of gunpowder and explosives. *Journal of Experimental and Theoretical Physics*, *12*(11-12), 498–524.

Zeldovich, Ya. B. (1948). On the theory of flame propagation. *Journal of Physical Chemistry*, *1*, 27–48.

Zeldovich, Ya. B. (1961). Chain reactions in gas flames, approximate theory of flame speed. *Kinetics and Catalysis*, *11*(3), 305–374.

Zeldovich, Ya. B., Barenblatt, G. I., Librovich, V. B., & Makhviladze, G. M. (1980). *Mathematical theory of combustion*. Nauka.

Zeldovich, Ya. B., & Frank-Kamenetsky, D. A. (1938). Theory of thermal propagation of the flame. *Journal of Physical Chemistry*, *12*(1), 100–105.

Zeldovich, Ya. B., & Kompaneets, A. S. (1955). *Theory of detonation*. Gostekhizdat.

Zeldovich, Ya. B., Leipunsky, O. I., & Librovich, V. B. (1975). *The theory of non-stationary combustion of gunpowder*. Nauka.

Zenin, A.A. (1966). Formal kinetic characteristics of reactions occurring during the combustion of gunpowder. *Physics of combustion and explosion*. 2, 28 - 32.

Zenin, A.A., & Nefedova, O.I. (1968). On the combustion of ballistic powder in a wide range of initial temperatures. *Physics of combustion and explosion*. 9(4), 343-349.

Zenin, A. A. (1973). On one model of the reaction layer of the condensed phase of ballistic powder. *Reports of the Academy of Sciences of the USSR*, *213*(6), 1357–1360.

Zenin, A. A. (1980). *Processes in the combustion zones of ballistic powders. Physical processes during combustion and explosion*. Atomizdat.

Zhizhin, G. V. (1972). *Qualitative study of one-dimensional steady flows*. Diss... cand. Phys.-Math. Sciences. Leningrad.

Zhizhin, G. V. (1972). *Qualitative study of one-dimensional steady flows*. Dissertation of the Candidate of Physical and Mathematical Sciences. Leningrad: Leningrad Polytechnic Institute.

Zhizhin, G. V. (1995). Polymerization method. RF patent 2044746 (BI 27).

Zhizhin, G. V. (1999). Semi-infinite reaction zone method in the theory of self-regulating waves of exothermic reactions and non-standard analysis. Mathematical methods in engineering and technology. *Proceedings of the 12th International Conference*. Veliky Novgorod, 14 - 17.

Zhizhin, G. V. (1999). Semi-infinite reaction zone method in the theory of self-regulating waves of exothermic reactions and non-standard analysis. Mathematical methods in engineering and technology. Proceedings of the 12th International Conference. Veliky Novgorod, 14 – 17.

Zhizhin, G. V. (2010a). Nonequilibrium processes in the standing wave of dusty combustible gas. *Scientific Abstract Collection of III International Scientific and Technical Conference "AEROENGINES ofXXI Century", Electronic data, Moscow: Federal State Unitary Enterprise "Central Institute of Aviation Motors named after P.I. Baranov".*

Zhizhin, G. V. (2010b). Mathematical model of a standing detonation wave in a jet of dusty gas. Jet, detached and unsteady flows. Anniversary seminar with international participation (collection of abstracts). St. Petersburg, 161 - 163.

Zhizhin, G. V. (2011). Mathematical model of a solitary wave of a population of bark beetles. *Proceedings of the St. Petersburg State Forest Engineering Academy*, 196, 153-162.

Zhizhin, G. V. (2011). Mathematical model of a solitary wave of a population of beetles - bark beetles. *Proceedings of the St. Petersburg State Forest Engineering Academy*, 196, 153 - 162.

Zhizhin, G. V. (2012). Mathematical model of a solitary wave of a population of beetles - bark beetles. *Proceedings of the St. Petersburg State Forestry Academy*, 196, 153 - 162.

Zhizhin, G. V. (2012). Supersonic motions of gas with condensed particles in the reaction zone of a standing detonation wave. International Scientific Conference on Mechanics: SIXTH POLYAKHOV`S READING. Book of abstract. Saint Petersburg: Saint Petersburg State University. 136.

Zhizhin, G. V., & Segal, A. S. (1988). Hydrodynamic stability of a spherical reaction front accompanied by a strong increase in viscosity. *Proceedings of the Academy of Sciences of the USSR. Mechanics of liquid and gas*. 3, 46-53.

Zhizhin, G. V., & Selikhovkin, A. V. (2012). *Mathematical modeling of the development and distribution of insect populations - stem pests in the forests of Russia*. S.-Petersburg: St. Petersburg State Forestry University.

Zhizhin, G.V. (1990). *Macrokinetics in frontal polymerization reactors*: Abstract of Doc. dissertations. St. Petersburg: LTI named after Lensoviet.

Zhizhin, G.V. (2004). Synergetic of flows. *Life and safety*. 2 - 3a, 365 - 385.

Zhizhin, G.V. (2004). Synergetics of flows. *Life and safety. Ecology*, 2 – 3 a, 366 – 385.

Zhizhin, G.V. (2004). Synergetics of flows. *Life and safety. Ecology*. 2 – 3 a, 366 – 385.

Zhizhin, G.V. (2009a). *Waves of detonation in nature and technology, their mathematical models and significance in violation of the ecological systems of the biosphere*. St. Petersburg: Preprint of the North -Western State Correspondence Technical University.

Zhizhin, G.V., & Poritskaya, I.Ya. (1994). Self-regulating waves of exothermic reactions of the n-th order in condensed media. *Physics of combustion and explosion*. 6, 61 - 68.

Zhizhin, G.V., & Selikhovkin, A.V. (2012). *Mathematical modeling of the development and distribution of insect populations - stem pests in the forests of Russia*. St. Petersburg: St. Petersburg State Forestry Technical University named after S.M. Kirov.

Zhizhin, G. V. (1982). The structure of the wave front of the polymerization. *Reports of the USSR Academy of Sciences*, *263*(6), 1399–1402.

Zhizhin, G. V. (1982). Wave structure of frontal polymerization. *Reports of the Academy of Sciences of the USSR*, *263*(6), 1399–1402.

Zhizhin, G. V. (1984). On the calculation of the wave structure of frontal polymerization. *Kinetics and Catalysis*, *2*, 292–298.

Zhizhin, G. V. (1985). Determination of the coordinates of the location of the front of radical polymerization. *Kinetics and Catalysis*, *4*, 1002–1005.

Zhizhin, G. V. (1988). Autowave processes of distribution of chemical reactions in a dispersion medium. *Journal of Applied Mechanics and Technical Physics*, *6*, 35–43.

Zhizhin, G. V. (1992). *Macrokinetics in Frontal – Polymerization Reactors*. Politechnika.

Zhizhin, G. V. (1992). *Macrokinetics in frontal polymerization reactors*. Polytechnic.

Zhizhin, G. V. (1992). *Macrokinetics of frontal polymerization reactors*. Polytechnica.

Zhizhin, G. V. (1997). *Mathematical Models for Radical- Polymerization Waves*. Izd. CZPI.

Zhizhin, G. V. (1997). *Mathematical models of waves of radical polymerization*.

Zhizhin, G. V. (1997a). *Mathematical models of waves radical polymerization*. Publishing Northwestern Correspondence Polytechnic Institute.

Zhizhin, G. V. (1997b). Stationary waves of reversible radical polymerization. *Chemical Physics, 16*(3), 114–123.

Zhizhin, G. V. (1997b). Stationary waves reversible radical polymerization. *Chemical Physics, 3*, 114–123.

Zhizhin, G. V. (2001). *Anthropogenic impact on the development of biological populations in the environment. The impact of anthropogenic impact on public health and environmental quality. Materials of the regional conference*. St. Petersburg State University.

Zhizhin, G. V. (2004). Model of an Ideal Solid-Flame Combustion Wave with a Variable Chemical -Reaction Surface. *Combustion, Explosion, and Shock Waves, 40*(1), 85–91. DOI: 10.1023/B:CESW.0000013671.95999.6c

Zhizhin, G. V. (2004). *Self-regulating waves of chemical reactions and biological populations*. Science.

Zhizhin, G. V. (2004a). *Self - regulating waves of chemical reactions and biological populations*. Science.

Zhizhin, G. V. (2004b). Model of a wave of ideal solid-flame combustion with a variable surface of chemical interaction. *Chemical Physics, 40*(1), 95–102.

Zhizhin, G. V. (2005). *Dissipative structures in chemical, geological and ecological systems*. Science.

Zhizhin, G. V. (2008). *Combustion waves with distributed zones of chemical reactions (Non -asymptotical theory of combustion)*. Werner Regen Publishing House.

Zhizhin, G. V. (2008). *Combustion waves with distributed zones of chemical reactions (nonasymptotic combustion theory)*. Publishing Werner Regen.

Zhizhin, G. V. (2008). *Combustion waves with distributed zones of chemical reactions (non-asymptotic theory of combustion)*. Werner Regen Publishing House.

Zhizhin, G. V. (2008). *Combustion waves with distributed zones of chemical reactions: (Non -asymptotic theory of combustion)*. Werner Regen Publishing House.

Zhizhin, G. V. (2009b). *Detonation wave in a gas with variable composition*. Polytechnic-service.

Zhizhin, G. V. (2014). Mathematical model of flight of the gregarious locust. *Biosphere, 6*(2), 112–117.

Zhizhin, G. V. (2014). Mathematical model of the flight of gregarious locusts. *Biosphere*, *6*(2), 112–117.

Zhizhin, G. V. (2021). Synergetics in Gas Streams. *Journal of Nature Science and Technology*, *15*(3), 127–144.

Zhizhin, G. V. (2021). Synergetics of Gas Streams. *Journal of Nature Science and Sustainable Technology*, *15*(3), 127–144.

Zhizhin, G. V.. (1986). On stationary states of a spherical reaction front in a fluid with variable viscosity. *Kinetics and Catalysis*, *6*, 1310–1314.

Zhizhin, G. V., & Bolshakova, N. N. (2000). Solitary waves in populations of multicellular animal organisms. *Mathematical Modelling*, *12*(12), 55–65.

Zhizhin, G. V., & Bolshakova, N. N. (2000a). Single waves in populations of multicellular animal organisms. *Mathematical Modelling*, *12*(12), 55–65.

Zhizhin, G. V., & Larina, T. I. (1994). Standing waves gas chemical reactions in porous inert media. *Combust*, *4*, 11–20.

Zhizhin, G. V., & Obukhova, I. A. (1997). Gel - effect at radical polymerization. *Mathematical Modelling*, *9*(11), 3–13.

Zhizhin, G. V., & Poritskaya, I. J. (1994). Self-regulating chemical reactions of n - th order in condensed matter. *Combust*, *6*, 61–68.

Zhizhin, G. V., & Poritskaya, I. Y. (1994). Self-regulating waves for order n exothermic reactions in condensed media. *Combustion, Explosion, and Shock Waves*, *30*(6), 781–787. DOI: 10.1007/BF00755250

Zhizhin, G. V., & Segal, A. S. (1983). Hydrodynamic stability of a cylindrical reaction front accompanied by a strong increase in viscosity. *Journal of Applied Mechanics and Technical Physics*, *2*, 62–71.

Zhizhin, G. V., & Segal, A. S. (1985). Stability and multiplicity of stationary states of the spherical and cylindrical reaction front in the condensed phase. *Abstracts of the All-Union Conference Dynamics of Processes and Apparatuses of Chemical Technology*. Cherkasy, 26 - 27.

Zhizhin, G. V., & Selikhovkin, A. V. (2012). *Mathematical modeling of the development and distribution of stem insect populations in the forests of Russia*. St. Petersburg State Forestry Technical University.

Zvonkov, A. K., & Shubin, M. A. (1984). Non-standard analysis and singular perturbations of ordinary differential equations. *Advances in Mathematical Sciences*, *39*(2), 77–127.

About the Author

Gennadiy Zhizhin, was Head of the Russian Federation Branch of European Society of Mathematical Chemistry and Academician of the Russian Academy of Natural Sciences, Russia in 2017. In 2006 Gennadiy was a Corresponding Member of the Russian Ecological Academy. Gennadiy held Academic rank of Professor in the Department of Mathematical modeling and optimization of chemical processes, Russia in 1999. In 1991 Dr. Zhizhin earned Academic degree of Doctor of Technical Sciences. Dr. Zhizhin earned the degree of Candidate of Physical and Mathematical Sciences in 1972, USSR. From 2014 – 2019, Dr. Zhizhin was Chief Researcher OOO "Adamant" ("Skolkovo"), Russia. Dr. Zhizhin was Senior Fellow, Institute of Silicate Chemistry, Russian Academy of Sciences, S.-Petersburg, Russia from 2012 – 2014. Gennadiy was Professor at North-Western State Technical University, S.- Petersburg, Russia from 1994 – 2012. From 1967 – 1976, Dr. Zhizhin was a Senior Fellow at the Naval Academy named N.G. Kuznetsova.

About the Author

Index

A

C

D

E

F

G

www.ingramcontent.com/pod-product-compliance
Lightning Source LLC
LaVergne TN
LVHW081350300926
849161LV00008B/2115